PRINCIPLES AND STANDARDS
FOR MEASURING PRIMARY PRODUCTION

LONG-TERM ECOLOGICAL RESEARCH NETWORK SERIES
LTER Publications Committee

*Grassland Dynamics: Long-Term Ecological Research in Tallgrass Prairie*
Editors: Alan K. Knapp, John M. Briggs, David C. Hartnett, and Scott L. Collins

*Standard Soil Methods for Long-Term Ecological Research*
Editors: G. Philip Robertson, David C. Coleman, Caroline S. Bledsoe, and Phillip Sollins

*Structure and Function of an Alpine Ecosystem: Niwot Ridge, Colorado*
Editors: William D. Bowman and Timothy R. Seastedt

*Climate Variability and Ecosystem Response at Long-Term Ecological Sites*
Editors: David Greenland, Douglas G. Goodin, and Raymond C. Smith

*Biodiversity in Drylands: Toward a Unified Framework*
Editors: Moshe Shachak, James R. Gosz, Steward T. A. Pickett, and Avi Perevolotsky

*Long-Term Dynamics of Lakes in the Landscape:*
*Long-Term Ecological Research on North Temperate Lakes*
Editors: John J. Magnuson, Timothy K. Kratz, and Barbara J. Benson

*Alaska's Changing Boreal Forest*
Editors: F. Stuart Chapin III, Mark W. Oswood, Keith Van Cleve, Leslie A. Viereck, and David L. Verbyla

*Structure and Function of a Chihuahuan Desert Ecosystem:*
*The Jornada Basin Long-Term Ecological Research Site*
Editors: Kris M. Havstad, Laura F. Huenneke, and William H. Schlesinger

*Principles and Standards for Measuring Primary Production*
Editors: Timothy J. Fahey and Alan K. Knapp

# PRINCIPLES AND STANDARDS FOR MEASURING PRIMARY PRODUCTION

*Edited by*

Timothy J. Fahey

Alan K. Knapp

2007

# OXFORD
## UNIVERSITY PRESS

Oxford University Press, Inc., publishes works that further
Oxford University's objective of excellence
in research, scholarship, and education.

Oxford   New York
Auckland   Cape Town   Dar es Salaam   Hong Kong   Karachi
Kuala Lumpur   Madrid   Melbourne   Mexico City   Nairobi
New Delhi   Shanghai   Taipei   Toronto

With offices in
Argentina   Austria   Brazil   Chile   Czech Republic   France   Greece
Guatemala   Hungary   Italy   Japan   Poland   Portugal   Singapore
South Korea   Switzerland   Thailand   Turkey   Ukraine   Vietnam

Published by Oxford University Press, Inc.
198 Madison Avenue, New York, New York 10016

www.oup.com

Oxford is a registered trademark of Oxford University Press

Library of Congress Cataloging-in-Publication Data
Principles and standards for measuring primary production/
Edited by Timothy J. Fahey, Alan K. Knapp.
    p. cm.—(Long-Term Ecological Research Network series)
Includes bibliographical references and index.
ISBN 978–0–19–516866–2
1. Ecology—Research.   2. Primary productivity (Biology)
I. Fahey, Timothy J., 1952–   II. Knapp, Alan K., 1956–   III. Series.
QH541.2.P75  2007
577'.15—dc22 2006015476

9 8 7 6 5 4 3 2 1

Printed in the United States of America
on acid-free paper

# Foreword

A common and enthusiastically viewed exhibit at many museums is a cross section of a large tree on which annual rings recording the growth of the tree are labeled so that they can be correlated with events in human history. The cross section is the visualization of a long ecological record that reflects both changing environmental conditions and changes in the tree itself. Perhaps the tree was struck by lightning, which affected its growth rate. A drought, flood, or insect infestation may have affected growth. All of these interacting factors are integrated and summarized in a single variable: the width of the annual rings. However, the complexities of comparison of things such as growth rate or primary productivity are not really reflected in the simple visualization of the tree's cross section. In particular, the growth represented by the rings in the tree is only one element of primary productivity even though it may correlate strongly with overall productivity. This book attempts to address the complexity that is inherent in collecting, interpreting, and presenting long-term data sets on primary productivity. It focuses on the principles underlying the specific details of collecting data and emphasizes the need for standardized methods. Standardization is key because the overall goal of any kind of ecological study or environmental monitoring is an improved understanding of the processes that are ongoing in the ecosystem. In order to fully appreciate these processes, we need to examine their dynamics over time and space. But to compare various measurements along these temporal or spatial dimensions, we need to collect measurements that are comparable as well as accurate and precise. While there are many ways to manipulate data to allow comparison, certainly the most productive approach is to start data collection with a preconceived idea of the methodology and intent of the study. By standardizing methodology among studies endeavoring to measure the same parameter, we ensure the maximum comparability in our data sets.

Humans are fascinated by change in the world. Environmental change is a constant topic of conversation, although much discussion is speculative and uninformed by data. We constantly compare conditions as we see them now with those in the past, and anticipate how things might look in the future. However, most of our mental comparisons suffer from the imprecision of faulty memory or the kind of exaggeration that the passage of time introduces. This is why we are powerfully drawn to tree rings, which enable us to check our own memories and recollections against unbiased information from the past. We are inveterate collectors of bits of information that help us remember and record a historical event. Innumerable measurements made by amateurs have been aimed at comparing the present with the past and forecasting the future. In most cases, measurements of this type are simple enough that we need not worry about standardized methods or changes in technique over time. For example, many places in the northern hemisphere have long records of the duration of ice cover on lakes during the winter. Some of these data sets are municipal records, and others are used to anticipate such important events as the start of the fishing season or the optimal time for agricultural activities. These kinds of information put our lives in the context of an environment that constantly changes, sometimes too subtly for us to appreciate within our usual time frame of reference. Long-term data provide an unbiased record at a temporal scale appropriate for understanding the nature and pace of environmental change.

Productivity is among the most fundamental characteristics of ecosystems, and the measurement of productivity is usually a central element of research programs focused on ecosystems. For example, the Long Term Ecological Research Network, a group of twenty-six sites funded by the U.S. National Science Foundation, has primary productivity as one of its five core research areas. This interest in productivity stems both from its crucial role in ecosystem dynamics as well as from its importance to human enterprise. Productivity is intimately related to the services that humans receive from natural ecosystems. In many systems, the entire productive output is harvested by human societies and used for their support and maintenance. The level of productivity often determines whether humans can exist in a particular location, as well as the structure and the dynamics of their social systems and cultures. Productivity is closely related to the governance of nations and the relationships between nations. The importance of productivity is acknowledged in the development of religious systems around the world and is a fundamental measure of the health of ecosystems for indigenous peoples. The desire to improve productivity has engendered complicated scientific enterprises focusing on mechanisms to increase food and fiber output of communities and has resulted in multibillion-dollar industries with the sole goal of maximizing production. Most recently, the role of productivity in the global carbon cycle has elicited particular interest because of the developing crisis of greenhouse-gas-driven climate change.

Measurement of primary productivity is a complex task. The elements that go into determining productivity are varied, and measurement schemes for each of these elements are diverse.

Because of the way our scientific enterprise is structured, measurements at different times or at different places are often conducted by different investigative teams. They may be funded by a variety of agencies, employ diverse methods that capture different time scales, and have unique goals. In most cases, the immediate need or objective of study is the one that drives the methodology used. However, for science to make efficient use of resources and to provide information that can be compared across ecosystems and across scales, we need to go beyond local planning as we collect data. In fact, each study should have a developed plan for integrating with other scales in addition to the focal scale of the investigation. Only in this way can we be sure that the questions we ask in our own backyards are relevant to broader regional and national issues. Moreover, as we change our focus from local to regional to landscape to national and global scales, we need to be aware that our measures may also change both in the details of methodology and in underlying principles. Clearly, studies with methodologies that are robust over a wide range of temporal and spatial intervals will be the most valuable.

Understanding the interactions of ecological processes acting at different temporal and spatial scales is one of the key challenges facing ecologists. Most ecological data are collected at small spatial and short temporal scales, generally less than one year on approximately $1-10,000$ m$^2$ plots. However, environmental changes act at scales larger than those at which field data are generally collected, and result in a mismatch or decoupling of important parameters that contribute to our understanding. Mechanisms for converting locally collected data into information that applies to larger landscape or regional scales are lacking, as is the firm theoretical basis for scaling up in most of ecology. Nonetheless, the need to understand ecosystem processes at multiple scales is real, and results in the use of an amalgam of different tools and methods. For example, measures of primary productivity at the plot or stand scale are often collected by field workers measuring individual organisms or extrapolating from stand-level simulation models. At larger scales, primary productivity may be estimated through remotely sensed data, aerial photography, land-use change analysis, or a variety of other large-scale techniques. The conversion or the homogenization of data collected by one technique with those collected by another technique is a serious problem for ecological science. The fine-scale spatial and temporal heterogeneity that leads to stand-level productivity is not measured by aircraft or satellite instruments. At the same time, the expense and time required for ground-based measurements do not permit widespread sampling, which results in the aforementioned problem with scaling up. Because of these issues, it is even more important that the utility of ground observations be maximized through a methodology which is comparable and scalable over large spatial extents. We cannot afford to lose data because of idiosyncratic methodology or mismatched methods.

Various obstacles confront research synthesis by acting as barriers to the standardization of ecological data. The training that ecologists undergo often involves considerable emphasis on individual effort and self-reliance. Research projects undertaken by undergraduates, dissertation studies, and postdoctoral research often involve single individuals who make a significant investment of their lives in these efforts. As a result, we are encouraged to think for ourselves through our academic development, and this training carries over into our later professional lives. However, the decisions on methodology made during this formative stage are often taken without regard to possible future expansion of the research horizon. Methodology is often chosen to address specific small-scale, relatively self-contained projects. Moreover, once we are locked into a particular method, we may find it difficult to change our approach because we have invested too much time and energy to risk a new method.

This problem affects more than just individuals. Although the U.S. LTER Network has focused on common core research areas (including primary production) since its inception, the lack of an initial emphasis on standardization of measurements has resulted in the development of varying methods at different sites even with the same biome type. Efforts to address this problem by adopting standard methods are hindered by the high cost of calibrating old and new methods over the long time frames of ongoing experiments and measurements. Hence, our reaction to new approaches tends to be conservative—unless, of course, our own chosen method is the basis for standardization.

If technology stood still, we would be more likely to reach common ground eventually. However, new technologies and approaches arise constantly, and may provide faster, cheaper, or more accurate measurements than our tried-and-true favorites. For example, the current revolution in wireless sensor networks holds enormous promise for ecology. However, early adopters of this new technology may find themselves out of step with practitioners of more traditional methods, and as a result their ability to compare results may suffer. Moreover, increased technology requires increased understanding of the processes and assumptions underlying the technology, adding an additional burden to our capability to conduct robust comparisons using different methods. One solution to this problem is to maintain a clear focus on common methodological principles and insist that new technology incorporate these principles. Finally, heterogeneity of sampling approaches is introduced by funding limitations and cost restrictions. Because the availability of resources varies among research sites, and the proportional allocation of funds to measurements of productivity differs across studies, there is inevitable heterogeneity in the precision and accuracy of our data sets. No simple solution exists that will resolve each and every issue that challenges us when we attempt to compare our results across sites, studies, or sampling periods. However, the establishment and adoption of underlying principles and standards to guide the selection of field methods will improve our ability to make these kinds of comparisons. With that in mind, the contributors to this book have endeavored to lay out a series of principles and standards for the measurement of primary productivity that, hopefully, will guide future investigators in choosing methods that are both effective and efficient, and at the same time lead to meaningful comparisons of data across studies, ecosystems, and times.

Robert B. Waide
LTER Network Office
Biology Department
University of New Mexico
Albuquerque, NM 87131–0001

# Contents

# Contributors

Jeanne E. Anderson, Complex Systems Research Center, Institute for the Study of Earth, Oceans, and Space, University of New Hampshire, Durham, NH 03824

John J. Battles, Department of Environmental Science, Policy, and Management, University of California, Berkeley, CA 94720–3110

Bobby H. Braswell, Complex Systems Research Center, Institute for the Study of Earth, Oceans, and Space, University of New Hampshire, Durham, NH 03824

John M. Briggs, School of Life Sciences, Arizona State University, Tempe, AZ 85287–4501

Daniel L. Childers, Department of Biological Sciences, Florida International University, Miami, FL 33199

Timothy J. Fahey, Department of Natural Resources, Cornell University, Ithaca, NY 14853

Evelyn E. Gaiser, Southeast Environmental Research Center and Department of Biology, Florida International University, Miami, FL 33199

Robert O. Hall, Jr., Department of Zoology and Physiology, University of Wyoming, Laramie, WY 82071

Mark E. Harmon, Department of Forest Science, Oregon State University, Corvallis, OR 97331–5752

Brian D. Kloeppel, Institute of Ecology, University of Georgia, Coweeta Hydrologic Laboratory, 3160 Coweeta Lab Road, Otto, NC 28763–9218

Alan K. Knapp, Department of Biology, Colorado State University, Fort Collins, CO 80523–1878

William K. Lauenroth, Department of Rangeland Ecosystem Science and Natural Resource Ecology Laboratory, Colorado State University, Fort Collins, CO 80523

Mary E. Martin, Complex Systems Research Center, Institute for the Study of Earth, Oceans, and Space, University of New Hampshire, Durham, NH 03824

William K. Michener, LTER Network Office, Department of Biology, University of New Mexico, Albuquerque, NM 87131–0001

James T. Morris, Department of Biological Sciences and Belle W. Baruch Institute for Marine and Coastal Sciences, University of South Carolina, Columbia, SC 29208

Scott V. Ollinger, Complex Systems Research Center, Institute for the Study of Earth, Oceans, and Space, University of New Hampshire, Durham, NH 03824

Donald L. Phillips, U.S. Environmental
Protection Agency, National Health &
Environmental Effects Research Labora-
tory, Corvallis, OR 97333

Andrew Rassweiler, Department of
Ecology, Evolution, and Marine Biology,
University of California, Santa Barbara,
CA 93106

Osvaldo E. Sala, Department of Ecology
and Evolutionary Biology, Center for
Environmental Studies, Brown University,
Providence, RI 02912

Marie-Louise Smith, USDA Forest Service
Northern Research Station, Durham, NH
03824

Raymond C. Smith, Institute for Computa-
tional Earth System Science and Department of Geography, University of
California-Santa Barbara, Santa Barbara,
CA 93016

Serge Thomas, Southeast Environmental
Research Center and Department of
Biology, Florida International University,
Miami, FL 33199

Geraldine L. Tierney, Department of
Environmental & Forest Biology, SUNY
College of Environmental Science &
Forestry, Syracuse, NY 13210

Robert N. Treuhaft, Jet Propulsion
Laboratory, California Institute of
Technology, Pasadena, CA 91109

Kristin L. Vanderbilt, Department of
Biology, University of New Mexico,
Albuquerque, NM 87131

Maria Vernet, Integrative Oceanographic
Division, Scripps Institution of Oceanog-
raphy, University of California–San
Diego, La Jolla, CA 92093–0218

Dale H. Vitt, Department of Plant Biology,
Southern Illinois University Carbondale,
Carbondale, IL 62901–6509

Donald R. Young, Department of Biology,
Virginia Commonwealth University,
Richmond, VA 23284

# PRINCIPLES AND STANDARDS
# FOR MEASURING PRIMARY PRODUCTION

# 1

# Primary Production: Guiding Principles and Standards for Measurement

Timothy J. Fahey
Alan K. Knapp

P rimary productivity is the rate at which energy is stored in the organic matter of plants per unit area of the earth's surface. It is often expressed in units of dry matter (e.g., grams of dry mass m$^{-2}$ year$^{-1}$) rather than energy because of the ease of determining mass and the relative constancy of the conversion from mass to energy (caloric) units for plant tissues. *Gross* primary productivity is the amount of energy fixed (or organic material created) by plants in photosynthesis per unit of ground area per unit of time. However, plants use a considerable amount of the organic matter that they produce for their own respiratory needs; *net* primary productivity (NPP) is the amount of organic matter that is left after respiration.

All heterotrophic organisms ultimately rely on the organic matter produced by green plants to meet their food requirements, except for the small amounts associated with chemoautotrophs. Hence, our understanding of the dynamics of ecosystems depends fundamentally upon knowledge of patterns and controls of NPP. Recent interest in obtaining accurate measurements of NPP has been stimulated by the problem of rising atmospheric concentrations of $CO_2$ and consequent greenhouse warming of Earth's climate. The balance between NPP and heterotrophic respiration in natural and human ecosystems largely determines the rate of change of atmospheric $CO_2$, and the interactions between global climatic change, human activity, and NPP will shape the future habitability of Earth. Both accurate measurements of NPP and a thorough understanding of the factors controlling this process in the world's biomes will be needed in the future to facilitate environmental management efforts to protect Earth's ecosystems (Geider et al. 2001).

A long history of the study of primary production has accompanied the needs of society to quantify the yield of managed ecosystems, especially agriculture and forestry. With advances in the concepts of ecology, more formalized study of primary production was undertaken first in aquatic systems, then in terrestrial biomes (McIntosh 1985), eventually leading to the development of handbooks to guide the measurement of NPP in forests (Newbould 1966) and grasslands (Milner and Hughes 1968) under the auspices of the International Biological Programme (IBP). These handbooks provided a valuable synthesis of methods and approaches for quantifying NPP, and they have been widely cited in production ecology studies and related literature. Research experience in the production ecology of terrestrial and aquatic ecosystems has advanced our understanding of primary production and some aspects of its measurement in ways that move us beyond the IBP handbooks; however, more recent overviews of methods in ecology and ecosystem science (e.g., Sala et al. 2000) have not provided detailed guidelines for NPP measurement. Moreover, recent attempts to synthesize the literature on NPP within (Clark et al. 2001a; Gower et al. 2001) and across biomes (Knapp and Smith 2001) have illustrated the need for improving the standardization of NPP measurements.

The demand for high-quality and standardized NPP data can be highlighted by several contrasting types of uses of these data. Numerous attempts have been made to better understand environmental controls and drivers of NPP within and among biomes (Rosenzweig 1968; Webb et al. 1983; Sala et al. 1988; Lauenroth and Sala 1992; Frank and Inouye 1994; Huxman et al. 2004) by relying on published or unpublished data that often are of unknown accuracy and precision. Knapp and Smith (2001) compiled long-term NPP data from 11 Long-Term Ecological Research (LTER) sites to assess controls of means and temporal variation in NPP in North American terrestrial ecosystems. Although NPP data quality in LTER research programs is high (because this process measurement is a core area of research within the LTER network), the limited spatial extent of available data highlights the need for additional data sets from sites that adopt those principles and standards required for meaningful comparative studies.

Accurate NPP measurements also are essential for the development and validation of simulation models that are capable of expanding the temporal and spatial scales of prediction of ecosystem dynamics. Moreover, field NPP measurements can provide a comparative basis for utilizing tower-based, aerodynamic measurements of ecosystem carbon (C) exchange (Goulden et al. 1996). Taken together, these field measurements will be needed to evaluate and credit the sources and sinks of C in the biosphere as society grapples with the problem of greenhouse gas accumulation in the atmosphere.

The principal audience for the present volume is ecological scientists who are directly involved in designing and implementing research programs that require accurate measurements of primary production. Although ecologists at established research institutions and sites (e.g., U.S. LTER) can draw upon extensive experience measuring NPP, this group also can profit from this volume as a reference work which they can recommend to students and technicians. In many areas of the world, new programs of ecological research and monitoring are being developed (e.g., international LTER). By adopting the principles and standards described in this volume, these programs will be better equipped to obtain meaningful data that will advance the purposes outlined above.

## Principles and Standards

### General Considerations

As pointed out by Newbould (1967) in the IBP *Handbook on Measurement of Forest NPP*, "Complete standardization of NPP measurements is both undesirable and impractical, but wide adoption of general principles will help make results comparable." Our experience since that time would reinforce this argument. Although it might seem reasonable that a particular suite of standard methods should exist that would optimize the measurement of NPP in any particular biome (e.g., grassland, deciduous forest), experience shows that the unique features of individual sites, the varying availability of funding and manpower, technological advances, and the differing objectives and time scales of each particular project dictate against method standardization for NPP. This assessment contrasts with measurements of climate and, to some extent, soils. Hence, previous LTER methods volumes (Robertson et al. 1999) have successfully described detailed procedures for standardizing climate and soil measurements across the U.S. LTER network.

The present volume provides an overview of the principles that should underlie a program of primary production measurement in any biome. By following these principles, a research group should be able to develop a set of procedures that will meet key objectives which ensure that the data collected are of high quality for their own purposes and of maximum value for use by others. These objectives are (1) accurate, field-based estimates of primary production at appropriate temporal and spatial scales for a range of biome types (chaps. 3–10, this volume); (2) application of remote sensing approaches to NPP (chap. 11, this volume); (3) an assessment of error, bias, and precision of the estimates (chap. 12, this volume); (4) documentation of the measurement program, including primary data, metadata, and synthetic use of the data (chap. 2, this volume). Although exact sampling procedures must be adapted to meet the local setting and financial constraints, uniform guiding principles should provide sufficient conformity of NPP measurement to permit reliable cross-site comparisons, model verification, and C accounting.

The working definition of NPP for purposes of field measurements must be adapted from the formal definition provided above, and in fact such a working definition will differ across biomes. For example, in evaluating methods for NPP measurement in tropical forest, Clark et al. (2001b, pp. 357–358) distinguished actual forest NPP from NPP (the measurement), "the sum of all materials that together are equivalent to (1) the amount of new organic matter that is retained by live plants at the end of the interval, and (2) the amount of organic matter that was both produced and lost by the plants during the same interval." The latter component is of particular importance in ecosystems with substantial herbivory and rapid turnover of organic matter. In contrast, for ecosystems in which annual changes in the standing stock of live biomass represent a negligible fraction of NPP (e.g., many terrestrial herbaceous communities) and biomass turnover or loss to herbivores ([2] in Clark et al.'s definition above) is minimal, the seasonal accumulation of biomass sampled at an appropriate temporal intensity is a reliable estimator of NPP (Sala et al. 1981; Briggs and Knapp 1995). Finally, because of tight coupling between photosynthetic and heterotrophic metabolism in most aquatic ecosystems, measurement of NPP is impractical, and the concept of NPP is of limited value (chap. 10, this volume).

A key consideration in any program of NPP measurement in terrestrial biomes will be what to do about belowground NPP. Field measurement of root growth is difficult and time-consuming; moreover, some of the losses of root organic matter that represent a substantial proportion of belowground NPP have so far defied most attempts at quantification: (1) root herbivory, (2) root exudation and rhizodeposition, and (3) root C allocation to mycorrhizal fungi. Hence, on one hand, measurement of NPP is incomplete without assessment of belowground NPP, but even the most intensive program of belowground NPP measurement may fail to provide estimates of belowground NPP that are comparable in accuracy and precision with aboveground NPP measurements.

## Temporal and Spatial Variation and Principles of Site Selection

The design of any program of primary production measurement must address the problems of scale with respect to temporal and spatial variability. Although interannual variation in NPP is often substantial (Knapp and Smith 2001), the annual cycle of climate provides a logical and fixed framework for defining the time scale of NPP measurements for terrestrial systems (one exception would be for evergreen vegetation, where there may be substantial incongruence between annual fine litterfall and shoot production). The spatial scale of NPP sampling will usually involve substantial subjectivity on the part of the researcher, and the range of applicability of the primary production measurements within and across sites must be carefully evaluated both before and after the sampling program. This principle is true for both terrestrial and aquatic ecosystems, but the difficulty of the problem probably depends mostly upon the magnitude of spatial variation in primary production within the study area.

Most study areas will present researchers with a complex landscape where the patterns of NPP are nonuniform, and even the pelagic zones of aquatic systems exhibit very high spatial variability in primary production. Although a purely random or regular sampling system may provide an unbiased estimate of the NPP for the study area, in many cases a stratified sampling program will be more efficient, especially when there are straightforward criteria for stratification. Also, the objective of the NPP measurement program may not be to estimate average NPP for the whole study area, but rather for some subset of the study area. For example, at the Konza Prairie LTER site, soil depth varies substantially with topographic position, and NPP sampling is stratified by upland, slope, and lowland locations. This sampling is further arrayed across fire frequencies (1–yr, 4–yr, and 20–yr intervals between fires). Thus, in both cases, the end points of important gradients are sampled (upland shallow soils vs. lowlands, and annual fires vs. rare fires), as well as an intermediate point. Nonetheless, often a single value for NPP is needed for the site, necessitating spatial interpolation of values. Another suitable approach to the problem of stratification of NPP sampling is vegetation classification, particularly where the environmental factors controlling NPP may be more obscure (e.g., wetlands).

In most NPP measurement programs it will be necessary to delineate areas for different sampling activities, especially to accommodate destructive sampling in ways that do not introduce bias into subsequent measurements. Also, where NPP measurement programs are integrated with other ongoing sampling programs (e.g., heterotroph populations) or with experimental treatments (e.g., fertilization trials), it will be important to coordinate the delineation of sampling areas in ways that will accommodate long-term measurements. Typically, within each vegetation stratum the delineation of an NPP sampling area will be necessary, with subplots for destructive measurements and sufficient buffer to avoid the introduction of sampling artifacts. The actual layout will vary, depending upon the procedures used for NPP sampling. Because root systems and belowground processes may be sensitive to trampling, careful attention to access to the sampling area is important.

## Field Sampling and Laboratory Procedures: Principles and Standards

Primary productivity can be regarded with equal validity as an energy flux or as a mass flux. Most studies of NPP have focused on the latter, presumably in part because it does not require the additional measurement of the caloric equivalent of biomass. Three different conventions exist for NPP units that express the process in terms of mass flux: dry weight, organic matter (or ash-free) dry weight, and mass of C. Strictly speaking, the latter two conventions better express the connection between the process of photosynthetic production and the synthesis of organic matter; however, most studies do not explicitly make the conversion from dry weight to organic matter or carbon content of biomass. The carbon concentration in plant tissues ranges from about 47% to 55% of dry weight, and a conventional conversion factor of 50% often has been used (Fahey et al. 2005). The concentration of minerals or nonvolatile elements or compounds in plant tissues is usually less than 2%; hence, the error introduced into NPP estimates by assuming equivalence between dry weight and organic matter usually will be small. For ease of comparison we recommend use of the convention of reporting oven-dry mass and a 50% carbon concentration conversion factor when these data are unavailable. Alternatively, careful reporting of actual ash and carbon concentrations is needed.

The standard units of area and time for expressing NPP are more straightforward. There is no particular reason to favor a larger or smaller area unit (e.g., ha vs. $m^2$) for direct interconversion; however, because the plots and areas in which NPP measurement is undertaken are almost always much smaller than 1 ha, some confusion could be avoided by standardizing to an $m^2$ basis. As noted earlier, the temporal unit of 1 yr will usually apply, but for some purposes, expression of NPP flux at some shorter or longer time scale may be appropriate.

The problem of determining the appropriate sample size for field measurements of NPP is complex. The intensity of the field sampling program will vary so much among biomes and study areas that it is not useful to provide general guidelines; rather, the question of sampling intensity is considered within individual chapters of this volume. However, two questions about field sampling that deserve general consideration can be posed. First, is there a minimum level of sampling below which field measurement of NPP is not worthwhile? This is not a trivial question, because primary production measurements in some biomes (e.g., tropical forest, open ocean) are so difficult that obtaining useful data may be very expensive and the generation of data from sub-minimal programs may be scientifically counterproductive. A second general consideration about NPP measurement is how to determine when additional sampling effort is not warranted. This problem is amenable to statistical analysis, as detailed in chapter 12; however, the answer to the question will depend upon both the resources available to the study and the questions and designs of the research. Hence, the applications of statistical criteria are always context specific.

## Statistical Analysis, Problems of Bias, and Quality Assurance/Quality Control

Although it would be desirable to develop a standard set of guidelines for expressing the uncertainty underlying NPP estimates, the varying nature of the field sampling programs will preclude a uniform approach. In principle, different sources of error and uncertainty can be distinguished: (1) spatial variability; (2) quantifiable sampling errors—for example, uncertainty associated with allometric conversions; and (3) unknown sampling errors—for example, uncertainty associated with the separation of live and dead tissues or difficulties in separating current-year senescent material from the previous year's senescent material. The nature and severity of these different sources of error will vary among biomes and study areas. Nevertheless, it should be possible for researchers to generate estimates of uncertainty in their NPP measurements. The general principle that must be followed in all NPP measurement programs is that the researchers must present their best estimate of error and document in detail how they arrived at their error estimates. Although the ideal of perfect intercomparability among data sets is not possible, if this principle is followed, then an objective basis for evaluating the probability of significant differences among sites and years should be available.

Problems of bias are common in field measurements of NPP. For example, Clark et al. (2001b) concluded that most of the procedural problems in field measurements of forest NPP result in underestimating NPP. In contrast, in many grassland systems where positive increments in biomass from sequential harvests during the growing season are summed, overestimation of NPP commonly results (Singh et al. 1984; Lauenroth et al. 1986; Biondini et al. 1991). Hence, methodological problems resulting in bias will be particular to the ecosystem under study, and the general principle is that researchers must be aware of the potential for introducing bias in NPP measurements. Often the most important sources of bias in NPP measurements for terrestrial vegetation include not accounting for (1) C lost to mycorrhizae, root exudation/rhizodeposition and volatile organic carbon (VOC) production; (2) C loss to foliage and root herbivory; and (3) turnover of biomass between sampling intervals. Most field NPP measurement programs lack sufficient resources to adequately address the contribution of unmeasured NPP components, but potential biases should be identified and considered explicitly when a sampling program is designed. Finally, quality assurance/quality control (QA/QC) protocols that ensure quality control during data collection and quality assurance during data analyses can reduce errors and increase precision (chap. 2, this volume).

## Organization of This Volume

Most of the chapters in this volume (chaps. 3–10) describe methods of NPP measurement specific to particular biomes or groups of biomes with similar physiognomy of the dominant plants. This organizational framework was chosen to minimize redundancy, since the methods of measuring NPP for all plants with similar physiognomy are about the same. Each of these chapters begins by identifying which biomes and community types are covered. However, many study sites support vegetation with a mixture of physiognomic classes (e.g., savanna with trees and perennial grasses). The most appropriate procedures for such sites will consist of a combination of the techniques for each of the "pure" biomes: forests (chap. 9) and grasslands (chap. 11) in the case of savannas. Sampling designs in these situations will likely be hybrids of the designs for the "pure" systems. The only extensive biome types that are not covered in the present volume are in the coastal ocean (e.g., coral reefs, sea grass beds, kelp forests).

The content of the chapters describing methods for each biome includes consideration of the question of scale specific to the biomes because of the profound variation among biome groups in spatial patterns of NPP. Also, for each biome, unique aspects of the ecosystem that may strongly influence the design of an NPP sampling program are identified. A table of representative data is provided in each chapter, not from a thorough review of the literature for purposes of scientific synthesis, but rather as a guide for the practitioner in the design, analysis, and interpretation of the sampling program. The suite of guiding principles and specific procedures for each biome has been developed with the objective of providing the information necessary to guide the practitioner in the design of an NPP measurement program. It is important to emphasize that the optimal sampling strategy will depend upon the question that is being addressed in the research program. For example, a strategy to quantify annual variation within a study site would likely differ markedly from one to quantify spatial variation or experimental treatment effects.

The science of production ecology is continually evolving as new techniques are developed and as new questions confront the ecologist. This volume provides current state-of-the-science descriptions of methodological principles and techniques as well as separate chapters addressing related topics of data and information management (chap. 2), error analysis (chap. 12), and remote sensing approaches for scaling up from plot-level measurements to larger areas (chap. 11). The hope is that this volume will contribute to the improved standardization of NPP measurement programs across the world.

References

Biondini, M. E., W. K. Lauenroth, and O. E. Sala. 1991. Correcting estimates of net primary production: Are we overestimating plant production in rangelands? Journal of Range Management 44:194–197.

Briggs, J. M., and A. K. Knapp. 1995. Interannual variability in primary production in tallgrass prairie: Climate, soil moisture, topographic position and fire as determinants of aboveground biomass. American Journal of Botany 82:1024–1030.

Clark, D. A., S. Brown, D. W. Kicklighter, J. Q. Chambers, J. R. Thomlinson, J. Ni, and E. A. Holland. 2001a. NPP in tropical forests: An evaluation and synthesis of existing field data. Ecological Applications 11:371–384.

Clark, D. A., S. Brown, D. W. Kicklighter, J. Q. Chambers, J. R. Thomlinson, and J. Ni. 2001b. Measuring net primary production in forests: Concepts and field methods. Ecological Applications 11:356–370.

Fahey, T. J., T. G. Siccama, C. T. Driscoll, G. E. Likens, J. Campbell, C. E. Johnson, J. D. Aber, J. J. Cole, M. C. Fisk, P. M. Groffman, S. P. Hamburg, R. T. Holmes, P. A. Schwarz, and R. D. Yanai. 2005. The biogeochemistry of carbon at Hubbard Brook. Biogeochemistry 75:109–176.

Frank, D. A., and R. S. Inouye. 1994. Temporal variation in actual evapotranspiration of terrestrial ecosystems: Patterns and ecological implications. Journal of Biogeography 21:401–411.

Geider, R. J., et al. 2001. Primary productivity of planet Earth: Biological determinants and physical constraints in terrestrial and aquatic habitats. Global Change Biology 7:849–882.

Goulden, M. L., J. W. Munger, S. Fan, B. C. Daube, and S. C. Wofsy. 1996. Measurement of carbon sequestration by long-term eddy covariance: Methods and a critical evaluation of accuracy. Global Change Biology 2:169–182.

Gower, S. J., O. Krankina, R. J. Olson, M. Apps, S. Linder, and C. Wang. 2001. Net primary production and carbon allocation patterns of boreal forest ecosystems. Ecological Applications 11:1395–1411.

Huxman, T. E., et al. 2004. Convergence across biomes to a common rain-use efficiency. Nature 429:651–654.

Knapp, A. K., and M. D. Smith. 2001. Variation among biomes in temporal dynamics of aboveground primary production. Science 291:481–484.

Lauenroth, W. K., H. W. Hunt, D. M. Swift, and J. S. Singh. 1986. Estimating aboveground net primary production in grasslands: A simulation approach. Ecological Modeling 33:297–314.

Lauenroth, W. K., and O. E. Sala. 1992. Long-term forage production of a North American shortgrass steppe. Ecological Applications 2:397–403.

McIntosh, R. P. 1985. The Background of Ecology. Cambridge University Press, Cambridge.

Milner, C., and R. E. Hughes. 1968. Methods for the Measurement of the Primary Production of Grassland. Blackwell Scientific, Oxford.

Newbould, P.J. 1967. Methods for Estimating the Primary Production of Forests. Blackwell Scientific, Oxford.

Robertson, G. P., D. C. Coleman, C. S. Bledsoe, and P. Sollins (eds.). 1999. Standard Soil Methods for Long-Term Ecological Research. Oxford University Press, New York.

Rosenzweig, M. L. 1968. Net primary productivity of terrestrial communities: Prediction from climatological data. American Naturalist 102:67–74.

Sala, O. E., V. Deraegibus, T. Schlichter, and H. Alippe. 1981. Productivity dynamics of a native temperate grassland in Argentina. Journal of Range Management 34:48–51.

Sala, O. E., R. B. Jackson, H. A. Mooney, and R. W. Howarth (eds.). 2000. Methods in Ecosystem Science. Springer, New York.

Sala, O. E., W. J. Parton, L. A. Joyce, and W. K. Lauenroth. 1988. Primary production of the central grassland region of the United States. Ecology 69:40–45.

Singh, J. S., W. K. Lauenroth, and R. K. Steinhorst. 1984. Bias and random errors in estimators of net root production: A simulation approach. Ecology 65:1760–1764.

Webb, W. L, W. K. Lauenroth, S. R. Szarek, R. S. Kinerson, and S. Russell. 1983. Primary production and abiotic controls in forests, grasslands, and desert ecosystems in the United States. Ecology 64:134–151.

# 2

# Information Management Standards and Strategies for Net Primary Production Data

Kristin L. Vanderbilt
William K. Michener

N et primary production (NPP) data can provide valuable insights into a variety of ecological patterns and processes at a particular site. Moreover, when such data are collected over long time periods and integrated with similar data from other sites spread across a region or biomes, long-term and broad-scale patterns and processes can be documented, enabling us to understand how changes in climate, land use, and other drivers affect net primary production.

All data, regardless of the purpose for which they were collected, are subject to "data entropy"—the gradual loss of information content over time that can be attributed to inadequate documentation (metadata), accidents, insecure storage, media degradation, and other factors (Michener et al. 1997). Sound information management practices help retard data entropy, protect our investment in science, and foster data-sharing now and in the future. In this chapter, a set of standards and best practices is provided that forms the basis for sound information management and facilitates data integration and synthesis.

## The Process of Managing NPP Data and Information

Information management is a critical component of the scientific process, and it is performed and directed by all scientists. The process of managing data and information begins with the conceptualization of a question and the documentation of the sampling design and protocols that will be used. The data management process continues through data collection, quality assurance and quality control, analyses, the publication cycle, and, ideally, the submission of the data to an archive center.

The sampling design for any study should be logically integrated with the experimental design, and both of these designs should inform the database design. All three (experimental, sampling, and database designs) influence the types of analyses that can be performed on the data and the ease with which they can be done. Consideration of experimental design is outside the scope of this chapter, so the reader is referred to excellent treatments of this subject by Green (1979), Snedecor and Cochran (1989), Sokal and Rohlf (1995), Underwood (1997), Resetarits and Bernardo (1998), Scheiner and Gurevitch (2001), and Gotelli and Ellison (2004). We focus attention in this chapter on standards and best practices associated with data acquisition, data design and management, quality assurance and quality control, data documentation (i.e., metadata), and the process of preparing data for submission to an archive.

## *Data Acquisition*

The importance of recording data in a format that reflects the statistical and sampling designs of the study cannot be overemphasized (Briggs et al. 1998). Mechanisms for manual collection of field data include paper data sheets, palmtop computers, and tape recorders. Each method has advantages and disadvantages. Regardless of which method is chosen, advance planning about how data will be captured and input into the computer can streamline the data management process and reduce errors.

A well-designed paper data sheet offers straightforward data entry as well as easy review and storage. Data sheets should include blanks for all data that need to be collected because visual prompts help ensure that some measurements are not omitted by mistake. Definitions of codes to be used by the data recorder should be included on the data sheet for easy reference, as should blanks for the initials of data collection and data entry personnel. Data sheets and data entry forms should be similarly structured (see fig. 2.1) to facilitate data entry. An advantage to using paper data forms is that they can be stored for decades and are easily referred to after the data have been entered. Primary disadvantages include the space required for paper storage and the potential for media degradation over time, although acid-free paper can be stored for centuries under proper conditions, far outliving most magnetic and optical media (Larsen 1999).

Using palmtop computers or personal digital assistants (PDAs) reduces errors made during data transcription by capturing data in digital form at the point of collection. Data entry software is commercially available for both systems. Palmtop computers are most useful in situations where two field technicians are working together and are not constantly moving around. For instance, one of the technicians can enter data while the other measures plant dimensions in an NPP quadrat. Data collected in the field on handheld units are downloaded onto a desktop computer as quickly as possible so that they can be backed up. A drawback to using palmtops is that it is not possible to see all data collected at once; one must scroll the screen up and down to look at data records. Battery life, unit durability in dusty and wet conditions, and screen glare in bright conditions are important considerations when selecting a computer or PDA.

**SEVILLETA LTER NPP FIELD QUAD DATA**

Recorder Name: _____          Reader Name: _____

| DATE(m/d/y) | SITE | WEB | PLOT | QD | SPECIES | OBS | COVER | HEIGHT | COUNT | COMMENTS |
|---|---|---|---|---|---|---|---|---|---|---|
| 2/3/2004 | B | 1 | N | 1 | SPCR | 1 | 5 | 10 | 3 | |
| | | | | | SPCR | 2 | 12 | 18 | 4 | grazed |
| | | | | | BOER4 | 1 | 11 | 15 | 1 | |
| | | | | 2 | SPCR | 1 | 3 | 2 | 3 | |
| | | | | | : | : | : | : | : | : |
| | | | | | | | | | | |
| | | | | | | | | | | |
| | | | | | | | | | | |
| 2/3/2004 | B | 2 | S | 1 | HOGL2 | 1 | 1 | 4 | 1 | seedling |

SITE: C,G,J,P,B          PLOT: N,S,E,W
WEB: 1,2,3,4,5          QD: 1,2,3,4, (clock-wise from northwest=1)

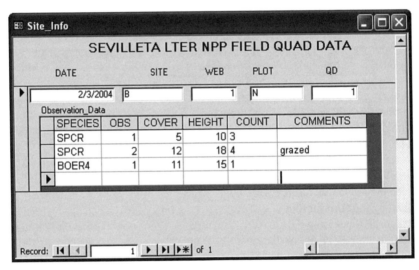

Figure 2.1.  The data sheet (upper form) contains blanks for every piece of information to be collected. The data entry form (lower form) has been designed to reduce redundant data entry; date, site, web, plot, and quad have to be entered only once for the many species on each quadrat.

Tape recorders are useful for fast and easy data collection that results in less field time but, typically, greater data loss. Tape recorder batteries die, pause buttons get stuck, tapes break, and voice data recorded under windy conditions can be inaudible. Tape-to-computer entry is done using a Dictaphone. Data entry and summarization should be done as quickly as possible after data collection, so that any garbled or missing data can be re-collected. Tapes are not a viable long-term data storage medium, and their shelf life is limited. A drawback of collecting data with tape recorders is that double-checking data after they have been entered into the computer is an onerous task.

   Automated environmental data collection can yield synchronous multiple observations at a high frequency over long periods of time. A typical automatic data collection scenario is a logger recording data from many meteorological or soil sensors at a single site, with the data being routinely manually downloaded or radi-oed to a computer for backup. To ensure data accuracy, scientists must be vigilant that instrumentation is functioning properly at all times. Sensors should be regularly serviced and calibrated, and periodic battery testing should be done, because instrument sensitivity may decrease as battery power declines. Results of maintenance checks should be entered into the data documentation in case any anomalies in the data can be related to the instrument. Programs that automatically graph or filter the data for out-of-range values allow quick detection of sensor malfunctions.

Data collected using handheld instruments such as spectral radiometers also should be documented with instrument servicing and calibration information. Instrument settings peculiar to the period of data collection should be recorded. Newly collected data should be downloaded as quickly as possible from the instrument and stored in a location where they can be backed up.

## Data Design and Management

The design of data (either simple tables or more complex databases) entails conceptualizing and implementing a logical structure within and among the data elements that can facilitate data acquisition, entry, storage, retrieval, and manipulation. For ease of data entry, table design should reflect the sampling strategy (and collection forms) used when acquiring the data. A table usually consists of a matrix where each row represents a complete record and the columns represent the parameters that make up the record. Cells in the table should contain information about a single parameter only. For example, a season variable could be created for which values might be Spring 2004, Fall 2004, and Winter 2004. The preferred way to store this information would be as two variables, season and year, which could be queried independently of one another. A single table should ideally contain a set of similar measurements taken for one study, using the same methods and instruments. One large data file is easier to manage than several smaller files defined by, for instance, month or site.

NPP and associated data may be managed in simple ASCII text files, spreadsheets, or relational databases. Many scientists prefer to manage their data in spreadsheets, which offer some limited analytical capabilities. This practice is not recommended, however, because spreadsheets do not preserve internal record consistency, meaning that each column can be manipulated independently of the other columns (Brunt 2000). Furthermore, the proprietary nature of spreadsheets does not lend itself to long-term data storage. Relational databases such as Microsoft Access and MySQL offer greater flexibility with respect to managing and manipulating data. Relational database software has many useful functions, such as sorting, querying, and indexing the data, and also preserves record consistency. Storing data in related tables also reduces the number of times that information is entered, which can greatly improve data integrity, as illustrated in figure 2.2. For instance, if site information is entered only once in one table and is related to 100 records in another table containing observations from that site, then all 100 records have identical site information. Repeated entry of site information for each record in a spread-

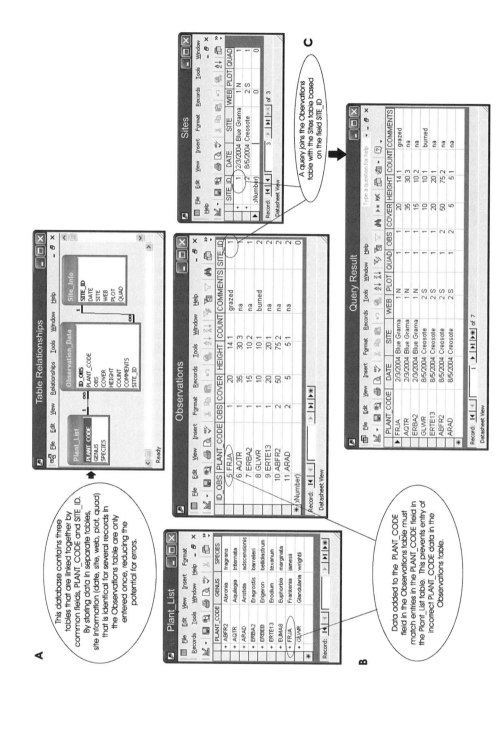

**A** This database contains three tables that are linked together by common fields, PLANT_CODE and SITE_ID. By storing data in separate tables, site information (date, site, web, plot, quad) that is identical for several records in the Observations table are only entered once, reducing the potential for errors.

**B** Data added to the PLANT_CODE field in the Observations table must match entries in the PLANT_CODE field in the Plant_List table. This prevents entry of incorrect PLANT_CODE data in the Observations table.

**C** A query joins the Observations table with the Sites table based on the field SITE_ID.

sheet could result in different spellings of the same location and other typos. Relational databases readily interface with analytical programs and WWW tools, making it easy to distribute data in a searchable format and download them as a text file or as a spreadsheet. ASCII flat files are a neutral and stable storage format, ensuring that data do not become unreadable as proprietary formats become obsolete, but flat files offer the fewest options for easy data manipulation.

When choosing how to manage data, the needs of the user community should be considered (Porter 2000). Users are likely to want to combine NPP data with other kinds of data (e.g., meteorology, biodiversity), and a relational database that facilitates aggregation at different spatial scales over varying temporal domains may be most useful. The types of questions the database needs to be able to address also are important to consider, so that the structure of the database can be designed to deal efficiently with common queries, or so that frequently requested representations of the data can be provided. An NPP research project, for example, might make all seasonally collected weight-per-species data available in one table, but also offer a table of annual NPP summed across all species in order to address most users' needs.

## Quality Assurance and Quality Control

Quality control (QC) and quality assurance (QA) are essential components of information management. QC refers to measures taken during data acquisition or data transcription to prevent data contamination, while QA refers to graphical and statistical methods that are applied in order to identify anomalous or missing values (Brunt 2000). Components of a typical QA/QC program for field data are described below.

Data entry programs that utilize validation rules, range checks, and lookup tables are standard quality control tools. Such programs, which limit input to specific values or ranges of values, can be custom written or created with many commercially available software packages. A lookup table, for instance, could contain codes for all species that may be encountered in a study, and the data entry program would constrain legal species code entries to codes in this table. A range check would report an error upon entry of a data point outside the maximum or minimum value allowable set by the researcher.

---

Figure 2.2. (A) This NPP relational database contains 3 tables. Each record in the Sites and Plant_List tables can be related to multiple records in the Observations table. (B) Data for each site and date combination have to be entered only once in the Sites table and can be related to multiple records in the Observations table. This reduces redundancy in data entry. The Plant_List table is used to constrain the possible entries into the Plant_Code field in the Observations table. Data are entered into the Observations and Sites tables, using an interface such as illustrated in figure 2.1, which conceals the database structure from the user. (C) A query is used to combine data from the Observations and Sites tables into a single table that can be exported to another application, such as a spreadsheet.

Once entered into a computer, data files can be further checked for quality by running scripts that identify illegal or missing data. Illegal data are combinations of values that cannot exist. A script written with statistical software such as SAS or R could be applied to long-term tree dimension data, for instance, to ensure that tree 94, identified as *Juniperus monosperma* in 2004, has not been misidentified as *Pinus edulis* in 2005. For studies with multiple sampling units, scripts that sum the number of plots per treatment are useful for assessing whether data have been entered for the correct number of plots or whether a plot may have been skipped. Both transcription errors and plots that were missed in the field can be identified in this way.

An excellent method for checking data is for two data entry technicians to enter the same data set and then compare the data. Few ecologists, however, have the luxury of funds for two data entry personnel. Another option is to use text-to-speech software, which allows the researcher to check data sheets while the computer reads the data aloud.

Data collected automatically via sensors located in the field need to be scrutinized for out-of-range values that may signal a malfunctioning sensor and questionable data. Some data, however, may fall within normal range limits and still be invalid. Graphing sensor data against time will help detect peculiar "blips" in the data and may reveal that the sensor readings are slowly drifting up or down. Statistical quality control charts (Dux 1986; Mullins 1994), which are commonly used to evaluate laboratory instrument function, are also a mechanism for determining if the variability observed is merely the variability inherent in any process or if it is due to a problem with the sensor.

In spite of the best QC efforts, contaminating data points may still be entered into the data set. Quality assurance methods may be applied to identify outliers (observations that have unusual values, given the rest of the data set). Outliers may result from measurement errors or data entry mistakes, or they may be valid extreme values. Graphical tools, such as box-and-whisker plots, stem-and-leaf diagrams, and normal probability plots, are used for detecting outliers in samples.

Most statistical analysis programs will readily produce summary statistics and graphics for a univariate data set. A glance at such an output (fig. 2.3) reveals if there are points that have abnormally small or large values with respect to the rest of the data. An observation that appears to be an outlier when modeled as a normal probability distribution may not be an outlier when modeled as a log-normal distribution, and exploratory visualization of the data helps determine the need for data transformations. Once a data set can be characterized as being normally distributed, Grubbs's test (Grubbs 1969; Edwards 2000) or Dixon's test (Snedecor and Cochran 1989; Barnett and Lewis 1994) can be applied to determine if a point is a statistical outlier. Details about the interpretation of box-and-whisker plots, normal probability plots, and histograms can be found in Cleveland (1993) and Edwards (2000).

Many NPP studies rely on regression analysis to relate a measure of plant cover or volume to biomass, and researchers should create scatter plots of their data to identify possible invalid observations with unusual values that may unduly influence regression coefficients. Plots of residuals against the explanatory variable and fitted values are also recommended for detecting outliers and ensuring that the

```
A.   SAS CODE:
/*input data*/
      data npp_data;
           infile 'nppgraphs.unx' DLM = ',';
           input volume weight;   run;
/*make normal probability plot, stem and leaf diagram, box and whisker plot*/
      proc univariate data = npp_data plot normal; run;
```

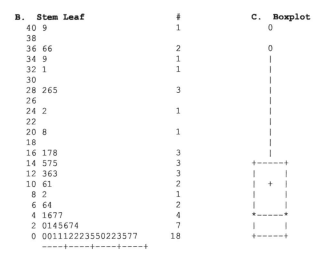

```
B.   Stem Leaf                        #            C.   Boxplot
     40 9                             1                 0
     38
     36 66                            2                 0
     34 9                             1                 |
     32 1                             1                 |
     30                                                 |
     28 265                           3                 |
     26                                                 |
     24 2                             1                 |
     22                                                 |
     20 8                             1                 |
     18                                                 |
     16 178                           3                 |
     14 575                           3            +-----+
     12 363                           3            |     |
     10 61                            2            |  +  |
      8 2                             1            |     |
      6 64                            2            |     |
      4 1677                          4            *-----*
      2 0145674                       7            |     |
      0 001112223550223577           18            +-----+
        ----+----+----+----+
```

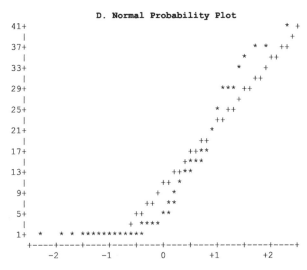

D. Normal Probability Plot

Figure 2.3. Selected output from the SAS Proc Univariate statement (A), which is used for exploratory data analysis. The stem-and-leaf diagram (B) and normal probability plot (D) indicate a right-skewed distribution. The box-and-whisker plot (C) indicates that there are two outliers.

assumptions of regression analysis are met (Chatterjee and Price 1977). Outliers in a regression context are points with Y values that do not fit the general X-Y trend. Other potentially influential observations, called leverage points, have an unusual X value, given the rest of the data. Both types of points can significantly impact the results of the analysis (fig. 2.4) by unduly influencing the slope of the regression line. Statistical packages will readily calculate *hat values* ($h_i$) or Cook's distance that measure leverage, and *studentized residuals* that identify outliers (Fox 1991; Neter et al. 1996; Draper and Smith 1998). Outliers and leverage points should be carefully examined to see if they represent contamination, in which case they should be removed from the analysis and documented accordingly.

## Metadata

Metadata, defined as "information about data," consist of information that describes the content, context, quality, structure, and accessibility of a specific data set (Michener et al. 1997). Metadata comprise all the essential information that allows the original investigator to reuse the data, and also enable use by scientists not directly involved in the data collection. A useful goal for metadata is for the data to be fully comprehensible to a user unfamiliar with the data or how they were obtained 20 years after the data were archived, based solely on the documentation provided (NRC 1991). Incomplete or inadequate metadata present a significant technical barrier to data integration efforts (Hale et al. 2003).

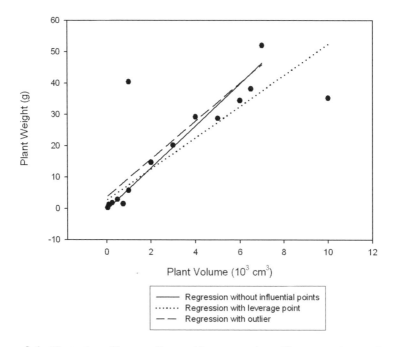

Figure 2.4.  Illustration of how outliers and leverage points affect regression results.

Metadata vary according to purpose, although several community standards have emerged in different scientific domains. Field notes represent a simple form of metadata that enable one to characterize observations made in the field, often including comments on the weather and other pertinent facts. They may be useful for refreshing one's memory about certain observations, but may be relatively meaningless to others (e.g., incomplete, much too abbreviated, illegible writing, or unintelligible codes). Much more comprehensive metadata may be necessary if one is providing a recipe for a colleague that would enable him or her to replicate an observation or measurement.

Numerous metadata standards have been developed to guide the documentation of data so that they can be discovered and reused. For example, the Dublin Core (Weibel 1997; http://dublincore.org/), Global Change Master Directory (http://gcmd.nasa.gov), and others focus on providing a limited number of descriptors that primarily support data discovery. The Federal Geographic Data Committee (FGDC 1998) and the International Standards Organization (http://iso.org) have developed more comprehensive metadata standards that are appropriate for many types of biological data, particularly those that have a large geospatial component. The Darwin Core metadata standard is used to document specimens in museums and herbaria (Bisby 2000).

Ecological Metadata Language (EML) was developed as a comprehensive metadata standard that is particularly applicable for a broad range of ecological and biodiversity data. It is organized into a suite of modules (table 2.1) that are based on a comprehensive list of metadata descriptors that were proposed for the ecological science community (Michener et al. 1997). EML is implemented in XML (eXtensible Markup Language), which defines the structure of the text file that contains the metadata. Unlike metadata that is written in an unstructured format (i.e., free-flowing textual descriptions), EML is machine-parsable and is therefore useful for a variety of research applications, from simple data discovery to advanced data processing. Importantly, EML contains a domain module for documenting the spatial, temporal, and taxonomic extents of the data set. Scientists are especially encouraged to document species names, including the relevant taxonomic authority and reference, so that search engines can identify taxon-specific data sets. Software tools have been developed to support EML-compliant metadata entry and management (e.g., Morpho and Metacat). More information on EML and metadata tools can be found at http://www.ecoinformatics.org, a community-based resource for ecoinformatics information and software.

## Data Archives

The goal of an ecological archive is to foster broader ecological objectives, such as regional and multidisciplinary analyses, through data-sharing. Data archives are permanent collections of data and metadata stored in such a way that they are easy to discover, acquire, understand, and use (Olson and McCord 2000). In addition to data and metadata, key components of a data archive include software (databases and user interfaces), storage media, a backup system, network connections (for security, data management, and accessibility), and staff (such as system adminis-

Table 2.1. Ecological metadata language (EML): Modules that describe characteristics of a data set

| Module | Descriptors | Purpose |
|---|---|---|
| EML data set | Data set title, abstract, keywords, maintenance history | Information that is used in a data catalog to alert secondary users of the theme and of spatial, temporal, and taxonomic domains of the data set and its accessibility |
| EML access | Access permissions to the data set (who can read, modify, and download the data set) | |
| EML party | Contact information for people and organizations associated with the data | |
| EML coverage | Spatial, temporal, and taxonomic extents of the data set | |
| EML project | Research context in which the data set was created, including a description of comprehensive research program the data set emerged from and its funding, personnel and description of the study area | Additional detail allowing a secondary data user to understand the overall goals of the research project, experimental design, sampling procedures, site selection, and data processing |
| EML methods | Methods followed in the creation of the data set, including description of field, laboratory, and processing steps, sampling methods and units, and quality control procedures | |
| EML physical | External (e.g. file name, file size) and internal (e.g. format (ASCII, RDBMS)) physical characteristics of the data set and distribution information | Structural attributes of the data set allowing a secondary user to understand what each data point represents and the format in which it is available. |
| EML attribute | Names of variables in the data set, variable definitions, range of values, missing value codes, and precision | |
| EML data table | Characteristics of each tabular unit in a data set (e.g. table name, number of records) | |

trators, information managers, and scientists) to provide user services (Scurlock et al. 2002).

Data contributed to a data archive are preserved so that they are available to and usable by researchers in the future; thus the longevity of the archive must be protected. Many data sets that were stored electronically in the early 1970s are no longer accessible because the storage media have become obsolete. Archives must have a plan and resources available to move data to new storage media as the old media degrade. Current technology has a life cycle of about 10 yr (Scurlock et al. 2002).

Formal data archives often house data related to a particular theme. NASA's distributed active archive centers (DAACs), for instance, store multidisciplinary data of interest to global change researchers and policymakers. Oak Ridge National Lab operates a DAAC that contains biogeochemical data, including a global NPP data set (Scurlock and Olson 2002). NASA's Global Change Master

Directory contains metadata that point to archives containing a wide range of data sets on themes including the atmosphere, biosphere, hydrosphere, oceans, land surface, and human dimensions. Many other organizations store and distribute environmental data (table 2.2). Archives are usually accessed through a Web-based search engine for browsing and viewing the data before ordering it via an electronic interface. Data may be made available in a variety ways, including FTP and CD.

Many research programs establish their own archives in order to share their data, and these project-based archives represent different stages of development toward the functionality of a formal archive such as the Oak Ridge DAAC. Long-Term Ecological Research (LTER) sites have Web sites from which data can be discovered and downloaded, and all are contributing metadata as EML to a centralized network-level database that is accessible via the WWW. Many LTER sites contributed to a centralized NPP database, which is housed in the Network Information System (NIS) (http://www.lternet.edu/data/). The NIS consists of a series of modules, each containing either research data or support data, such as site or personnel information (Baker et al. 2000).

Cook et al. (2000) describe several practices that researchers should follow as they prepare to archive and share their data as part of a formal or a project-based archive. Following these guidelines will yield data that is better organized, more easily understood, easier to locate, and most likely to persist for long periods.

1. Assign descriptive file names. File names should be unique and reflect the file contents. File names such as "Mydata" or "2005_data" lack key identifiers such as a project acronym, study location, and abbreviated study title. A better file name is "Sevilleta_LTER_2005_NPP.asc," where "Sevilleta_LTER" is the project name, "2005" is the calendar year, "NPP" represents Net Primary Productivity study, and "asc" stands for the file type (ASCII).

2. Use consistent and stable file formats. ASCII file formats, or other generic formats, should be used rather than proprietary formats that often become obsolete over time. Data tables should be consistently formatted, ensuring that the number and order of columns are identical throughout. Ideally, the file name, data set title, author, date, and companion file names should be included at the beginning of the file. Column headers should describe the content of each column, including both parameter names and parameter units. Within an ASCII file, fields may be delimited by commas, tabs, pipes (|), spaces, or semicolons, preferably in that order.

3. Define the parameters. Use commonly accepted parameter names that describe what the parameter is, and denote the parameter name consistently throughout the data file. Format a parameter consistently throughout the data file (e.g., significant digits should be the same for all records for a given parameter).

4. Assign descriptive data set titles. Ideally, data set titles should describe the type of data, time period, location, and instruments used. Titles should be restricted to 80 characters, and should be similar to names of data files.

Table 2.2. Ecological data archives

| Data Center | Web URL | Supporting Agency* |
|---|---|---|
| Atmospheric Radiation Measurement (ARM) Archive | http://www.archive.arm.gov | DOE |
| Carbon Dioxide Information Analysis Center (CDIAC) | http://cdiac.esd.ornl.gov | DOE |
| Databases at the Swedish Museum of Natural History | http://www.nrm.se/databas.html.en | Swedish Museum of Natural History |
| GEO Data Portal | http://gridca.grid.unep.ch/geoportal/index2.php | UNEP |
| Center for Ecology and Hydrology | http://www.ceh.ac.uk/ | NERC |
| National Climatic Data Center (NCDC) | http://www.ncdc.noaa.gov/wdcamet.html | NOAA |
| Forest Inventory and Analysis (FIA) | http://fia.fs.fed.us/tools-data/data/ | USFS |
| Global Population Dynamics Database | http://cpbnts1.bio.ic.ac.uk/gpdd/ | Center for Population Biology (CPB) and NSF |
| ISRIC Soil Information System | http://www.isric.nl | ISRIC |
| Long Term Ecological Research Network (LTER) | http://lternet.edu/ | NSF |
| EROS Data Center | http://edc.usgs.gov/ | USGS |
| National Environmental Satellite, Data, and Information Service (NESDIS) | http://www.nesdis.noaa.gov/satellites.html | NOAA |
| National Geophysical Data Center (NGDC) | http://www.ngdc.noaa.gov/ | NOAA |
| National Ocean Data Center (NODC) | http://www.nodc.noaa.gov/ | NOAA |
| Soil Data Mart | http://soildatamart.nrcs.usda.gov/ | NRCS |
| National Water Information System (NWIS) | http://waterdata.usgs.gov/usa/nwis/ | USGS |
| Oak Ridge National Laboratory Distributed Active Archive Center (ORNL DAAC) for biogeochemical dynamics | http://www.eosdis.ornl.gov/ | NASA |
| World Conservation Monitoring Centre | http://www.unep-wcmc.org/ | IUCN/ WWF/ UNEP |
| World Lakes Database | http://www.ilec.or.jp/database/database.html | ILEC, UNEP, Japanese Environment Agency |
| World Ozone and Ultraviolet Radiation Data Center (WOUDC) | http://www.woudc.org/ | Environment Canada |

*DOE: U.S. Department of Energy; ILEC: International Lake Environment Committee Foundation; ISRIC: International Soil Reference and Information Center; IUCN: International Union for the Conservation of Nature/ World Conservation Union; NASA: U.S. National Aeronautics and Space Administration; NOAA: U.S. National Oceanic and Atmospheric Administration, NERC: U.K. Natural Environment Research Council; NSF: U.S. National Science Foundation; NRCS: U.S. Natural Resources Conservation Service; UNEP: United Nations Environment Program; USFS: U.S. Forest Service; USGS: U.S. Geological Survey; WWF: Worldwide Fund for Nature/ World Wildlife Fund.

The title "Net Primary Production in Grasslands at the Sevilleta LTER, New Mexico, 1999–2005" is, for example, preferable to the title "NPP."
5. Provide comprehensive documentation (metadata). Record what data were collected, why the data were collected, when and where they were collected, and how they were collected. Describe the physical structure of the data file, and note any changes that have been made to the data.

## Conclusion

The discussion in this chapter offers suggestions for managing NPP data to achieve the goals of sharing and reusing high-quality, well-documented data; minimizing data entropy; and accumulating equity from research investments as others use and cite archived NPP data. Ecological data, once treated as a consumable, are now regarded as a valuable resource. As the focus of ecological research has shifted from isolated plot- or population-level studies to larger-scale, more interdisciplinary studies, scientists must more frequently synthesize data collected by others. NPP data are useful in many fields of research, and making high-quality, well-documented NPP data discoverable, accessible, and usable by other researchers should be an important component of NPP research programs.

References

Baker, K. S., B. J. Benson, D. L. Henshaw, D. Blodgett, J. H. Porter, and S. G. Stafford. 2000. Evolution of a multisite network information system: The LTER information management paradigm. BioScience 50:963–978.
Barnett, V., and T. Lewis. 1994. Outliers in Statistical Data. Wiley, New York.
Bisby, F. A. 2000. The quiet revolution: Biodiversity informatics and the Internet. Science 289:2309–2312.
Briggs, J. M., B. J. Benson, M. Hartman, and R. Ingersoll. 1998. Data entry. Pages 29–31 in W. K. Michener, J. H. Porter, and S. G. Stafford (eds.), Data and Information Management in the Ecological Sciences: A Resource Guide. LTER Network Office, University of New Mexico, Albuquerque, NM.
Brunt, J. W. 2000. Data management principles, implementation and administration. Pages 25–47 in W. K. Michener and J. W. Brunt (eds.), Ecological Data: Design, Management, and Processing. Blackwell Science, Oxford.
Chatterjee, S., and B. Price. 1977. Regression Analysis by Example. Wiley, New York.
Cleveland, W. S. 1993. Visualizing Data. Hobart Press, Summit, NJ.
Cook, R. B., R. J. Olson, P. Kanciruk, and L. A. Hook. 2000. Best practices for preparing ecological data sets to share and archive. Bulletin of the Ecological Society of America 82:138–141.
Draper, N. R., and H. Smith. 1998. Applied Regression Analysis. Wiley, New York.
Dux, J. P. 1986. Handbook of Quality Assurance for the Analytical Chemistry Laboratory. Van Nostrand Reinhold, New York.
Edwards, D. 2000. Data quality assurance. Pages 25–47 in W. K. Michener and J. W. Brunt (eds.), Ecological Data: Design, Management, and Processing. Blackwell Science, Oxford.
Federal Geographic Data Committee. FGDC-STD-001-1998. Content standard for digital

geospatial metadata (revised June 1998). Federal Geographic Data Committee. Washington, D.C.

Fox, J. 1991. Regression Diagnostics: An Introduction. Sage University Papers on Quantitative Applications in the Social Sciences, 07–079. Sage, Newbury Park, CA.

Gotelli, N. J., and A. M. Ellison. 2004. A Primer of Ecological Statistics. Sinauer Associates, Sunderland, MA.

Green, R. H. 1979. Sampling Design and Statistical Methods for Environmental Biologists. Wiley, New York.

Grubbs, F. 1969. Procedures for detecting outlying observations in samples. Technometrics 11:1–21.

Hale, S. S., A. H. Miglarese, M. P. Bradley, T. J. Belton, L. D. Cooper, M. T. Frame, C. A. Friel, L. M. Harwell, R. E. King, W. K. Michener, D. T. Nicolson, and B. G. Peterjohn. 2003. Managing troubled data: Coastal data partnerships smooth data integration. Environmental Monitoring and Assessment 81:133–148.

Larsen, P. S. 1999. Books and bytes: Preserving documents for posterity. Journal of the American Society for Information Science 50:1020–1027.

Michener, W. K., J. W. Brunt, J. Helly, T. B. Kirchner, and S. G. Stafford. 1997. Nongeospatial metadata for the ecological sciences. Ecological Applications 7:330–342.

Mullins, E. 1994. Introduction to control charts in the analytical laboratory: Tutorial review. Analyst 119:369–375.

Neter, J., M. Kutner, C. Nachtsheim, and W. Wasserman. 1996. Applied Linear Statistical Models. Irwin, Chicago.

NRC [National Research Council]. 1991. Solving the Global Change Puzzle: A U.S. Strategy for Managing Data and Information. National Academy Press, Washington, DC.

Olson, R. J., and K. A. McCord. 2000. Archiving ecological data and information. Pages 117–141 in W. K. Michener and J. W. Brunt (eds.), Ecological Data: Design, Management, and Processing. Blackwell Science, Oxford.

Porter, J. H. 2000. Scientific databases. Pages 48–69 in W. K. Michener and J. W. Brunt (eds.), Ecological Data: Design, Management, and Processing. Blackwell Science, Oxford.

Resetarits, W. J., Jr., and J. Bernardo (eds.). 1998. Experimental Ecology. Oxford University Press, New York.

Scheiner, S. M., and J. Gurevitch (eds.). 2001. Design and Analysis of Ecological Experiments. Oxford University Press, New York.

Scurlock, J. M. O., and R. J. Olson. 2002. Terrestrial net primary productivity—A brief history and a new worldwide database. Environmental Reviews 10:91–109.

Scurlock, J. M. O., R. J. Olson, R. A. McCord, and W. K. Michener. 2002. Data banks: Archiving ecological data and information. Pages 248–259 in E. T. Mun (ed.), Encyclopedia of Global Environmental Change. Wiley, New York.

Snedecor, George W., and William G. Cochran. 1989. Statistical Methods. Iowa State University Press, Ames, IA.

Sokal, R. R., and F. J. Rohlf. 1995. Biometry. W. H. Freeman, New York.

Underwood, A. J. 1997. Experiments in Ecology: Their Logical Design and Interpretation Using Analysis of Variance. Cambridge University Press, Cambridge.

Weibel, S. 1997. The Dublin Core: A Simple Content Description Model for Electronic Resources. Bulletin of the American Society for Information Sciences. October/November, 9–11.

# 3

# Estimating Aboveground Net Primary Production in Grassland- and Herbaceous-Dominated Ecosystems

Alan K. Knapp

John M. Briggs

Daniel L. Childers

Osvaldo E. Sala

E stimating aboveground net primary production (NPP) in grasslands, particularly those in which woody plants are rare, can be accomplished with relatively straightforward procedures compared with those required in other biome types (for shrublands, see chapter 4; for forests, see chapter 5; for aquatic systems, see chapters 9–10). However, the apparent structural simplicity of grasslands, and other systems dominated by herbaceous plants, belies a host of unique and complicating factors that can introduce significant errors into aboveground NPP estimates (figure 3.1). For example, most grasslands were historically home to large migratory ungulates (Stebbins 1981; Axelrod 1985) that today have been replaced by domesticated large grazers. Accounting for the intermittent and cumulative amounts of foliage consumed by herbivores (both large and small) and potential compensatory regrowth responses by plants (McNaughton 1983; Coughenour 1985) represents a significant challenge to accurate aboveground NPP estimates in these systems.

Grasslands also have been, and still are, subject to fire at a much greater frequency than most other biomes (Whelan 1995). Fire can occur in the dormant or the growing season and, unlike grazing, rapidly and completely removes biomass. Fire can either simplify or complicate estimates of aboveground NPP, and is recognized as an important determinant of aboveground NPP in many grasslands (Knapp and Seastedt 1986). Complex interactions between the behaviors of grazers and fire (Coppedge and Shaw 1998; Knapp et al. 1999) further complicate estimates.

Typically, grasslands occupy the interior of large landmasses, and the continental climates associated with these locations accentuate extremes in temperature and water availability. Droughts in particular can alter the expected phenology of plants

Figure 3.1. Structurally, grasslands are relatively simple compared to many other biomes (top left), but inherent features of grasslands such as fire (top right) and grazing (bottom left) can influence the methods chosen for quantifying aboveground net primary production. Furthermore, many grasslands have a significant woody plant component (bottom right), which is rapidly increasing around the world (Briggs et al. 2005). Thus, methods may need to be selected to accommodate this structural component. Photo credits: (top left and right): A. K. Knapp; (bottom left): M. D. Smith (Yale University); (bottom right): J. M. Briggs.

and lead to premature tissue senescence and decomposition, thus affecting aboveground NPP sampling strategies. Fire and grazing also strongly interact with drought in ways that must be recognized in order to estimate aboveground NPP accurately.

Many herbaceous wetlands are essentially grasslands, and many of the aboveground NPP issues listed above are also challenges in these systems (chapter 7, this volume). Herbivory can be an important process in herbaceous wetlands (Kreeland and Young 1997; Evers et al. 1998; Gough and Grace 1998), particularly those of high nutrient status. Fire is also a controlling factor in many large grassland wetlands (Schmalzer and Hinkle 1992; Ford and Grace 1998; Rheinhardt and Faser 2001). Interannual variation in hydrologic drivers, including hydroperiod, salinity, and depth of inundation, may directly affect aboveground NPP rates. These controls may also be manifest indirectly, such as through changes in plant phenology (Hooper and Vitousek 1998; Din et al. 2002; Brewer 2003). In oligotrophic herbaceous wetlands, such as the Florida Everglades, dramatic and rapid shifts in plant community composition in response to subtle changes in nutrient status

(Childers et al. 2003) also can complicate aboveground NPP measurements in both space and time.

Despite these challenges, consistent and accurate estimates of aboveground NPP in these different systems can be made if a few key principles and guidelines are incorporated into sampling procedures. Because grasslands are among the ecosystems most responsive to climate variability (Knapp and Smith 2001; Knapp et al. 2002), accurate estimates of aboveground NPP can be key for detecting global changes in energy flow through ecosystems.

The purpose of this chapter is to briefly review past and currently accepted methods of estimating aboveground NPP in grass and herb-dominated ecosystems, provide some guiding principles and recommendations to facilitate accurate determinations of aboveground NPP, and discuss biases and errors and sampling adequacy. For these types of ecosystems, aboveground NPP is operationally defined as all aboveground plant biomass produced during a specified interval (typically the growing season, but usually expressed on an annual basis), accounting for losses due to herbivory and decomposition when appropriate. We will focus on annual and perennial grasslands, but these principles should apply to most herbaceous-dominated systems such as old fields, tundra, and many agroecosystems. In savanna, woodland, or wetland communities with significant woody plant cover (fig. 3.1), combining methods for the herbaceous strata with those recommended for shrubs or trees (chapters 4 and 5, this volume) should permit aboveground NPP to be estimated in proportion to the growth forms present.

## Key Determinants and Representative Values of Grassland Aboveground NPP

Grassland climates vary widely, and, depending on the fire and grazing regime, any of several resources (water, temperature, nutrients, light) may limit aboveground NPP (Borchert 1950; Collins and Wallace 1990; Knapp et al. 1998; Ni 2004). In most arid and semi-arid grasslands, soil moisture is primarily limiting to aboveground NPP, but even in more mesic grasslands, water can limit production in most years (Lauenroth and Sala 1992; Knapp et al. 2001). At a continental scale, annual precipitation is strongly correlated with aboveground NPP (Sala et al. 1988), whereas in many temperate and tropical grasslands, nutrients or light also can be limiting (Knapp and Medina 1999). Hydrologic conditions often exert strong control on aboveground NPP in both freshwater and estuarine herbaceous wetlands. In some cases, all of these factors can co-limit aboveground NPP simultaneously or in sequence throughout the growing season (Knapp et al. 1998). Because of the high interannual variability in climate inherent to grasslands (Borchert 1950), different controls can be the primary limiting factors in different years. As a result of variability in climate, fire frequency, and grazing pressures, estimates of aboveground NPP for grassland ecosystems can vary over an order of magnitude, ranging from <100 g m$^{-2}$ yr$^{-1}$ in desert grasslands and <200 g m$^{-2}$ yr$^{-1}$ in oligotrophic freshwater wetlands to >1500 g m$^{-2}$ yr$^{-1}$ in tropical grasslands, with an incredible estimate of >9000 g m$^{-2}$ yr$^{-1}$ in a perennially wet tropical grassland (Long et al. 1989, table 3.1). Moreover, grasslands exhibit more extreme temporal variability in aboveground NPP than other biomes (Knapp and Smith 2001), and spatial variability can also be substantial (Briggs and Knapp 1995; Lauenroth et al. 1999), despite topographic gradients that are often more subtle than in other ecosystems. Because of this wide range in aboveground NPP across grasslands and the fundamental differences among determinants of this variation, sampling strategies and methods must be customized for each type of grassland.

As in most ecosystems, there are far fewer reliable estimates of belowground NPP (BNPP) in grasslands (Milchunas and Lauenroth 2001). Although there are exceptional grasslands where aboveground biomass and productivity account for 95% of the total (Long et al. 1989), it is generally accepted that a significant fraction of productivity in most grasslands occurs belowground (Sims and Singh 1978; Rice et al. 1998). Indeed, the high organic matter content of most grassland soils reflects this allocation pattern. Scurlock et al. (2002) estimated that BNPP accounted for between 40% and 90% of total NPP in grasslands globally, with BNPP greater than aboveground NPP in most grassland types. Moreover, responses and dynamics in BNPP do not necessarily mirror those of aboveground NPP; thus, aboveground NPP:BNPP ratios may not be constant (Milchunas and Lauenroth 2001; Ni 2004). Given the magnitude of BNPP in grasslands, and the direct (allocation strategies, resource uptake, etc.) and indirect (soil properties, microbial processes, etc.) effects

Table 3.1. Estimates of aboveground net primary production (NPP) for grassland sites globally

| Grassland Type | Mean (g m$^{-2}$ yr$^{-1}$) | Range | References |
|---|---|---|---|
| Hot desert | 94, 148, 184, 229 | 16–292 | Webb et al. 1983 Knapp and Smith 2001 |
| Cold desert steppe | 109, 188 | 69–330 | Scurlock et al. 2002 Webb et al. 1983 |
| Temperate steppe | 94, 116, 189, 388 | 18–986 | Scurlock et al. 2002 Webb et al. 1983 Knapp and Smith 2001 Lauenroth and Sala 1982 |
| Temperate mesic | 277, 354, 443, 508 | 197–1072 | Scurlock et al. 2002 Webb et al. 1983 Knapp and Smith 2001 |
| Subtropical savanna | 316, 518, 553 | 80–1121 | Scurlock et al. 2002 Knapp and Medina 1999 |
| Tropical wet | 732, 3223 | 331–9425 | Scurlock et al. 2002 Long et al. 1989 |
| Herbaceous wetlands | | 900–5500 | Mitsch and Gosselink 2000 |
| Herbaceous oligotrophic wetlands | 900, 2900, 300 | 150–2900 | Davis 1989 Daoust and Childers 1998 Childers et al. 2003 |

*Notes:* These values were derived from a wide variety of studies employing different methods, some of which likely underestimate and others of which overestimate aboveground NPP. Data for means and ranges are based on both temporal (long-term data from one or a few sites) and spatial sampling (combining sites on different continents), and should be used only as guides for the expected magnitude of aboveground NPP encountered in different grassland types.

of BNPP on aboveground NPP, measuring BNPP is critically important for understanding ecological interactions in these ecosystems. A review of techniques and recommendations for estimating BNPP in grasslands can be found in chapter 8 of this volume.

## Guiding Principles and Recommendations for Grasslands

### Review of Methods

Despite the relative ease with which most grasslands can be sampled for aboveground production, temporal and spatial variability in aboveground NPP is significant, and field sampling and lab processing time can be substantial. As a result, numerous methods have been proposed, many having the goal of reducing sampling effort while optimizing the information gained from the time and resources expended to estimate aboveground NPP (Wiegert 1962; Briggs and Knapp 1991; Brummer et al. 1994). These can be divided into two general approaches to estimating aboveground NPP in grasslands: direct harvest methods and indirect or "nondestructive" techniques. Although numerous variations have been proposed for each of these approaches, we will review only a few here.

Harvest methods require the direct removal of aboveground plant biomass from plots of a specified size (usually <1 m$^2$), separation of this biomass into components (live vs. dead, by growth form, by species), drying to a constant mass, and weighing. Variations in this method involve plot size, shape, and number; sampling frequency; pairing plots in grazed systems; and the mode of biomass harvest (Wiegert 1962; Van Dyne et al. 1963; Kelly et al. 1974; Singh et al. 1975; Dickerman et al. 1986; Brummer et al. 1994). Harvests are typically accomplished with handheld scissors, but because harvesting of biomass in this way can be quite time-consuming, alternative means ranging from the use of handheld electric clippers to large mowers to vacuum devices have been proposed to speed the process (Van Dyne et al. 1963; Milner and Hughes 1968). Error due to spatial variability is inherent in harvest methods, and this problem is typically addressed by harvesting many plots at a time. The major labor cost is thus in the lab—in sorting and drying many samples. There are few alternatives that can substantially reduce this labor cost. A number of double sampling protocols and indirect techniques (see below) have been developed in which easily measured or estimated parameters (plant height, cover, etc.) are correlated with harvest data (Catchpole and Wheeler 1992; Daoust and Childers 1998; Vermeire and Gillen 2001). We provide more details on the harvest method below as the recommended technique for estimating aboveground NPP in most grasslands.

A second general approach, using indirect or nondestructive techniques, includes numerous variations for estimating aboveground NPP in grasslands. Among them are the use of electrical capacitance and beta attenuation devices for estimating leaf area and canopy volume as correlates of aboveground NPP (Mitchell 1972; Knapp et al. 1985; Sala and Austin 2000), point intercept methods (Catchpole and Wheeler 1992), disk pasture meters (Trollope and Potgieter 1986; Dorgeloh 2002), visual obstruction methods (Vermeire and Gillen 2001), the measurement of canopy optical properties with handheld devices or via remote sensing (Tucker 1980; Turner et al. 1992, 2005), and simulation modeling (Roxburgh et al. 2004). Indirect techniques, in which easily measured attributes are correlated with biomass, can be a preferred alternative to harvest methods in grasslands where the number of plots that must be harvested is prohibitively large due to continuous growth of the vegetation (e.g., tropical systems), where a high sampling frequency is required to account for rapid turnover of biomass, or where large harvests are difficult (e.g., in national parks). In addition, many of these indirect methods can be useful for coarse estimates of standing crop biomass and fuel loads, and it is important to emphasize that some indirect methods, such as remotely sensed "greenness indices" or NDVI, can be quite valuable for estimates across large spatial scales. This is particularly true at the regional and global scales, where standing crop is often coarsely correlated with aboveground NPP (Prince and Goward 1995). However, the uncertainty and error inherent in predicting the rate of a process such as aboveground NPP from a pool size (standing crop) are often too high for site-based ecological studies in grasslands (Turner et al. 1992, 2005). Recent advances in combining remote sensing output with process-based models may improve predictions of aboveground NPP with satellite/airborne sensors (chapter 11, this volume).

### Recommended Methods

### General

Approaches based on the harvest method are recommended for estimating aboveground NPP in most grasslands, with a series of modifications depending on the accuracy required, the important drivers of aboveground NPP, and the inherent attributes of a given grassland type. However, the harvest method may not be best for tropical grassland systems or any herbaceous system in which destructive harvesting is a problem. For these situations, allometric techniques that are regularly validated with harvests (e.g., Daoust and Childers 1998; chapter 7, this volume) are recommended. General principles of these methods are outlined below for different types of grasslands. These are presented from the simplest to the most complex situations.

   Key to the success of most harvest methods is the ability to accurately recognize and partition aboveground biomass into three pools: green (living) biomass ($b_g$); senesced material produced during the current year, often referred to as current year's dead ($b_{sc}$) or standing dead because it usually, though not always, is elevated above the surface litter and is typically not in contact with the soil surface (however, this material need not be attached to the living plant); and dead biomass from previous years ($b_l$) in the form of litter on the soil surface or as standing dead material. The latter two pools may be readily distinguished on the basis of their color/appearance in many ecosystems, but may be more difficult to separate in others (Singh et al. 1975). In tropical grasslands or herbaceous wetlands, for instance, there is often little remaining of the previous year's dead material. Despite this difficulty, it is critical for investigators to be able to distinguish $b_{sc}$ from $b_l$ because accurate aboveground NPP estimates depend on quantifying these pools. In contrast, the simplest harvest methods are those that require measuring just green or living biomass, which is the most easily distinguished aboveground component (Singh et al. 1975; Ni 2004). Those methods that require measuring only living biomass are not appropriate for most grasslands, however. This is because there are virtually no natural grasslands where plant growth phenologies and patterns of senescence are so uniform and temporally distinct that senescent biomass can be ignored without introducing significant errors (Singh et al. 1975; Sala and Austin 2000). Intensively managed artificial grasslands, such as wheat fields and other agroecosystems, would be the exception to this rule. Thus, the harvest methods recommended all include some level of accounting for plant senescence (i.e., mortality, as per Wiegert and Evans [1964], or turnover).

## Peak Standing Biomass Harvest

As the name implies, this method bases estimates of aboveground NPP on aboveground biomass harvested once, usually near the end of the growing season, at or just after the time of peak biomass. This method is recommended for grasslands that meet the following criteria: (1) there is little carryover of living biomass from previous years due to a distinct dormant season or fire during the dormant season, or the previous year's biomass can be easily recognized and separated from the current year's biomass (living and dead); (2) the growing season is sufficiently short or plant material is of such low quality that decomposition of biomass produced during the growing season can be ignored; (3) consumption of plants by herbivores is minimal (i.e., large grazers are absent and small vertebrates and invertebrates can be ignored). If these criteria are met, or if the errors associated with relaxing them are acceptable, then green and current year's standing dead biomass at the time of harvest can be summed to estimate aboveground NPP. Hence,

$$\text{Aboveground NPP} = b_g + b_{sc.}$$

This method has been used extensively at the Konza Prairie long-term ecological research site in the central United States, where $C_4$ grasses dominate productivity, and fire in the dormant season is frequent (Briggs and Knapp 1995). In this grassland, there are early-season $C_3$ forb species that are not entirely accounted for during a single end-of-season biomass harvest, but early-season sampling of their productivity indicate that they comprise <5% of total aboveground NPP (Briggs and Knapp, unpubl. data). Thus, the effort required to include this component was deemed excessive relative to the increase in accuracy gained.

## Sequential Biomass Harvests

This more labor-intensive method requires that aboveground biomass be harvested at two or more times during the growing season, typically coinciding with peaks in productivity of species with distinct phenologies. For example, if a distinct $C_3$ cool-season flora dominates aboveground NPP in the spring and a $C_4$ warm-season flora dominates in the summer, then the positive differences in green biomass are summed.

An initial harvest may also be made at the beginning of the growing season to estimate beginning biomass if some carryover of living biomass occurs. This method is recommended for grasslands that meet the following criteria: (1) biomass produced in previous years is present but cannot be distinguished from current-year production later in the season, and thus an initial standing biomass value may be needed to correct for this carryover biomass; (2) the grassland is composed of species with substantially different seasonal patterns of growth (cool- and warm-season species, $C_3$ vs. $C_4$ plants, etc.), and each contributes significantly to aboveground NPP; (3) consumption of plants by herbivores is minimal (i.e., large grazers are absent and small vertebrates and invertebrates can be ignored). In this method, green pools are measured at each sampling period, and these are summed over the season to estimate aboveground NPP. Hence,

$$\text{Aboveground NPP} = (b_{g2} - b_{g1}) + (b_{g3} - b_{g2}),$$

where 1, 2 . . . refer to sampling periods. ANPP is the sum of the positive values of these differences when $b_{gn} - b_{gn-1} > 0$.

In some cases, investigators may want to sample numerous times during the growing season (Singh et al. 1975), but care must be taken with such frequent sampling because both under- and overestimates of aboveground NPP can result (see below and Sala and Austin 2000 for excellent analyses of these potential errors).

## Aboveground Biomass Harvest(s): Accounting for Decomposition

If production and subsequent disappearance (decomposition) of plant material are likely to be substantial during the sampling interval, then an estimate of the dynamics of all three biomass pools, as well as decomposition losses, must be made and used to account for the turnover of biomass. Wiegert and Evans (1964) proposed a method to account for decomposition in old fields dominated by grasses by harvesting biomass at frequent intervals and sorting it into live ($b_g$) and dead ($b_{sc}$ and $b_l$) components. They also measured decomposition either with litterbags or through the disappearance of dead biomass in paired plots where all living biomass had been removed (plots were paired with those in which all biomass was harvested). By accounting for changes in live biomass between intervals and the mortality of live material (defined by changes in the standing crop of dead material, adjusted for decomposition), they calculated aboveground NPP by summing positive growth increments. In practice, pairing plots with identical characteristics (required for this technique) is almost impossible, and other approaches have been favored (Singh et al. 1975). Lomnicki et al. (1968) suggested a simplifying modification to the Wiegert-Evans method. They proposed accounting for the mortality (senescence) of live plant material by removing all dead material at the beginning of the sampling interval and then, for each sampling interval, summing living and dead material (which could have been produced only during the current growing season) in harvested plots. Although they argued that for their grassland, removing the previous year's litter had no effect on the current year's production, the key role that litter (detritus) can play in affecting microclimate, nutrient cycling, water relations, and, ultimately, aboveground NPP has been well documented in other grasslands (Knapp and Seastedt 1986). Thus, this method is not recommended.

Long et al. (1989) reviewed the extensive studies of Singh et al. (1975) and many others, and concluded that in grasslands with long growing seasons (i.e., some temperate and most tropical grasslands) and where there are long time intervals between harvests, so that losses due to decomposition will occur, the best method for estimating aboveground NPP (with the fewest assumptions) involves simultaneous measurements of changes in all plant biomass components and decomposition. This approach, which includes the basic elements of the Wiegert and Evans (1964) method, has substantial merit. Changes in mass of both living and dead pools are measured at intervals appropriate for the grassland under study, and losses due to decomposition are added to the net change in biomass. Estimating and correcting for losses due to decomposition can be time-consuming and introduce additional uncertainty (see Harmon et al. 1999 for a review of methods). However, in some wet tropical grasslands, decomposition during the growing season can be so rapid and of such a large magnitude that failure to account for this process can lead to significant underestimates in aboveground NPP (Long et al. 1989). Hence,

$$ANPP = \sum_{i=1}^{n} \Delta b_g + d,$$

where $d$ = loss due to decomposition during the sampling interval and aboveground NPP is summed over $i$ sampling intervals. The decision of whether or not to include nonstatistically significant changes in any component between samples should be based on the length of the sampling interval (and hence the amount of change expected) and the sample size (relatively small changes require very large sample sizes if spatial heterogeneity is high [Scurlock et al. 2002]).

## Aboveground Biomass Harvest(s): Accounting for Grazing

Herbivory (mostly by large mammals in terrestrial systems and mostly by small mammals or insects in wetlands) is a widespread determinant of grassland aboveground NPP. Unfortunately, this key biotic factor complicates estimates of aboveground NPP more than any other factor discussed thus far. This is because the act of grazing (consumption) can be a continuous or intermittent process, and is almost always spatially heterogeneous. Of course, regrowth responses of the plants mirror the activities of the grazers, and both consumption and regrowth must be accounted for in aboveground NPP estimates (McNaughton et al. 1996). It is well established that plants have numerous compensatory responses to herbivory (McNaughton 1983), and simply measuring aboveground NPP in permanent or season-long grazing exclosures will not capture alterations in productivity manifest under grazed conditions. Thus, estimating aboveground NPP in grazed grasslands requires a substantial number of temporary, movable exclosures that allow for estimates of consumption by herbivores and regrowth responses of grazed plants.

This method is recommended with either of two variations. In grasslands actively grazed by large herbivores, a large number of temporary exclosures are randomly placed for a relatively short period of time (1–2 weeks). At the end of this period, all living and current year's senesced biomass within the exclosures is harvested. If a similar harvest is made at the same time outside the exclosures, an estimate of consumption can be made; these estimates can be summed for all intervals and added to residual biomass at the end of the season to estimate aboveground NPP. Hence,

$$\text{Aboveground NPP} = \Sigma\, C + (b_g + b_{sc})_{final},$$

where C = consumption and "final" refers to standing crop biomass at the end of the year (or growth season). Decomposition of current-year growth is assumed to be unimportant if grazing intensity is high or the growth season is relatively short (Frank and McNaughton 1993).

A variation of this method involves harvesting plots outside the exclosures at the beginning of the interval rather than at the end, when exclosed biomass is harvested. This approach measures regrowth within the exclosures during the interval rather than consumption. The sum of these estimates comprises aboveground NPP for the season. Exclosures should be placed prior to growth at the beginning of the growing season (such that the initial measures of $b_g$ and $b_{sc}$ outside of exclosures is zero), and continue to be harvested and moved at regular intervals until the end of the season or year. Hence,

$$\text{Aboveground NPP} = \Sigma\, [(b_g + b_{sc})_{ex} - (b_g + b_{sc})_{outside}],$$

where exclosure (ex) biomass is harvested at the end of intervals and outside biomass is harvested at the beginning of the interval.

There are also unique situations that may require combinations of methods or additional alterations (Cox and Waithaka 1989). For example, in lightly and patchily grazed grasslands, ungrazed areas may require methods different from those used in grazed areas. Indeed, it is advisable to estimate grazing pressure prior to selecting a sampling method and estimating the number of plots needed. Heavily and uniformly grazed grasslands will require fewer exclosures and plots than patchily grazed systems (McNaughton et al. 1996) but, as noted earlier, each grassland will require a unique sampling scheme. Table 3.2 summarizes some of the characteristics of grasslands that are important to consider when choosing among the four primary methods for estimating aboveground NPP reviewed above.

## Additional Methodological Issues

This chapter does not provide a detailed methodological discourse on field harvesting, processing biomass from plots, the selection of the optimal plot size and shape, or the best temporal and spatial sampling strategies for estimating aboveground NPP in grasslands. This is because there are other works that provide such detail (Van Dyne et al. 1963; Milner and Hughes 1968; Dickerman et al. 1986; Brummer et al. 1994; Wiegert 1962) and because grasslands vary in so many different ways (e.g., dominance by rhizomatous vs. caespitose grasses, annual vs. perennial, desert vs. tropical wet grasslands vs. herbaceous wetlands) that specific techniques must be customized for each grassland type. Instead, the focus is on a few methodological

Table 3.2. General guide for determining the most appropriate method of estimating aboveground NPP in grassland ecosystems based on key attributes of the site

| Grassland Criteria | Peak Biomass | Sequential Harvest | Sequential Harvest + Decomposition | Temporary, Movable Exclosures | Nondestructive Phenometric |
|---|---|---|---|---|---|
| 1. Relatively short, distinct growing season with no carryover of live biomass | ✓ | | | | |
| 2. Decomposition not important during sampling interval | ✓ | ✓ | | | |
| 3. Grazing not important | ✓ | ✓ | ✓ | | |
| 4. Current year senesced biomass $(b_{sc})^*$ can be distinguished from previous year's litter $(b_l)^{**}$ | ✓ | | | | |
| 5. Long growing season with plants having distinct phonologies, >1 peak in biomass, or carryover of live biomass from the previous year | | ✓ | ✓ | | ✓ |
| 6. Long growing season and $b_{sc}$ cannot be distinguished from $b_l$ by season's end | | ✓ | ✓ | | |
| 7. Decomposition of current year's production must be accounted for during the sampling interval | | | ✓ | | |
| 8. Continuous growth, low species diversity, perennial forms | | | | | ✓ |
| 9. Consumption of current year's production by grazers is substantial | | | | ✓ | |

*Note*: If grazing is not important in a given grassland site, any of three methods may be selected, depending on the other criteria listed. If decomposition must also be accounted for in this site, then only one method is appropriate.
*$b_{sc}$ = biomass produced during the current year that has senesced
**$b_l$ = litter or dead biomass produced in years prior to the current year

details that, if overlooked, are likely to lead to systematic errors in aboveground NPP estimates. In addition, a case study is provided as a guide for how to determine an appropriate sampling effort for ecological studies that include estimates of aboveground NPP. Although the results of this case study are specific to one grassland, it serves as a relatively detailed model for quantifying and reducing sampling error in aboveground NPP estimates.

As noted earlier, correctly partitioning harvested biomass into green ($b_g$), senescent ($b_{sc}$), and previous year's dead ($b_l$) components can be critical for accurate aboveground NPP estimates. Operationally, defining green biomass as any plant material (foliage or stems) that contains visible chlorophyll is recommended, even if the majority of the tissue is senesced and brown. There is no loss of accuracy with including leaves that are 95% brown and 5% green in the $b_g$ category because $b_{sc}$ and $b_g$ are subsequently combined to estimate aboveground NPP. Further, the presence of green tissue usually ensures that the biomass was produced that year. An exception to this rule must be made for grasses that form aerial tillers in subtropical and tropical climates. In this case, care must be taken to separate these tillers from the senescent stalk that was produced the previous year. As noted earlier, $b_{sc}$ and $b_l$ can usually be distinguished by color, with $b_l$ darker brown or gray, depending on the state of decay. Sorting biomass into these categories, at least coarsely, while harvesting in the field is also recommended. It is often much easier to distinguish $b_{sc}$ from $b_l$ in the field than in the lab. A secondary check of the accuracy of this rough field sorting should be performed in the lab.

Greig-Smith (1983) and others (Sala and Austin 2000) have emphasized the need for sampling plot (and hence quadrat) sizes to be larger than the average plant size, which is not difficult in grasslands. However, when harvesting biomass, strict rules must be followed in defining the edge of the sampled plot and for determining if plant material is to be included or excluded from harvest. At the edge of quadrats, it is recommended that the basal portions of plants, not canopy position, be used to determine if material is to be harvested or excluded. Thus, when locating the quadrat, care should be taken to ensure that it rests on the soil surface as much as possible, and that plants are harvested from their base. Thus, canopy foliage that occurs within the vertical projection of the quadrat, but whose basal contact with the soil is outside the quadrat, would not be included. Conversely, the foliage of plants whose bases are within the quadrat would be harvested even if this foliage extends outside the vertical projection of the plot. Quadrat edges that fall across the bases of large caespitose grasses will require that only those portions of the individual bunch within the plot be harvested. In other biomes or when sampling other growth forms, alternative rules for determining portions of plants to harvest may be employed (see chapter 4, this volume). Clearly, the use of consistent methodology by all field personnel is crucial to avoid unnecessarily high variances in aboveground NPP estimates in any biome.

Long-term studies of aboveground NPP present additional challenges due to the potential cumulative effect of sampling (biomass removal) and investigator trampling on the plant community. When a site is considered for long-term sampling, sampled plots need to be marked with flags or metal tags to ensure that no further sampling takes place in that year and for at least two additional years. Thus, the size of the overall sampling area must be sufficient to accommodate harvests more than once during the season and for several years without resampling the same plot.

## Errors in Estimating Aboveground Net Primary Production

Estimates of aboveground NPP (ANPP) have several sources of error that can be classified in two types: errors leading to underestimation (ELUs) of ANPP and errors leading to overestimation (ELOs) of ANPP (Sala et al. 1988). These two types of errors are different in nature and will be discussed separately.

### Errors Leading to Underestimation

Errors leading to underestimation of aboveground NPP result from two sources: missing peaks of biomass and the simultaneous nature of production, senescence, and decomposition. The first source of error can be reduced by a high frequency of sampling that reduces the possibility of missing peaks of biomass, and the consequent underestimation of annual aboveground NPP. The second source of error is associated with the simultaneous nature of production and senescence, and is also conceptually very clear. Solutions to address this issue have been to simultaneously estimate biomass of different species and different categories, such as green, current years and previous years dead biomass. The combination of high frequency and sampling several species and types of biomass should result in a reduction of ELUs.

### Errors Leading to Overestimation

Errors leading to overestimation were first recognized by Singh et al. (1984) and Lauenroth et al. (1986). ELOs result from the fact that random errors in estimates of biomass inevitably accumulate but may not compensate for each other, leading to a positive bias in the estimates of ANPP. Sala et al. (1988) analytically determined that aboveground NPP, defined as the increase in positive biomass increments during periods of time, is a biased estimator of the true net primary production. When biomass at time t+1 is larger than biomass at $t_0$, we consider that $B_{t+1} - B_{t_0}$ is the estimate of NPP. When $B_{t+1}$ is less than $B_{t_0}$, we consider that production and growth have been zero. Biomass estimates are random variables, but this error accumulates in the estimate of NPP. For example, assume that the true value $B_{t+1} = B_{t_0}$ and consequently the true NPP = 0. However, because B is a random variable, in some instances $B_{t+1}$ will be higher than $B_{t_0}$, and in other instances $B_{t+1}$ will be lower than $B_{t_0}$. The NPP bias occurs as a result of the fact that most sampling protocols require that all the negative values ($B_{t+1} < B_{t_0}$) be discarded, but that all positive values ($B_{t+1} > B_{t_0}$) be included.

Singh et al. (1984) and Lauenroth et al. (1986) performed modeling experiments in which they simulated a true value of production and calculated the magnitude of the ELOs by randomly sampling from a true distribution. These experiments indicated that the magnitude of the ELOs could be quite large. For belowground productivity, estimates of NPP were 5 times higher than true NPP, and for aboveground NPP the estimated value was 33% higher than the true value.

Sala et al. (1988) analytically derived the distribution function of the estimator of NPP, which is nonnormal, and concluded that the estimator is a biased estimate of productivity. From the distribution function of NPP, it was possible to derive the equation to calculate the magnitude of the overestimation error (OE):

$$OE = (\sigma / \sqrt{2\pi})\, e^{-\frac{1}{2}(\mu/\sigma)^2} - q\mu$$

This equation yields interesting conceptual results. The magnitude of the overestimation error is a function of (1) the magnitude of the standard deviation ($\sigma$) of NPP that is directly related to the magnitude of the error in estimating biomass ($B_{t+1}$ and $B_{t_0}$), and (2) the true value of NPP ($\mu$). The magnitude of the overestimation error increases as the error increases and as the true value of NPP decreases. Consequently, methods that try to reduce ELUs by increasing the sampling frequency will necessarily increase the ELOs because as the sampling frequency increases, the true value of productivity decreases. The magnitude of the true increase in biomass decreases as the sampling dates get close to each other. Similarly, methods that choose to estimate biomass by species result in larger ELOs because the true value of $B_{t+1}$ – $B_{t_0}$ is lower for subunits such as a species than for functional groups or total biomass. ELOs are lower when aboveground NPP is estimated from differences in total biomass than when it is estimated from differences in biomass per species that then are added.

The paradox is that those techniques devised to reduce one kind of error (ELUs) inevitably result in the increase of the other kind of error (ELOs). The same is true of efforts aimed at reducing ELOs that result in high ELUs. Biondini et al. (1991) developed an algorithm that can be used to estimate the magnitude of the ELOs based on the observed mean of aboveground NPP and the observed standard deviation. This algorithm allows for correction of the OE and for focusing on reducing the ELUs.

This analysis of ELOs and ELUs shows that methods that are more complicated and conceptually more complete may not necessarily yield results closer to the true value of aboveground NPP than simpler methods. In many cases, elaborate and expensive methods based on high frequency of sampling and estimates per species yield results farther from the true aboveground NPP than those yielded by simpler methods.

## Determining Sample Adequacy: A Case Study

Below, a case study is presented that determined the appropriate level of sampling needed to reliably estimate aboveground NPP in tallgrass prairie. This analysis focused on determining the effect of varying sample sizes on aboveground NPP estimates and the impact of sample size on the statistical determination of the response of this ecosystem to fire. The case study, a summary of Briggs and Knapp (1991), used a combination of jackknifing and Monte Carlo simulations based on sampling of aboveground NPP over a 14-yr period. As mentioned earlier, Wiegert (1962) discussed the trade-off between many small and few large quadrats when calculating variance for biomass estimates. This analysis did not include variables such as plot size, shape, and area, and sampling protocol, but instead focused on a single quadrat size ($0.1m^2$) used extensively in estimating aboveground NPP in tallgrass prairie (Hulbert 1969; Barnes et al. 1983; Abrams et al. 1986; Steuter 1987; Briggs et al. 1989; Briggs and Knapp 1995; Briggs and Knapp 2001). This size quadrat fits the criteria of Greig-Smith (1983) regarding the relationship between quadrat dimension and average plant size. A three-sided metal frame (a rectangular quadrat with one open end) was used that allowed the frame to be easily inserted into dense vegetation.

Research was conducted at the Konza Prairie Biological Station, in a $C_4$-dominated grassland with a typical Midwestern continental climate characterized by warm, wet summers and dry, cold winters (Knapp et al. 1998). Fire is critical to the maintenance and functioning of tallgrass prairie (Collins and Wallace 1990), and since 1981 (and in some areas on Konza since 1971) entire watersheds have been subjected to late spring (April $10 \pm 20$ days) fires at intervals of 1, 2, 4, 10, and 20 yrs.

Although estimates of aboveground productivity are made in numerous watersheds on Konza Prairie, analyses for this study were limited to a unique data set from two intensively studied watersheds. One watershed (hereafter referred to as the high-fire-frequency site) had been burned annually for > 20 years, and the adjacent watershed (hereafter referred to as the low-fire-frequency site) had been burned only once in 20 years (1994). Beginning in 1984 and continuing until 1997, 20 $0.1$-$m^2$ plots were harvested at about 2-wk intervals from May to September in these two adjacent watersheds. Soil type and topographic position were similar at both sites of biomass harvest. Detailed methodology for sampling the aboveground components are given in Abrams et al. (1986), Briggs and Knapp (1995), and Knapp et al. (1998). Briefly, all harvested plant material was first separated into live (at least partially green), current-year senescent, and previous-year(s) dead biomass (on unburned sites). Live plant material was further separated into graminoids (dominated by $C_4$ plants) and forbs (primarily $C_3$ plants, including a minor woody plant component (less < 5%). All material was oven-dried at $60°$ C for 48 hr, and weighed to the nearest $0.1$g, with values expressed as $g/m^2$. Although permanent sites were used in this study, the plots sampled biweekly were marked so that resampling of specific locations could be avoided for several years.

Sample adequacy was determined in two ways. First, a running mean and the standard error of the mean were plotted for sequentially sampled quadrats. When the standard error of the mean was reduced to <10% of the mean, it was concluded that the sample size was sufficient (National Academy of Sciences 1962). In addition, jackknifing techniques were used to randomly select data with samples sizes of 2 to 18, and again a standard error of <10% was used as the criterion for sample adequacy. This analysis was replicated 20 times, with a maximum and a minimum standard error (SE) for each sample size determined.

To examine the effect of sample size on the statistical detection of treatment effects (fire, in this case), data sets were used from this jackknifing analysis, and t-statistics were computed for each sample size (total of 720 t-tests). Since all variables (total aboveground, grass and forb biomass) were significantly different at a sample size of 20, the sample size was deemed adequate if all 20 randomly generated comparisons were also significantly different at $P = 0.05$. Finally, since the effect of fire on biomass in tallgrass prairie can vary by over 60% (Knapp et al. 1998), the analysis was extended using Monte Carlo simulations. Simulations were begun with mean biomass values from a long-term record for both upland and lowland soil types under high fire frequency and low fire frequency. These mean values were adjusted to obtain a 10%, 20%, 30%, and 40% increase in biomass in burned relative to unburned sites. Variances were also adjusted on the basis of long-term data sets. Similarly, when the sample size was decreased to determine its effect on statistical comparisons, the sample variance was also increased. This was accomplished with data from the jackknifing exercise. Maximum variances were used to ensure that sample adequacy estimates were as conservative as possible.

When the running mean was calculated for sequential quadrats, the standard error of the mean decreased to <10% of the mean at 10 quadrats for the high-fire- frequency site and 18 for the low-fire-frequency site. Based on the jackknifing analysis, 14 and 16 quadrats were deemed adequate (maximum SE <10% of the mean) for estimating aboveground NPP in the high- and low-fire-frequency sites, respectively. The random ordering of the plots and the large number of repetitions generated via this jackknifing procedure provided greater confidence in this recommended sample size.

Results of analyses on the effect of sample size on the statistical detection of treatment effects were specific to the particular components of aboveground NPP that were being estimated. For example, if only the grass component was of interest, a sample size of 14 was deemed adequate. But if differences in total aboveground NPP were of interest, a sample size of 18 was necessary, with 30 quadrats required for reliable statistical comparisons of the forb component. Overall, results indicated that to detect a treatment effect on aboveground NPP with a magnitude of 20%, a sample size of 20 quadrats (0.1 m²) was required (fig. 3.2). Since a sample size of 20 is near the inflection point of the relationship in figure 3.2, it may represent the optimal sample size (for 0.1 m² quadrats) for assessing fire effects in this tallgrass prairie.

Based on these analyses, it is recommended that to estimate aboveground NPP with an SE of the mean of <10% of the mean, 14 and 16 quadrats should be harvested from burned and unburned sites in tallgrass prairie, respectively. If the goal of aboveground NPP estimates is to detect treatment effects due to fire, at least 20 quadrats (0.1 m²) would have to be harvested per site. Finally, these results suggest that a treatment effect of <20% would be very difficult to detect using aboveground NPP as the response variable in this grassland. Moreover, because forbs are patchily distributed and are a relatively small component of aboveground NPP, only large changes in this growth form component could be detected with a sampling effort deemed adequate for total aboveground NPP.

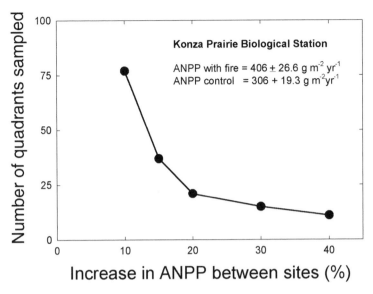

Figure 3.2. The relationship of the magnitude of change in aboveground net primary production (ANPP) between burned and unburned sites in a mesic grassland (Konza Prairie,
Kansas) and the number of quadrats that would need to be sampled to statistically detect
this change. This analysis was based on long-term field estimates of ANPP in response to
fire and on Monte Carlo simulations in which sample size and variances were adjusted (see
Briggs and Knapp 1991 for more details).

## Final Comments

The goal of this chapter was to provide guiding principles for estimating aboveground NPP in grasslands, not detailed methods. Even cursory consideration of the
varied attributes of grasslands and the number of potential determinants of productivity should convince the reader of the difficulty in providing detailed recommendations that would be useful for more than a few types of grasslands. Alternatively,
brief coverage of those factors (herbivory, decomposition, fire, phenology, etc.) are
included that, if not considered, are likely to lead to substantial errors in aboveground
NPP estimates. Then, by providing guiding principles for coping with potential errors
in aboveground NPP estimates, some unique to grasslands and some more general,
investigators can make informed decisions when selecting the best method to adopt
for their system. A summary of the primary methods recommended, and a general
guide for selecting among them, are presented in table 3.2. For each of these methods, corresponding grassland attributes are indicated.

*Acknowledgments*   Thanks to Linda Wallace (University of Oklahoma) and Priscilla Baker
(Colorado State University) for reviewing and providing helpful comments on an earlier
version of this chapter.

References

Abrams, M. D., A. K. Knapp, and L. C. Hulbert. 1986. A ten-year record of aboveground biomass in a Kansas tallgrass prairie: Effect of fire and topographic position. American Journal of Botany 73:1509–1515.

Axelrod, D. I. 1985. Rise of the grassland biome, central North America. The Botanical Review 51:163–201.

Barnes, P. W., L. L. Tieszen, and D. J. Ode. 1983. Distribution, production, and diversity of $C_3$ and $C_4$-dominated communities in a mixed prairie. Canadian Journal of Botany 61:741–751.

Biondini, M. E., W. K. Lauenroth, and O. E. Sala. 1991. Correcting estimates of net primary production: Are we overestimating plant production in rangelands? Journal of Range Management 44:194–198.

Borchert, J. R. 1950. The climate of the central North American grassland. Annals of the Association of American Geographers 40:1–39.

Brewer, S. J. 2003. Why don't carnivorous pitcher plants compete with non-carnivorous plants for nutrients? Ecology 84(2):451–462.

Briggs, J. M., J. T. Fahnestock, L. E. Fischer, and A. K. Knapp. 1994. Aboveground biomass production in tallgrass prairie: Effect of time since fire. Pages 165–170 in R. G. Wickett, P. D. Lewis, A. Woodliffe, and P. Pratt (eds.), Proceedings of the 13th North American Prairie Conference, Windsor, Ontario: Dept. of Parks and Recreation.

Briggs, J. M., and A. K. Knapp. 1991. Estimating aboveground biomass in tallgrass prairie with the harvest method: Determining proper sample size using jackknifing and Monte Carlo simulations. Southwestern Naturalist 36:1–6.

Briggs, J. M., and A. K. Knapp. 1995. Interannual variability in primary production in tallgrass prairie: Climate, soil moisture, topographic position and fire as determinants of aboveground biomass. American Journal of Botany 82:1024–1030.

Briggs, J. M., and A. Knapp. 2001. Determinants of $C_3$ forb growth and production in a $C_4$-dominated grassland. Plant Ecology 152:93–100.

Briggs, J. M., A. K. Knapp, J. M. Blair, J. L. Heisler, G. A. Hoch, M. S. Lett, and J. K. McCarron. 2005. An ecosystem in transition: Causes and consequences of the conversion of mesic grassland to shrubland. BioScience 55:243–254.

Briggs, J. M., T. R. Seastedt, and D. J. Gibson. 1989. Comparative analysis of temporal and spatial variability in aboveground production in a deciduous forest and prairie. Holarctic Ecology 12: 130–136.

Brummer, J. E., J. T. Nichols, R. K. Engel, and K. M. Eskridge. 1994. Efficiency of different quadrat sizes and shapes for sampling standing crop. Journal of Range Management 47:84–89.

Catchpole, W. R., and C. J. Wheeler. 1992. Estimating plant biomass: A review of techniques. Australian Journal of Ecology 17:121–131.

Childers, D. L., R. F. Doren, R. Jones, G. B. Noe, M. Rugge, and L. J. Scinto. 2003. Decadal change in vegetation and soil phosphorus patterns across the Everglades landscape. Journal of Environmental Quality 32:344–362.

Collins, S. L., and L. L. Wallace (eds.). 1990. Fire in North American Tallgrass Prairie. University of Oklahoma Press, Norman, OK.

Coppedge, B. R., and J. H. Shaw. 1998. Bison grazing patterns on seasonally burned tallgrass prairie. Journal of Range Management 51:258–264.

Coughenour, M. B. 1985. Graminoid responses to grazing by large herbivores: Adaptations, exaptations, and interacting processes. Annals of the Missouri Botanical Garden 72:852–863.

Cox, G. W., and J. M. Waithaka. 1989. Estimating aboveground net production and grazing harvest by wildlife on tropical grassland range. Oikos 54:60–66.

Daoust, R., and D. L. Childers. 1998. Quantifying aboveground biomass and estimating productivity in nine Everglades wetland macrophytes using a non-destructive allometric approach. Aquatic Botany 62:115–133.

Davis, S. M. 1989. Sawgrass and cattail production in relation to nutrient supply in the Everglades. Pages 325–341 in R. R. Sharitz and J. W. Gibbons (eds.), Freshwater Wetlands and Wildlife. CONF-8603101. USDOE, Office of Scientific and Technical Information, Oak Ridge, TN.

Dickerman, J. A., A. J. Stewart, and R. G. Wetzel. 1986. Estimates of net annual aboveground production: Sensitivity to sampling frequency. Ecology 67:650–659.

Din, N., R. J. Priso, M. Kenne, D. E. Ngollo, and F. Blasco. 2002. Early growth stages and natural regeneration of *Avicennia germinanas* (L.) Stearn in the Wouri estuarine mangroves (Douala-Cameroon). Wetlands Ecology and Management 10(6): 461–472.

Dorgeloh, W. G. 2002. Calibrating a disc pasture meter to estimate above-ground standing biomass in mixed bushveld, South Africa. African Journal of Ecology 40:100–102.

Evers, D. E., C. E. Sasser, J. G. Gosselink, D. A. Fuller, and J. M. Visser. 1998. The impact of vertebrate herbivores on wetland vegetation in Atchafalaya Bay, Louisiana. Estuaries 21:1–13.

Ford, M. A., and J. B. Grace. 1998. The interactive effects of fire and herbivory on a coastal marsh in Louisiana. Wetlands 18:1–8.

Frank, D. A., and S. J. McNaughton. 1993. Evidence for the promotion of aboveground grassland production by native large herbivores in Yellowstone National Park. Oecologia 96:157–161.

Gough, L., and J. B. Grace. 1998. Herbivore effects on plant species density at varying productivity levels. Ecology 79:1586–1594.

Greig-Smith, P. 1983. Quantitative Plant Ecology. 3rd ed. Blackwell Scientific Publications, Oxford.

Harmon, M. E., K. J. Nadelhoffer, and J. M. Blair. 1999. Measuring decomposition, nutrient turnover, and stores in plant litter. Pages 202–240 in G. P. Robertson, D. C. Coleman, C. S. Bledsoe, and P. Sollins (eds.), Standard Soil Methods for Long-Term Ecological Research. Oxford University Press, New York.

Hooper, D. U., and P. M. Vitousek. 1998. Effects of plant competition and diversity on nutrient cycling. Ecological Monographs 68:121–149.

Hulbert, L. C. 1969. Fire and litter effects in undisturbed bluestem prairie in Kansas. Ecology 50:874–877.

Kelly, J. M., G. M. van Dyne, and W. F. Harris. 1974. Comparison of three methods of assessing grassland productivity and biomass dynamics. American Midland Naturalist 92:357–369.

Knapp, A. K., M. D. Abrams, and L. C. Hulbert. 1985. An evaluation of beta attenuation for estimating aboveground biomass in a tallgrass prairie. Journal of Range Management 38:556–558.

Knapp, A. K., J. M. Blair, J. M. Briggs, S. L. Collins, D. C. Hartnett, L. C. Johnson, and E. G. Towne. 1999. The keystone role of bison in North American tallgrass prairie. BioScience 49:39–50.

Knapp, A. K., J. M. Briggs, J. M. Blair, and C. Turner. 1998. Patterns and controls of aboveground net primary production in tallgrass prairie. Pages 193–221 in A. K. Knapp,

J. M. Briggs, D. C. Hartnett, and S. L. Collins (eds.), Grassland Dynamics: Long-Term Ecological Research in Tallgrass Prairie. Oxford University Press, New York.

Knapp, A. K., J. M. Briggs, and J. K. Koelliker. 2001. Frequency and extent of water limitation to primary production in a mesic temperate grassland. Ecosystems 4:19–28.

Knapp, A.K., P.A. Fay, J.M. Blair, S. L. Collins, M. D. Smith, J. D. Carlisle, C. W. Harper, B. T. Danner, M.S. Lett and J. K. McCarron. 2002. Rainfall variability, carbon cycling and plant species diversity in a mesic grassland. Science 298: 2202-2205.

Knapp, A. K., and E. Medina. 1999. Success of $C_4$ photosynthesis in the field: Lessons from communities dominated by $C_4$ plants. Pages 251–283 in R. F. Sage and R. K. Monson (eds.), $C_4$ Plant Biology. Academic Press, San Diego.

Knapp, A. K., and T. R. Seastedt. 1986. Detritus accumulation limits productivity in tallgrass prairie. BioScience 36:662–668.

Knapp, A. K., and M. D. Smith. 2001. Variation among biomes in temporal dynamics of aboveground primary production. Science 291:481–484.

Kreeland, B. D., and P. J. Young. 1997. Long-term growth trends of bald cypress (*Taxodium distichum* (L.) Rich.) at Caddo Lake, Texas. Wetlands 17:559–566.

Laurenroth, W. K., I. C. Burke, and M. P. Gutmann. 1999. The structure and function of ecosystems in the central North American grassland region. Great Plains Research 9: 223–259.

Lauenroth, W. K., H. W. Hunt, D. M. Swift, and J. S. Singh. 1986. Estimating aboveground net primary production in grasslands: A simulation approach. Ecological Modelling 33:297–314.

Lauenroth, W. K., and O. E. Sala. 1992. Long-term forage production of North American shortgrass steppe. Ecological Applications 2:397–403.

Lomnicki, A., E. Bandola, and K. Jankowska. 1968. Modification of the Wiegert-Evans method for estimation of net primary production. Ecology 49:147–149.

Long, S. P., E. Garcia Moya, S. K. Imbamba, A. Kamnalrut, M. T. F. Piedade, J. M. O. Scurlock, Y. K. Shen, and D. O. Hall. 1989. Primary productivity of natural grass ecosystems of the tropics: A reappraisal. Plant and Soil 115:155–166.

McNaughton, S. J. 1983. Compensatory plant growth as a response to herbivory. Oikos 40:329–336.

McNaughton, S. J. 1985. Ecology of a grazing ecosystem: The Serengeti. Ecological Monographs 55:259–294.

McNaughton, S. J., D. G. Milchunas, and D. A. Frank. 1996. How can net primary productivity be measured in grazing ecosystems? Ecology 77:974–977.

Milchunas, D. G., and W. K. Lauenroth. 2001. Belowground primary production by carbon isotope decay and long-term root biomass dynamics. Ecosystems 4:139–150.

Milner, C., and R. E. Hughes. 1968. Methods for the Measurement of the Primary Production of Grassland. Blackwell Scientific, Oxford.

Mitchell, J. E. 1972. An analysis of the beta attenuation technique for estimating standing crop of prairie range. Journal of Range Management 71:220–227.

Mitsch, W. J., and J. G. Gosselink. 2000. Wetlands, 3rd ed. Wiley, New York.

National Academy of Sciences, National Research Council, and American Society of Range Management. 1962. Basic Problems and Techniques in Range Research. Natural Resource Council Publication 890. National Academy of Sciences, Washington, D.C.

Ni, J. 2004. Estimating net primary productivity of grasslands from field biomass measurements in temperate northern China. Plant Ecology 174:217–234.

Prince, S. D., and S. N. Goward. 1995. Global primary production: A remote sensing approach. Journal of Biogeography 22:815–835.

Rheinhardt, R. D., and K. Faser. 2001. Relationship between hydrology and zonation of freshwater swale wetlands on Lower Hatteras Island, North Carolina, USA . Wetlands 21:265–273.

Rice, C. W., T. C. Todd, J. M. Blair, T. R. Seastedt, R. A. Ramundo, and G. W. T. Wilson. 1998. Belowground biology and processes. Pages 244–264 in A. K. Knapp, J. M. Briggs, D. C. Hartnett, and S. L. Collins (eds.), Grassland Dynamics: Long-Term Ecological Research in Tallgrass Prairie. Oxford University Press, New York.

Roxburgh, S. H., D. J. Barrett, S. L. Berry, J. O. Carter, I. D. Davies, R. M. Gifford, M. U. F. Kirschbaum, B. P. McBeth, I. R. Noble, W. G. Parton, M. R. Raupach, and M. L. Roderick. 2004. A critical overview of model estimates of net primary productivity for the Australian continent. Functional Plant Biology 31:1043–1059.

Sala, O., A. Deregibus, T. Schlichter, and H. Alippe. 1981. Productivity dynamics of a native temperate grassland in Argentina. Journal of Range Management 43:48–51.

Sala, O. E., and A. T. Austin. 2000. Methods of estimating aboveground net primary productivity. Pages 31–43 in O. E. Sala, R. B. Jackson, H. A. Mooney, and R. W. Howarth (eds.), Methods in Ecosystem Science. Springer, New York.

Sala, O. E., M. E. Biondini, and W. K. Lauenroth. 1988. Bias in estimates of primary production: An analytical solution. Ecological Modelling 44:43–55.

Sala, O. E., W. J. Parton, W. K. Lauenroth, and L. A. Joyce. 1988. Primary production of the central grassland region of the United States. Ecology 69:40–45.

Schmalzer, P. A., and C. R. Hinkle. 1992. Soil dynamics following fire in *Juncus* and *Spartina* marshes. Wetlands 12:8–21.

Scurlock, J. M. O., K. Johnson, and R. J. Olson. 2002. Estimating net primary productivity from grassland biomass dynamics measurements. Global Change Biology 8:736–753.

Sims, P. L., and J. S. Singh. 1978. The structure and function of ten western North American grasslands. III. Net primary production, turnover, and efficiencies of energy capture and water use. Journal of Ecology 66:573–597.

Singh, J. S., W. K. Lauenroth, H. W. Hunt, and D. M. Swift. 1984. Bias and random errors in estimators of net root production: A simulation approach. Ecology 65:1760–1764.

Singh, J. S., W. K. Lauenroth, and R. K. Steinhorst. 1975. Review and assessment of various techniques for estimating net aerial primary production in grasslands from harvest data. The Botanical Review 41:181–232.

Stebbins, G. L. 1981. Coevolution of grasses and herbivores. Annals of the Missouri Botanical Garden 68:75–86.

Steuter, A. A. 1987. $C_3/C_4$ production shift on season burns—northern mixed prairie. Journal of Range Management 37:392–397.

Trollope, W. S. W., and A. L. F. Potgieter. 1986. Estimating grass fuel loads with the disc pasture meter in Kruger National Park. Journal of the Grassland Society of South Africa 3:148–152.

Tucker, C. J. 1980. A critical review of remote sensing and other methods for non-destructive estimation of standing crop biomass. Grass and Forage Science 35:177–182.

Turner, C. L., T. R. Seastedt, M. I. Dyer, T. G. F. Kittel, and D. S. Schimel. 1992. Effects of management and topography on the radiometric response of a tallgrass prairie. Journal of Geophysical Research 97:18,855–18,666.

Turner, D. P., W. D. Ritts, W. B. Cohen, T. K. Maeirsperger, S. T. Gower, A. A. Kirschbaum, S. W. Running, M. Zhao, S. W. Wofsy, A. L. Dunn, B. E. Law, J. L. Campbell, W. C. Oechel, H. J. Kwon, T. P. Meyers, E. E. Small, S. A. Kurc, and J. A. Gamon. 2005. Site-level evaluation of satellite-based global terrestrial gross primary production and net primary production monitoring. Global Change Biology 11:666–684.

Van Dyne, G. M., W. G. Vogel, and H. G. Fisser. 1963. Influence of small plot size and shape on range herbage production estimates. Ecology 44:746–759.

Vermeire, L. T., and R. L. Gillen. 2001. Estimating herbage standing crop with visual obstruction in tallgrass prairie. Journal of Range Management 42:57–60.

Webb, W. L., W. K. Lauenroth, S. R. Szarek, and R. S. Kinerson. 1983. Primary production and abiotic controls in forests, grasslands, and desert ecosystems in the United States. Ecology 64:134–151.

Whelan, R.J. 1995. The Ecology of Fire. Cambridge University Press, Cambridge, UK.

Wiegert, R. G. 1962. The selection of an optimum quadrat size for sampling the standing crop of grasses and forbs. Ecology 43:125–129.

Wiegert, R. G., and F. C. Evans. 1964. Primary production and the disappearance of dead vegetation on an old field in southeastern Michigan. Ecology 45:49–63.

# 4

## Estimating Aboveground Net Primary Production in Shrub-Dominated Ecosystems

Donald R. Young

Measurement of production in shrub communities may be difficult, but it is not impossible (Whittaker 1961). Estimating aboveground net primary production (ANPP) in shrub-dominated ecosystems can be quite challenging relative to other biome types. The inherent structural complexity of shrubs, especially those with multiple stemmed bases and a high degree of vegetative propagation, frequently leads to dense communities and sampling challenges. Perhaps most reflective of these statements is the dearth of published estimates for shrub ANPP.

Shrub-dominated ecosystems span a variety of climates, including the chaparral and matorral of Mediterranean climates; the floristically diverse fynbos and Renosterveld vegetation groups; arid communities associated with deserts, steppes, and lower montane zones; maritime shrub thickets; and riparian communities including many at high elevations and high latitudes. The shrub growth form is also an important component of many biomes, such as forests and grasslands, and frequently represents a midsuccessional seral stage.

The purpose of this chapter is to review methods for estimating ANPP in shrub-dominated ecosystems and to provide guiding principles and recommendations to facilitate accurate determinations of ANPP. For these systems, ANPP is defined as all aboveground plant biomass produced per unit area during a year, accounting for losses due to herbivory and decomposition when appropriate. In most systems, production may be restricted to intervals of favorable climate during the year; nevertheless, ANPP is expressed on an annual basis. The focus of this chapter will be ecosystems in which shrubs are the dominant growth forms, but these principles should apply to the shrub component of all ecosystems. In prairie or forested communities, combining shrub methods with those for grasslands and forests should permit ANPP to be estimated in proportion to the growth forms present.

## Representative Values for ANPP and Key Determinants

ANPP for shrub-dominated ecosystems varies by more than two orders of magnitude, from less than 10 to more than 1300 g m$^{-2}$ yr$^{-1}$ (table 4.1). Numerous environmental factors may affect ANPP in shrub-dominated ecosystems. Variations are primarily related to water and temperature limitations, but light and soil nutrient levels must also be considered. Most shrub species and shrub-dominated communities occur in relatively high-light environments, and include early- to mid-successional communities that are replaced by forests. The limitations of a low-light environment are evident in the low ANPP values for understory shrubs (table 4.1). Moisture availability relative to favorable temperatures is perhaps the most important combination of physical factors. Seasonal drought reduces ANPP components in tropical forest shrubs (Wright 1991; Wright et al. 1992). Both temperature and moisture are important for controlling shrub ANPP in the Patagonian steppe (Jobbagy and Sala 2000). The more moderate values of arid environments may indicate that temperature is more limiting than water to shrub ANPP, especially for deeply rooted species. The highest values for ANPP are warm desert phreatophytes (table 4.1). The lowest estimates for ANPP are from low- temperature-limited environments at high elevation or latitude. Increased temperatures enhance alpine shrub production in northern Japan (Kudo and Suzuki 2003).

Nutrient availability often affects shrub ANPP. For example, leaf turnover and shrub production in heathland communities of The Netherlands increase as nutrient availability is enhanced (Aerts 1989). Fire increases nutrient availability, and numerous studies have investigated shrub response to fire. Shrub production in response to fire has been quantified in prairies (McCarron and Knapp 2003; Heisler et al. 2004; Lett et al. 2004), desert riparian systems (Busch and Smith 1993), heathlands of The Netherlands (Diemont and Oude Voshaar 1994), and arctic dwarf shrub communities in Sweden (Wardle et al. 2003). Effects on ANPP are related to the magnitude of the fire, damage to aboveground structures, and interspecific differences in mortality or the ability to resprout after a fire.

Biotic interactions, as affected by fire, also may influence shrub production. For example, the clonal shrub *Cornus drummondii* had greater post-fire productivity for new shoots compared with unburned areas of prairie due to the post-fire transient period of high resource availability (McCarron and Knapp 2003; Heisler et al. 2004). However, Lett et al. (2004) found increased shrub cover and density when fire was excluded in the prairie and the growth of dominant grasses was suppressed. Overall, fire had a negative effect on aboveground biomass for *C. drummondii* because the dominant grasses responded more quickly to the post-fire increase in resources. Rapid growth by the grasses inhibited resource acquisition by the resprouting shrubs.

Herbivory, both leaf herbivory by insects and browsing by mammals, may influence shrub ANPP, depending on the extent of damage to the shrub at the individual and community levels. The importance of herbivory also depends on the

Table 4.1. Estimates of aboveground net primary production (ANPP) for shrub communities

| Community | ANPP (g m$^{-2}$ yr$^{-1}$) | References |
|---|---|---|
| Desert | 50–180 | Rodin et al. (1972); Szarek (1979); West (1983); Nobel (1987) |
| Desert phreatophytes | 600–1300 | Sharifi et al. (1982); Salinas and Sanchez (1971) |
| African savanna | 317–634 | Menaut and Cesar (1979) |
| California chaparral | 205–1395 | Mooney et al. (1977) |
| Chile matorral | 180–504 | Mooney et al. (1977) |
| Heath | | |
| The Netherlands | 42–263 | Aerts (1989); Diemont and Oude |
| Great Smoky Mts., GA | 12–983 | Whittaker (1963) |
| Steppe | | |
| N. American intermountain | 80–250 | West (1983) |
| Patagonian | 11–46 | Jobbagy and Sala (2000) |
| Dwarf shrubs | | |
| Arctic | 9–400 | Wielgolaski et al. (1981); Bliss (2000) Wardle et al. (2003) |
| Alpine | 88–503 | Wielgolaski et al. (1981) |
| Prairie | | |
| Kansas | 34–1266 | Heisler et al. (2004); Lett et al. (2004) |
| Wisconsin | 8–161 | Harrington et al. (1989) |
| Forest understory | | |
| European mixed oak | 22 | Duvigneaud and Denaeyer-De Smet (1970) |
| Wisconsin mixed oak | 6–14 | Harrington et al. (1989) |

*Notes*: Values were derived from Archibold (1995) and the primary literature. ANPP estimates are based on both temporal and spatial sampling, may include a herbaceous component, and should be used only as a guide for the expected magnitude of ANPP encountered in different shrub communities.

ability of shrubs to respond to or recover from damage or loss of aboveground structures. However, herbivory effects are difficult to quantify for shrubs (Pitt and Schwab 1988), especially when distinguishing new growth from reallocation. Thus, determining the net positive or negative effects of herbivory on shrub ANPP is challenging (McNaughton et al. 1996). Additional discussion of herbivory is provided later in this chapter.

Belowground net primary production (BNPP) may be affected by the same physical and biotic factors that influence ANPP (Wardle et al. 2004). Direct measurement of shrub BNPP is difficult and often impossible. In addition to the challenges of soil physical properties for root excavation and measurement, shrub root systems can be well below the soil surface—more than 100 m in arid environments. A review of techniques and recommendations for estimating BNPP in shrub-dominated ecosystems can be found in chapter 8 of this volume.

Due to the inherent difficulty in measuring belowground structures, there are very few estimates of BNPP in shrub-dominated ecosystems. This is unfortunate, because belowground biomass represents a significant component of total shrub biomass in most ecosystems, and BNPP would, therefore, also be a significant portion of total shrub production. Root:shoot ratios for shrubs vary from 0.6 to 6.8 in desert environments, 0.3 to 2.9 in Mediterranean systems, and 2.0 to 3.1 for dwarf species in alpine tundra (Archibold 1995). Whittaker (1962) determined that shrub root:shoot ratios in Great Smoky Mountain communities vary from 0.5 to 2.2, depending on species and location. Whittaker (1961) also estimated BNPP as 38%–74% of ANPP for several shrubs in this system. For dwarf shrubs of the alpine tundra, estimated BNPP was 75%–114% of ANPP (Wielgolaski et al. 1981). In arctic tundra, shrub BNPP values were twice as high as ANPP values (Bliss 2000).

## Guiding Principles and Recommendations for Shrub-Dominated Ecosystems

### Review of Methods

A casual search of the primary scientific literature will find a variety of studies that have measured shrub production. Many, however, are really measures of growth characteristics and not of ANPP. Leaf production is often determined by counting the number of new leaves on branches (e.g., Wright 1991; Busch and Smith 1993) or dry mass per year (Kudo and Suzuki 2003). New-shoot dry mass is considered as production or at least a relative measure of production (e.g., Young et al. 1995; Jobbagy and Sala 2000; McCarron and Knapp 2003; Wardle et al. 2003; Gibson et al. 2004), but reallocation from belowground reserves may not be considered, and woody increment to preexisting branches and stems often is not included. Seed or fruit production is often expressed as the number of seeds per branch or shrub (e.g., Zammit and Zedler 1992; Young et al. 1995) or total seed or fruit weight (Pierce and Cowling 1991). In many studies, these production components are expressed per shrub or per branch, not per unit of ground surface area; thus, they are growth estimates and not production. In large part, this may be due to sampling logistics and the time-consuming work of developing the dimension analysis relationships needed for production estimates.

The accurate determination of ANPP for shrubs requires a combination of destructive sampling, dimension analysis, and extrapolation to a unit area of the landscape. There are two approaches: classic dimension analysis, described in detail by Whittaker (1961), and the difference method summarized by Gimingham and Miller (1968). Both methods require the development of allometric relationships (see Sprugel 1983 for discussion of potential bias in allometric equations).

### Dimension Analysis

This method requires sampling and subsampling of individual shrubs to develop allometric relationships between easily measured parameters (e.g., stem basal area or diameter) and woody, leaf, or fruit dry mass. Gray and Schlesinger (1981) used dimension analysis to quantify production in sage scrub along the coast of California. Specific details are discussed below.

Shrubs spanning a representative range of size (i.e., diameter at base of main stem) or age should be selected for destructive sampling. For an individual shrub, the main stem is sectioned. Branches are removed from each stem section, and for each branch, height above the main stem base, basal diameter, length, age, number of twigs, and mass of twigs, leaves, and fruits (if available) are measured. Basal and midstem diameters are measured for each section. Basal age and radial wood increment are also determined. For some of the stem sections, live:dry mass ratios should be determined so that all sample components can be converted to dry mass.

Given the above measurements, dry mass production can be determined for the different fractions of the shrub. Current twig and leaf production is measured directly by weighing the clipped twigs. Stem wood growth for each shrub fraction can be determined from the ratio of the cross-section area increase for the current year to the cross-section area of the wood, times the wood dry weight for that section. To account for taper, the cross-section area of the midpoint of the section should be used. (See Whittaker [1961] if separation of bark growth and wood growth is desirable.) Quantifying the growth of individual branches that were attached to the main stem sections requires the development of multiple regression relationships of branch weight and age. The production values are then summed for all twigs, branches, and main stem sections.

The ANPP of each sampled shrub is regressed against shrub size or age. This relationship can then be used to estimate annual production for all individual shrubs' stems within a defined plot or area to estimate ANPP.

The above procedure is followed for all shrub species within the community. In addition, allometric relationships may vary spatially and, therefore, should be developed for each field site or location of interest unless prior experience indicates negligible differences. Further, these relationships may vary with climatic fluctuations, so at least several shrubs should be sampled each year to determine if development of new relationships is warranted when comparing ANPP estimates among years.

## Difference Method

This approach requires destructive sampling to determine relationships between diameter at the stem base and shrub dry mass parameters. By measuring a change in stem diameter over a time period, biomass increase can be estimated and expressed as new production. Lett et al. (2004) used the difference method to determine shrub production in tallgrass prairie of Kansas. Specific details are presented below.

As described for dimension analysis, shrubs spanning a representative range of size (i.e., diameter at base of main stem) or age should be selected for destructive sampling. For each shrub, basal stem diameter should be determined. The shrub can then be divided into woody, leaf, and, if available, fruit components, with each weighed. Subsamples of each component should be dried to develop live:dry mass ratios, so that all sample components can be converted to dry mass. The relationships between stem basal diameter and woody, leaf, and fruit dry mass are determined through regression analyses. Allometric relationships must be developed for all shrub species within the community. In addition, there may be spatial variation in the relationships among plots or specific locations.

Using the allometric relationships, individual shrub stem diameters can be measured to estimate the dry mass of the three components which are summed to estimate total shrub dry mass. An increase in shrub dry mass can be estimated by measuring stem basal diameter at two points in time. During the initial measurement, the position on each stem must be carefully marked for use in the second measurement. For ANPP estimates, the measurement interval should be one year, and the difference in estimated dry mass represents annual production per shrub. If all shrub stems within a plot or specified area are measured, then their combined increase in dry mass per unit area is considered ANPP for the community.

### Spatially Explicit Sampling of Shrub Systems

A slightly different approach was developed for use at the Jornada Basin LTER site in southern New Mexico, where shrubs constitute a minor component of the community in some ecosystems (e.g., *Bouteloua* grasslands) and a dominant component in others (e.g., *Larrea* or *Prosopis* shrublands). The objective was to develop a method for estimating ANPP and its variation in space (both within and among study sites) and over time (between seasons and years). Permanent quadrats (1 m²) were established in multiple sites of a given ecosystem type. All plants or plant parts contained within the rectangular volume above a quadrat were measured in a non-destructive way. That is, if only a portion of a branch extended into that volume (from a plant rooted outside the quadrat), only that portion of the branch would be measured. Conversely, if a plant rooted inside the quadrat had branches extending outside, only the central portion of the plant (contained inside the volume above the quadrat) was measured. Measurements taken of a plant part were the vertical height or vertical dimension of the portion inside the volume, and the horizontal area or cover. Cover was quantified as percentage, or number of squares of cover, in a grid of 100 squares that were 10 cm × 10 cm, using a portable frame of PVC "legs" supporting 1 m × 1 m square gridded with twine. These nondestructive measurements were repeated at 3 sample dates per year on the same quadrats at each site, to allow estimation of seasonal contributions to ANPP and to minimize the chances of missing significant but temporary gains in plant volume.

During the first 4 yrs of sampling, measurements were made on plants or plant parts of similar range of size, for each species found at a site at each date, and those plant parts were then harvested. Harvested material was sorted, and live material was oven-dried and weighed. Resulting harvest data were used to calculate regressions between volume (a rectangular index calculated from cover or area and height) and biomass. Multiple seasons and years of harvests allowed a statistical determination of the degree to which regressions varied among seasons or among sites for a given species (Huenneke et al. 2001). Application of regressions allowed the estimation of biomass of each portion of each species within a quadrat in a given season. The sum of all positive increments of biomass for each species, from one sample date to the next, was used as the estimate of ANPP for that square meter. The result was a spatially explicit estimate of ANPP that could be used to assess variability within and between sites of a given ecosystem type, and to compare both spatial and temporal variability of ANPP among ecosystem types.

The general method described above was applied to both herbaceous and woody species at all Jornada sites. Shrubs usually were measured as multiple individual branch systems rather than as a single large mass. Shrubs were among those species requiring different seasonal volume-biomass regressions, because a partially leafed-out shrub (in spring) and the same individual, fully leafed-out in late summer, would yield different biomass estimates, even if the area and height measurements had not changed. Even evergreen species, such as *Larrea tridentata*, required seasonally distinct regressions. The ability to measure all vascular plant species with a consistent approach, and to measure enough quadrats within a site to allow statistical characterization of spatial variability, allowed a more powerful assessment of patterns of ANPP than had been available previously (Huenneke et al. 2002).

## Additional Methodological Considerations

### Herbivory

Depending on the ecosystem and the palatability of the species, herbivory may represent a significant loss of ANPP (McNaughton et al. 1996). Quantifying production lost to herbivores can be challenging in shrub-dominated systems because losses may be due to leaf grazing (by insects and mammals) and/or to browsing (by mammals). As discussed in chapter 3 of this volume, the use of movable exclosures or cages is the recommended method to provide both a measure of consumption and the detection of compensatory growth. Gimingham and Miller (1968) recommended this method for dwarf shrub heaths; however, this may be impractical where shrubs are taller.

Another approach is to tag randomly selected shoots and quantify the extent of leaf loss and/or shoot length loss to herbivory throughout the year. Leaf and twig losses throughout the year should be summed per shoot, and can be compared against intact shoots to estimate production loss due to herbivory. Controlled clipping experiments can be used to estimate compensatory growth, avoiding the time and effort associated with the construction and movement of exclosures. In addition, potential confounding effects of exclosures (e.g., microclimate alteration) are eliminated. Frequency of shoot monitoring should be a function of the extent and nature of herbivory pressures and shrub growth patterns. Leaves lost or damaged by herbivory can be counted, and current twig growth lost to browsing can be estimated from the diameter at the point of browsing. Twig loss due to browsing requires the regression of twig diameter against twig dry mass. Pitt and Schwab (1988) review methods and corresponding accuracies for estimating loss due to browsing.

An alternative approach developed by Jensen and Urness (1981) accounts for loss of leaves and twigs to herbivory. Mean twig diameter at the point of browsing

($D_p$), mean diameter of unbrowsed twig tips assessed as the first measurable internode below the terminal bud ($D_t$), and mean basal diameter of browsed twigs ($D_b$) are measured for a randomly selected number of shrubs/branches. Shrub utilization (U) or loss to herbivory is then calculated as a percentage of production:

$$U = 100 \; (D_p - D_t)/(D_b - D_t).$$

## Deciduous vs. Evergreen Leaf Habit

Leaf production is usually quantified by harvesting all leaves on individual shoots and then extrapolating to an appropriate area basis. Sampling in this manner at the conclusion of the growing season and prior to leaf fall is appropriate for determining annual leaf production for most deciduous species. Shrubs with an evergreen leaf habit may require an alternative approach, especially if leaves remain for several growth seasons. If current-season branch growth is identifiable, then current-year and previous-year leaf production can be distinguished. For species whose leaves persist for more than one season and current branch growth may not be readily identified, or where significant leaf fall occurs throughout the year, leaf litter traps may be a more accurate method for estimating annual leaf production. Although they are used to quantify litter fall, annual totals may be considered as estimates of leaf production for the previous year. Traps must be designed to ensure complete litter collection while draining precipitation. Considerable attention must be devoted to the number of traps, their placement throughout the shrub community, and frequency of collection in order to adequately estimate litter production. Litter sampling is beyond the scope of this discussion; however, numerous references describe trap construction and sampling procedures (e.g., Martinez-Yrizar et al. 1999).

Most deciduous species and many evergreens produce distinct annual growth rings. If growth is indeterminate, often starting and stopping several times during a year in response to temperature and/or moisture fluctuations, then false rings may be produced and estimates of age or annual growth may be erroneous. This may be especially prevalent in evergreen species. For example, stem cross sections for the evergreen shrub *Myrica cerifera* were examined at the Virginia Coast Reserve and related to aerial photographs and anecdotal information to determine maximum shrub age. For nearly all stems sampled, the number of growth rings exceeded the maximum age by as much as 40%, indicating the presence of numerous false growth rings (Young unpubl. data). For shrub species or habitats where growth rings do not correspond to years, the dimension analysis method described above may be inappropriate for determining radial growth increment and branch age. The difference method, which uses changes in stem diameter over time, should be used to provide a more accurate estimate of ANPP growth.

## Sampling Considerations

The number, size, shape, and placement of sample plots to reliably estimate ANPP for shrubs and shrub-dominated ecosystems is a function of the individual species growth characteristics, effects of local climate on growth, and heterogeneity of both shrub structure and the landscape. Plot size, shape, and placement in the shrub community must consider potential edge effects, and a procedure for shrub overlap, both stem and especially canopy, must be implemented. Pitt and Schwab (1988) include an excellent review of shrub sampling strategies. They recommend that initial measurements in shrub plots follow Zamora (1981), who concluded that 1 m × 1 m plots provide adequate sampling of widely spaced shrubs while permitting easy observation and measurement in dense thickets. Plot size and shape (square, rectangular, or circular) should be adjusted for each ecological system according to shrub characteristics, sampling methodology, and statistical variability revealed in preliminary sampling.

Statistical procedures for determining sample adequacy can be found in most biostatistical texts. Also, see chapter 12 in this volume for discussion of bias, error, and uncertainty in the measurement of production. Placement of plots may be especially important where shrub structure and/or the landscape varies. In addition, spatial variations in the age of the shrub community, whether successional or due to disturbance, should be considered when positioning sample plots. A case study based on research on Hog Island at the Virginia Coast Reserve (VCR) Long-Term Ecological Research (LTER) site will serve to illustrate several of these points.

Variations in new shoot growth for *Myrica cerifera* as a result of year-to-year variations in temperature and precipitation have been monitored as one of the long-term studies ongoing at the VCR LTER site. *Myrica* is an evergreen shrub that is the dominant woody species across the VCR and on most barrier islands along the southeastern Atlantic coast (Young et al. 1995). Research focused on Hog Island, where accretion on the ocean side of the island over the past hundred years has formed a chronosequence of soils (Hayden et al. 1991). The soils become progressively older from the ocean beach toward the island interior. *Myrica* shrub thickets form in the protected swales along the chronosequence; however, seedling establishment and thicket formation are delayed until exposure to salt spray and storm-related flooding are reduced. Thus, there is considerable variation in shrub structural characteristics across the Hog Island landscape.

Four sites, representing different stages in shrub thicket development and landscape position, have been monitored since 1992. These include an area where shrub seedlings are just beginning to establish, a second site where *Myrica* seedlings have emerged above the grass canopy and have formed a new thicket since a third site with a mature, stable thicket, and the oldest site, farthest inland from the beach, where *Myrica* shrubs are senescing and recruitment of new stems is apparent (fig. 4.1). Each site differs from the others with regard to shrub structural characteristics and exposure to salinity. Depending on sampling needs, both transects and plots (5 m × 10m) have been used.

Young et al. (1995) determined that frequency distributions of *Myrica* stem diameters differed significantly among the 3 sites with thickets. As expected, the developing thicket was composed primarily of small-diameter stems, reflecting recruitment. The stable thicket included a broad range of stem diameters and the largest size classes accounted for the majority of stems in the senescing thicket. In addition, slopes for the relationship between growth rings and stem diameter differed

Figure 4.1. Understory of developing (top panel) and senescing (bottom panel) *Myrica cerifera* shrub thickets on Hog Island in the Virginia Coast Reserve. The sign in the top panel marks the location of the shoreline in 1967; by 2005 it was more than 300 m to the east.

significantly among all three sites. *Myrica* stem density decreased at the older sites, with 1.7 stems m⁻² for the young thicket, 1.4 stems m⁻² for the stable thicket, and 1.1 stems m⁻² for the senescing thicket (Brantley and Young, in press). Fruit production was greatest for the stable *Myrica* thickets, and new shoot growth was lowest where *Myrica* stems were senescing (Young et al. 1995). Thus, sampling across the landscape is essential for capturing the range of structural and growth characteristics used for determining production.

To quantify *Myrica* shrub growth in response to annual variations in summer temperature and precipitation patterns, all 3 thicket sites have been sampled for more than a decade. To ensure that all portions of the thickets were sampled (i.e., edge and interior), new shoots were harvested at random locations along transects that completely cross each thicket. Shoots were harvested in mid-October, after growth had ceased and before significant leaf fall. Over the 12 years of monitoring, all 3 thicket sites showed variations in shoot growth related to the timing and amount of precipitation during the summer months (fig. 4.2). Although differences in annual shoot growth were usually apparent in most years, extreme summer drought reduced these differences. Thus, determining the extent of spatial variations for proper sampling of growth characteristics and ANPP should include several years of measurements.

Annual shoot growth for the senescing *Myrica* thicket has been consistently lower than for the others (fig. 4.2). There was greater fluctuation in new shoot growth over the first 7 years for the young, developing thicket. Thereafter, there was no discernible difference in shoot growth between the developing and stable thickets (fig. 4.2). The earlier difference between the two sites was related to a more open

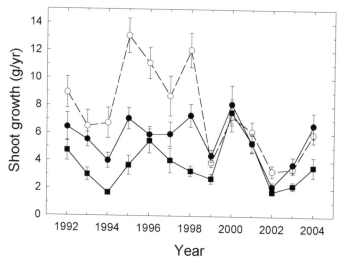

Figure 4.2. Annual variations in new shoot dry mass for developing (open circles), stable (closed circles), and senescing (closed squares) for *Myrica cerifera* shrub thickets on Hog Island in the Virginia Coast Reserve.

*Myrica* canopy in which individual shrubs were recognizable. As young shrubs continued to increase in size, individual canopies merged into a thicket similar to the stable thicket. Thus, age and/or location affected *Myrica* shrub shoot growth on Hog Island, and must be considered when determining ANPP estimates for shrub thickets over the island landscape.

In the Jornada Basin study described earlier, Huenneke et al. (2001) specifically assessed the adequacy of sample size, in terms of number of 1 m$^2$ quadrats per site, in estimating aboveground biomass and ANPP. The substantial patchiness of plant cover and biomass in the semi-arid systems under study, including open shrublands and grassland systems with scattered shrubs, meant that estimates of biomass and ANPP were quite unreliable for sample sizes of less than 20 quadrats per site. However, 40 quadrats appeared to be sufficient to characterize mean biomass and mean ANPP in most of the 15 study sites. The labor-intensive nature of most methods of determining ANPP places a substantial constraint on the ability to characterize variability within or between sites.

## Final Comments

Shrub-dominated ecosystems occur in a very broad range of environments from tropical to arctic, from arid to riparian; and a significant shrub component occurs in many forests and grasslands. In addition, the diversity of shrub growth forms, leaf habits, and physiologies makes the recommendation of a uniform method impractical. Two methods for estimating ANPP are provided, and depend on shrub growth characteristics. In addition, recommendations to reduce errors are discussed. Proper sampling, or placement of samples to account for spatial variations in shrub age, edaphic factors, fire history, herbivory intensity, and other factors is essential for the accurate estimate of shrub ANPP.

*Acknowledgments*   Laura Huenneke provided the descriptions of shrub sampling and ANPP measurements at the Jornada Basin LTER site. Contributions and suggestions from Neil West, Phil Rundel, Rasoul Sharifi, and John Briggs also enriched this chapter. The hard work of former students Ed Crawford, Dave Martin, Kathryn Tolliver, and Steven Brantley significantly contributed to the understanding of shrub dynamics at the Virginia Coast Reserve LTER site. Much of the shrub thicket research presented in this chapter was partially funded by the National Science Foundation through the LTER program.

### References

Aerts, R. 1989. The effect of increased nutrient availability on leaf turnover and aboveground productivity of two evergreen ericaceous shrubs. Oecologia 78:115–120.

Archibold, O. W. 1995. Ecology of World Vegetation. Chapman & Hall, New York.

Bliss, L. C. 2000. Arctic tundra and polar desert biome. Pages 1–40 in M. G. Barbour and W. D. Billings (eds.), North American Terrestrial Vegetation, 2nd ed. Cambridge University Press, Cambridge.

Brantley, S. T., and D. R. Young. 2007. Leaf area index and light attenuation in rapidly expanding shrub thickets. Ecology, in press.

Busch, D. E., and S. D. Smith. 1993. Effects of fire on water and salinity relations of riparian woody taxa. Oecologia 94:186–194.

Diemont, W. H., and J. H. Oude Voshaar. 1994. Effects of climate and management on the productivity of Dutch heathlands. Journal of Applied Ecology 31:709–716.

Duvigneaud, P., and S. Denaeyer-De Smet. 1970. Biological cycling of minerals in temperate deciduous forests. Pages 199–225 in D. E. Reichle (ed.), Analysis of Temperate Forest Ecosystems. Springer, Berlin.

Gibson, A. C., M. R. Sharifi, and P. W. Rundel. 2004. Resprout characteristics of creosote bush (*Larrea tridentata*) when subjected to repeated vehicle damage. Journal of Arid Environments. 57:411–429.

Gimingham, C. H., and G. R. Miller. 1968. Measurement of primary production of dwarf shrub heaths. In C. Milner and R. E. Hughes (eds.), IBP Handbook No. 6. Methods of the Measurement of the Primary Production of Grassland. Blackwell Scientific, Oxford.

Gray, J. T., and W. H. Schlesinger. 1981. Biomass, production, and litterfall in the coastal sage scrub of southern California. American Journal of Botany 68:24–33.

Harrington, R. A., B. J. Brown, P. B. Reich, and J. H. Fownes. 1989. Ecophysiology of exotic and native shrubs in Southern Wisconsin. II. Annual growth and carbon gain. Oecologia 80:368–373.

Hayden, B. P, R. D. Dueser, J. T. Callahan, and H. H. Shugart. 1991. Long-term research at the Virginia Coast Reserve. BioScience 41:310–318.

Heisler, J. L., J. M. Briggs, A. K. Knapp, J. M. Blair, and A. Sherry. 2004. Direct and indirect effects of fire on shrub density and aboveground productivity in a mesic grassland. Ecology 85:2245–2257.

Huenneke, L. F., J. P. Anderson, M. Remmenga, and W. H. Schlesinger. 2002. Desertification alters patterns of aboveground net primary production in Chihuahuan ecosystems. Global Change Biology 8:247–264.

Huenneke, L. F., D. Clason, and E. Muldavin. 2001. Spatial heterogeneity in Chihuahuan Desert vegetation: Implications for sampling methods in semi-arid ecosystems. Journal of Arid Environments 47:257–270.

Jensen, C. H., and P. J. Urness. 1981. Establishing browse utilization from twig diameters. Journal of Range Management 34:113–116.

Jobbagy, E. G., and O. E. Sala. 2000. Controls of grass and shrub aboveground production in the Patagonian steppe. Ecological Applications 10:541–549.

Kudo, G., and S. Suzuki. 2003. Warming effects on growth, production, and vegetation structure of alpine shrubs: A five-year experiment in northern Japan. Oecologia 135: 280–287.

Lett, M. S., A. K. Knapp, J. M. Briggs, and J. M. Blair. 2004. Influence of shrub encroachment on aboveground net primary productivity and carbon and nitrogen pools in a mesic grassland. Canadian Journal of Botany 82:1363–1370.

Martinez-Yrizar, A., S. Nunez, H. Mirando, and A. Burquez. 1999. Temporal and spatial variation in litter production in Sonoran Desert communities. Plant Ecology 145:37–48.

McCarron, J. K., and A. K. Knapp. 2003. $C_3$ shrub expansion in a $C_4$ grassland: Positive post-fire responses in resources and shoot growth. American Journal of Botany 90:1496–1501.

McNaughton, S. J., D. G. Milchunas, and D. A. Frank. 1996. How can net primary productivity be measured in grazing ecosystems? Ecology 77:974–977.

Menaut, J. C., and J. Cesar. 1979. Structure and primary productivity of Lamto savannas, Ivory Coast. Ecology 60:1197–1210.

Mooney, H. A., J. Kummerow, A. W. Johnson, et al. 1977. The producers—their resources and adaptive responses. Pages 85–143 in H. A. Mooney (ed.), Convergent Evolution in Chile and California. Dowden, Hutchinson & Ross, Stroudsburg, PA.

Nobel, P.S. 1987. Photosynthesis and productivity of desert plants. Pages 41–66 in L. Berkofsky and M. G. Wurtele (eds.), Progress in Desert Research. Rowman & Littlefield, Totowa, NJ.

Pierce, S. M., and R. M. Cowling. 1991. Dynamics of soil-stored seed banks of six shrubs in fire-prone dune fynbos. Journal of Ecology 79:731–747.

Pitt, M. D., and F. E. Schwab. 1988. Quantitative Determinations of Shrub Biomass and Production: A Problem Analysis. Land Management Report 54. Research Branch, British Columbia Ministry of Forests and Lands, Vancouver, BC.

Rodin, L. E., N. I. Bazilevich, and Y. M. Miroshnichenko. 1972. Productivity and biogeochemistry of *Artemisieta* in the Mediterranean area. Pages 193–198 in L. E. Rodin (ed.), Eco-physiological Foundation of Ecosystem Productivity in Arid Zones. Nauka, Leningrad, USSR.

Sharifi, M. R., E. T. Nilsen, and P. W. Rundel. 1982. Biomass and net primary production of *Prosopis glandulosa* (Fabaceae) in the Sonoran Desert of California. American Journal of Botany 69:760–767.

Sprugel, D. G. 1983. Correcting for bias in log-transformed allometric equations. Ecology 64:209–210.

Szarek, S. R. 1979. Primary production in four North American deserts: Indices of efficiency. Journal of Arid Environments 2:187–209.

Wardle, D. A., R. D. Bardgett, J. N. Klironomos, H. Setala, W. H. van der Putten, and D. H. Wall. 2004. Ecological linkages between aboveground and belowground biota. Science 304:1629–1633.

Wardle, D. A., G. Hornberg, O. Zackrisson, M. Kalela-Brundin, and D. A. Coomes. 2003. Long-term effects of wildfire on ecosystem properties across an island area gradient. Science 300:972–975.

West, N. E. 1983. Overview of North American temperate deserts and semi-deserts. Pages 321–374 in N. E. West (ed.), Ecosystems of the World, vol. 5, Temperate Deserts and Semi-deserts. Elsevier Scientific, New York.

Whittaker, R. H. 1961. Estimation of net primary production of forest and shrub communities. Ecology 42:177–180.

Whittaker, R. H. 1962. Net production relations of shrubs in the Great Smoky Mountains. Ecology 43:358–377.

Whittaker, R. H. 1963. Net production of heath balds and forest heaths in the Great Smoky Mountains. Ecology 44:176–182.

Wielgolaski, F. E., L. C. Bliss, J. Svoboda, and G. Doyle. 1981. Primary production of tundra. Pages 187–225 in L. C. Bliss, O. W. Heal, and J. J. Moore (eds.), Tundra Ecosytems: A Comparative Analysis. Cambridge University Press, Cambridge.

Wright, J. S. 1991. Seasonal drought and the phenology of understory shrubs in a tropical moist forest. Ecology 72:1643–1657.

Wright, J. S., J. L. Machado, S. S. Mulkey, and A. P. Smith. 1992. Drought acclimation among tropical forest shrubs (*Psychotria*, Rubiaceae). Oecologia 89:457–463.

Young, D. R., G. Shao, and J. H. Porter. 1995. Spatial and temporal growth dynamics of barrier island shrub thickets. American Journal of Botany 82:638–645.

Zammit, C. A., and P. H. Zedler. 1992. Size, structure and seed production in even-aged populations of *Ceanothus greggii* in mixed chaparral. Journal of Ecology 81:499–511.

Zamora, B. A. 1981. An approach to plot sampling for canopy volume in shrub communities. Journal of Range Management 34:155–156.

# 5

# Estimating Aboveground Net Primary Productivity in Forest-Dominated Ecosystems

Brian D. Kloeppel

Mark E. Harmon

Timothy J. Fahey

The measurement of net primary productivity (NPP) in forest ecosystems presents a variety of challenges because of the large and complex dimensions of trees and the difficulties of quantifying several components of NPP. As summarized by Clark et al. (2001a), these methodological challenges can be overcome, and more reliable spatial and temporal comparisons can be provided, only if greater conceptual clarity and more standardized approaches to the problem are achieved. The objective of this chapter is to contribute to correction of these limitations in forest NPP measurement. Because Clark et al. (2001a) did an exemplary job with this topic, our task is made somewhat easier. We focus our attention on a variety of practical matters concerning field measurements and calculations for aboveground NPP in broadleaf deciduous, evergreen coniferous, and tropical forest biomes. We evaluate the advantages and disadvantages of contrasting approaches to key measurements and provide recommendations that should aid researchers in designing field campaigns.

In general, field measurement of NPP involves quantifying two distinct sets of organic matter: (1) that which was added and retained by the plants through the measurement interval (net biomass increment) and (2) that which was produced, but lost by the plants during the same interval:

$$NPP = \Delta B + M + H + L + V,$$

where $\Delta B$ is net biomass increment, and M, H, L, and V are losses owing to mortality, herbivory, leaching, and volatilization, respectively. Clark et al. (2001a) thoroughly reviewed the likely magnitudes and some approaches for H, L, and V, all of which can be significant in certain situations, and we refer readers to that paper for details. Here we focus our attention on approaches to accurate and precise mea-

63

surement of ΔB (live biomass increment) and M (mortality of living tissues, including litterfall, pruning, stem rot, and tree death). The reason that loss terms such as M must be added to ΔB to calculate NPP can be seen in the case where ΔB is zero: if the live biomass is constant over time, and losses of living organic matter are occurring, then the plants must have replaced this material in the form of new tissue production. These loss terms are therefore equivalent to this new production. Finally, we consider only ANPP in this chapter, since the approaches for belowground NPP are detailed in Tierney and Fahey (chap. 8, this volume). The allocation of carbon (C) to ANPP and BNPP in forests has been the subject of several reviews (Nadelhoffer and Raich 1992; Gower et al. 1996), and remains an important topic for future research because global changes in climate and pollution loading are intimately tied to this balance.

In estimating ANPP as the sum of ΔB and M over some time interval, it is important to maintain consistent and internally complementary definitions of these components. Because of the large size of trees, ΔB is usually estimated by applying allometric biomass equations (developed from carefully harvested trees) to stand survey data, particularly for production of wood and other perennial tissues. Mixed approaches to quantifying M are necessary because of the varied nature of the components of this term. For example, forest ecosystems contain aboveground tissues in the vegetation canopy with both short persistence (less than 1 yr) and long persistence (greater than 1 yr). Tissues with short persistence include deciduous leaves (life span less than 1 yr), flowers, and seeds (from nonserotinous cones). Aboveground tissues with long life spans may include evergreen leaves (life span greater than 1 yr), serotinous cones, branches, bark, and stem wood. The production of short life span tissues is often best estimated with fine litterfall collections (as a component of M), while long life span tissues (especially woody tissues) are best estimated with tree allometric relationships (as a component of ΔB). This is because the former tend to reach a "steady-state" biomass relatively early in stand development. The key point is to avoid double counting or omission of these components in ANPP calculations. Similarly, stand survey and allometric estimates of tree mortality during a measurement interval must be added to ΔB to obtain an accurate estimate of ANPP, but if tissues of dead trees are collected as woody litterfall, then these components of M could be doubled counted. Errors of this sort have been common in the forest NPP literature.

## Representative Values and Key Determinants of Forest ANPP

Large projects have been undertaken to summarize and synthesize NPP data to assist in global model evaluation, such as the Global Primary Production Data Initiative described in Scurlock et al. (1999) and Olson et al. (2001). For the present chapter we synthesize a representative suite of these data arranged by forest biome types (table 5.1). There is a tremendous range of aboveground biomass and ANPP for forested biomes. Clark et al. (2001b) have summarized data from 38 tropical forest study sites. We selected 7 sites to include in table 5.1, among them the L'Anguédédou site in the Ivory Coast that had an estimated ANPP of 1430 g C m$^{-2}$ year$^{-1}$. We also in-

Table 5.1. Estimates of forest aboveground biomass and aboveground net primary productivity

| Forest Type | Estimated Aboveground Biomass (g C m$^{-2}$) | | Estimated Aboveground Net Primary Productivity (g C m$^{-2}$ year$^{-1}$) | | Summary Data Sources |
|---|---|---|---|---|---|
| | Minimum | Maximum | Minimum | Maximum | |
| Boreal evergreen | 572 | 8656 | 120 | 439 | Gower et al. 2001 |
| Boreal deciduous | 2644 | 9334 | 169 | 635 | Gower et al. 2001 |
| Temperate evergreen | 540 | 49126 | 60 | 1555 | Grier and Logan 1977; Runyon et al. 1994; Gower et al. 1996 |
| Temperate deciduous | 4085 | 4845 | 230 | 555 | Gower et al. 1996; Reich et al. 1997; Elliott et al. 2002 |
| Temperate subalpine | | 12860 | 36 | 190 | Arthur and Fahey 1992; Hansen et al. 2000 |
| Tropical evergreen | 2380 | 32450 | 140 | 1505 | Gower et al. 1996; Clark et al. 2001a |
| Tropical seasonal | 15055 | 22580 | | | Jaramillo et al. 2003 |
| Tropical montane | 4696 | 21850 | 390 | 956 | Kitayama and Aiba 2002; Jaramillo et al. 2003 |
| Mangrove | 1115 | 35550 | 305 | 1350 | Sherman et al. 2003 |
| Overall | 540 | 49126 | 36 | 1555 | This summary |

*Note*: These estimates should be used as a guide for the expected magnitude of ANPP in different forest ecosystems.

cluded data from several studies of temperate forest systems that had ANPP ranging from a low of 105 g C m$^{-2}$ year$^{-1}$ (Ryan and Waring 1992) to a high of 1030 g C m$^{-2}$ year$^{-1}$ (Runyon et al. 1994). Gower et al. (2001) summarized boreal ANPP from 9 sites ranging across North America and Eurasia. The boreal systems exhibited a moderate range of ANPP from 129 g C$^{-2}$ year$^{-1}$ (Gower et al. 1997) to 635 g C m$^{-2}$ year$^{-1}$ (Ruess et al. 1996).

The primary drivers (climatic and biotic) behind the exhibited range in ANPP vary widely. Predominant drivers include soil moisture as influenced by soil water-holding capacity and annual precipitation (Knapp and Smith 2001), N availability (Reich et al. 1997), temperature (Schuur 2003), and light (Runyon et al. 1994). While ANPP generally increases as the magnitude of these drivers increases, in some cases complex interactions can lead to negative relationships; for example, Schuur (2003) noted that NPP declined where mean annual precipitation increased more than approximately 2200 mm year$^{-1}$ because of reduced light. Canopy trees usually comprise the great majority of NPP in closed-canopy forests because they utilize most of the light resource. However, in tropical systems, vine production can be an important contributor and controller of ANPP. Increased vine production as a result of increased atmospheric $CO_2$ concentrations can cause increased mortality of over-story trees, resulting in a net decline of forest ANPP (Fearnside 1995). As the over-story canopy becomes more open, an increasing proportion of NPP is contributed by understory vegetation. As noted by Vitt (chap. 6, this volume), in boreal forest ecosystems, moss and lichen ground cover is a large contributor to ANPP. O'Connell et al. (2003a) summarized ANPP in black spruce ecosystems with a feather moss understory and a sphagnum understory where 19.7% and 78.2% of ANPP were contributed by the bryophyte and understory layers, respectively.

## Guiding Principles and Recommendations for Measuring Forest ANPP

The 3-dimensional structure of forest ecosystems and the perennial age structure of tree woody tissues pose several challenges to accurately measuring forest ANPP. Because the field methods for measuring $\Delta B$ and M in forests are typically applied separately to the production of woody tissues, foliage (and other ephemeral tissues), and tree mortality, we organize our detailed review of methods and their limitations in these three categories.

### Wood Production

The determination of wood production or woody biomass increment typically involves repeated measurement of tree diameters and the application of allometric equations to estimate changes in biomass from these diameter measurements. We focus our attention in this section on the accurate and precise measurement of tree diameters in fixed-area permanent plots. While variable radius and other "plotless" methods have been used, they are problematical for tree ingrowth, defined as the growth of small trees into the minimum size class used in the forest survey. Neither this size class nor the time interval of remeasurement can be strictly specified, being dependent on the structure and growth rate of the particular forest. Ingrowth will become a significant proportion of the estimated $\Delta B$ only when the measurement interval is relatively long (e.g., more than 5 yr). In cases of slower-growing forests, repeated measures are not made annually; rather, the measurements may be made at the start and end of some longer time span and divided by the number of years to determine average annual diameter increment. It is best to make repeated tree diameter measurements during the dormant season, so that the entire annual growth increment can be captured. Consistent measurement season becomes most critical when the measurement interval is short, although it can influence estimates for intervals of up to 10 years by as much as 5%. Tree diameters usually are measured at breast height (1.37 m [4.5 feet] above the soil surface), a common forestry definition (Avery and Burkhart 1983) that avoids swollen tree bases in most, but not all, tree species. For example, special approaches are needed for buttressed trees; Sherman et al. (2003) measured tree diameters above the highest prop root in a tidal mangrove ecosystem to maintain a biologically consistent measurement location on the bases of the tapered stems. On sloping ground it is best to use the uphill side of the tree as the point to determine breast height.

Tree diameters can be measured with a tape that is placed around the circumference of the tree stem. The absolute value of precision for diameter measures is an inverse function of diameter, although as a proportion, precision appears to be relatively constant (fig. 5.1). For example, over a range in DBH from 17 to 125 cm, 2 standard deviations of 7 measurements varied between 0.3% and 1.7% of the mean. The average of 2 standard deviations was 0.8%, indicating that a precision of 1% should be expected for most measurements. Bark sloughing between measurements will result in an underestimate of tree diameter increment or even in a negative growth increment. It may be best to remove any obviously loose bark at breast height before the first diameter measurement is made. Heavy epiphyte growth on stems can also cause errors in diameter measurement; as with loose bark, it may be best to remove these plants before the first measurement and before subsequent measurements. Closely growing trees may eventually grow into each other, causing difficulties in determining individual diameters. One may either measure the combined diameters and partition them according to relative size, or measure the portion of circumference that is exposed. Regardless, it is important to have some sense of relative growth of the trees before they merge. Diameter tapes are easy to use, but care should be taken to make repeated measurements at the same height on the tree stem and at same orientation relative to the axis of the stem. Diameter measurements should be oriented to be perpendicular to the long axis of the stem. This can be ensured by painting or otherwise marking the tree stem at breast height. For many studies, it is necessary to permanently tag trees for repeated census. In this case, the tree tag can be held with an aluminum nail, with plenty of room to grow at breast height, to mark the location of future tree measurements. It is best to avoid using steel or galvanized nails that can corrode over time and cause serious safety issues decades later, when the tree may be cut for future studies or in forest harvesting operations. Galvanized nails can also cause bark necrosis in some species, particularly angiosperms. In plots on very steep ground or in cases where trees are very large, have significant butt swell, or are not growing vertically, it is helpful to have several points marked to assure that diameter measurements are repeatable.

An important issue in the repeated measurement of tree diameters in permanent plots is whether and how to use previous measurements while making current measurements. Often diameter increments will be small relative to the precision of diameter measurements. Checking current diameter measurements against previous values for the same tree will help to reveal obvious inaccurate values (e.g., shrinkage or unrealistic high growth). Cases of unlikely shrinkage would be indicated if the DBH "shrinks" by more than 1%. Cases of unlikely growth would be indicated if the DBH increases more than the expected growth plus 1%. We recommend the following procedure both to reduce measurement error and to avoid possible bias owing to greater circumspection of apparently low rather than apparently high measurements (i.e., it is discomfiting to observe trees shrinking but less so to observe trees growing slightly fast): (1) when first remeasuring a tree, the person with the DBH tape makes the measurement without knowledge of the previous value; (2) the data recorder checks the new value against previous measurements and asks for a remeasurement if there is an obvious discrepancy; (3) if the discrepancy persists, further work may be necessary to resolve the problem.

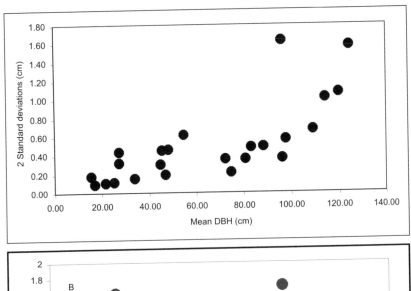

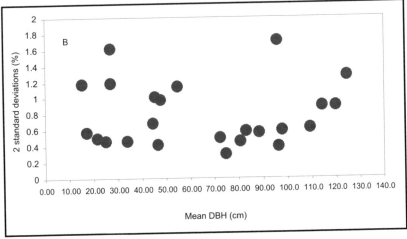

Figure 5.1. Precision of DBH measurements for 24 trees ranging in DBH from 17 to 125 cm. Seven independent sets of measurements were taken. Top panel: absolute difference from the mean expressed as 2 standard deviations as a function of mean diameter. Bottom panel: relative difference expressed as 2 standard deviations as a function of mean diameter.

An alternative to using diameter tapes is to install dendrometer bands on the tree at breast height. These are metal bands that expand as the tree grows. This expansion in circumference can be measured with a digital micrometer, adhesive labels installed with vernier scales, or dial gauges (Heinrich and Banks 2005). The circumference increment can be easily converted to diameter increment. Dendrometer bands provide greater measurement precision than using a diameter tape, but the costs can be high. Dendrometer bands are advantageous when diameter increments are extremely small (for trees growing on extremely cold, infertile, or dry sites), when remeasurements of tree diameter are taken more frequently than once a year, or in forests without distinct growing seasons. They have the disadvantage of expansion and contraction with changes in temperature as well as of detecting moisture- related changes that are not part of the growth signal.

In some studies, a reconstruction of historical tree diameter increment may facilitate the analysis of wood biomass productivity. To accomplish this, the annual growth rings are measured from extracted wood cores. Annual rings are distinguishable in most temperate and boreal tree species; however, many tropical species do not produce tree rings that represent annual growth cycles (Clark et al. 2001a). Therefore, this method is not useful in most tropical ecosystems. The reconstruction of past tree diameter increments facilitates analysis of several years of wood increment during one sampling campaign, rather than having to wait for many years of wood growth after the initial plot establishment. To reconstruct the wood increment, the current tree diameter needs to be measured with a diameter tape. Then a wood core should be extracted from the tree, using an increment borer. One needs to account for the bark as well. As discussed below, most allometric biomass equations are developed for outside bark diameter. Hence, bark thickness should be measured when the core is extracted. Since tree stems are rarely perfectly round in cross section, it is best to extract two wood cores from the tree at a 90° angle to one another. Given the pattern of compression and reaction wood, orienting these cores up and across slope is a good strategy. Because of the often high variation of individual tree growth, at least 50 trees typically need to be measured in each plot sampled. Subsampling in plots with larger numbers of trees is possible with a stratification by size and species. After cores are extracted, they should be mounted in plastic or wood, sanded or trimmed with a sharp blade to clearly expose the annual growth rings, and, if possible, kept moist and cold. Any wood shrinkage due to drying will result in an underestimate of diameter increment of approximately 10%. Alternatively, one can estimate radial shrinkage and make an adjustment to the increments measured. The increments can be measured with an optical dendrometer, dissecting microscope, precision ruler, or scanned and analyzed with a bed scanner and software. After measurement, the 2 growth increments for each year (one from each core) are added to generate the wood diameter increment. Due to internal decay or subsampling, some trees will not be cored. The diameter increment of these trees and the error introduced by this substitution can be estimated using Monte Carlo methods.

The approaches for converting diameter increment measurements into woody biomass production are detailed later in this chapter.

## Production of Ephemeral Tissues

Foliage usually comprises a high proportion of the ephemeral tissues produced by trees, although in "mast" years fruits and seeds may be significant in some species (Barnes et al. 1998). Foliage productivity often is easier to estimate than wood productivity, although in cases of young forests with increasing leaf biomass, the same challenges may exist as for wood. The same basic foliage method can be used whether the tree foliage phenology is broadleaf or needleleaf, or the leaf life span is less or greater than 1 yr. Assuming ΔB for foliage is zero, the equivalent of foliage production in forest stands can be measured (as a component of M) with a leaf littertrap. Trees that have a deciduous leaf habit will excise all foliage by the end of the growing season, whereas trees with an evergreen leaf habit contain more than 1 yr of foliage. Evergreen foliage can live for many years; in extreme cases, over 20 yrs of foliage may be retained in the forest canopy (Schoettle and Fahey 1994). In mature evergreen forests the collection of foliage over 1 yr is approximately equivalent to the new foliage produced in the canopy. This approximation is most meaningful over the long term, and at an annual interval is not necessarily related to the actual foliage production for that year.

Numerous approaches have been used for forest litter collection, and the most appropriate approach will depend upon a variety of site-specific factors. In particular, the size and nature of the collection device, the number and spatial arrangement of collectors, and the frequency of collection must be chosen carefully to avoid bias and error. Collectors with a fine-mesh bottom allow drainage of rainwater while retaining most of the smallest litter fragments. Spatial variation in litterfall depends mostly on the canopy structure; for example, higher variation typically occurs for forests of excurrent canopy form (e.g., conifers) than for decurrent form. A sample size of 4 to 6 traps of 0.25 m² or larger is usually sufficient to measure litterfall with precision better than ± 5%. The frequency of litter collection must be chosen to minimize losses owing to leaching and litter decomposition; hence, the optimum interval depends upon environmental and biotic conditions affecting these processes as well as the timing of leaf abscission. For example, in the lowland tropics, collecting at 2-wk intervals year-round is often deemed necessary (Clark et al. 2001a), whereas in cold, temperate deciduous forest, collection can be focused on the autumn leaf fall period, with a few additional collections in other seasons.

An alternative to using litter traps for foliage production estimates is to utilize the tree diameter data described above in conjunction with allometric relationships. For deciduous trees, the annual tree foliage production is roughly equal to the estimated foliage biomass from the allometric equation. For evergreen trees, the annual tree foliage production is equal to the estimated foliage biomass from the allometric equation divided by the median leaf longevity in years. If significant amounts of organic matter are resorbed from foliage prior to abscission (Fahey et al. 2005), then this allometric approach will yield systematically higher estimates of foliage production than the litterfall method; the latter may be closer to the actual foliage production, assuming that resorbed organic matter is subsequently utilized in leaf growth. Although the allometric method may not be as reliable as the collection of excised leaf litter, it does provide an estimate of foliage production when other data may not be available. For forests in the early stages of recovery from disturbance, allometric equations can also be used to estimate ΔB of foliage.

### Tree Mortality

As mentioned under "Woody Production," depending on the method used to estimate ΔB, neglecting tree mortality will lead to underestimation of woody production. Tree mortality is a crucial component of ANPP calculations that often is not included (Gower et al. 2001). The magnitude of this error will depend mostly upon the time interval between plot censuses. For example, biomass mortality in mature forests typically is in the range of 1%–2% per year; hence, if annual increments from tree cores are used to estimate $\Delta B$, the error from mortality will be small, whereas if repeated measures of diameters of tagged trees at decadal intervals are used, the error will be substantial. Also, in the latter case, the growth of any trees that died during the measurement interval should be added to the mortality estimates described below. Tree mortality estimates can be determined from the same tree census and diameter data collection mentioned in the "Woody Production" section above. Mortality is best determined on trees that have been marked or mapped, as it is extremely difficult to determine the exact year of tree death. It is easy to overlook new mortality unless each tagged or marked tree is individually checked. If mortality data are to be used to determine forest carbon balance, it is helpful to know the proportion of mortality that remains standing versus falling to the ground. Tree mortality is highly variable in time and space. To provide adequate estimates, one needs to sample a suitably sized area over a number of years. Assuming that the death of at least 10–20 trees would form an adequate sample (if all trees were similar in size, this would provide an estimate within 5% to 10% if one tree was added or deleted), a minimum of 5 ha-years (the product of area and time) would be required in many forests.

### Understory Production

While understory plants do not comprise a large fraction of forest biomass, sometimes they can form a significant share of the ANPP. For herbaceous, shrub, and moss growth forms, the reader is referred to chapters 3, 4, and 6 in this volume, respectively. For understory trees the methods must be adapted from those for larger trees. Typically a nested subplot can be employed to manage the higher stem density in smaller size classes. Because tagging trees is difficult below about 10 cm DBH and the diameter growth rates of suppressed stems often are very small, estimation of $\Delta B$ from resurveys, as described for larger trees, is impractical. Moreover, because these trees normally constitute less than 5% of forest ANPP, the precision of estimates does not need to be as high as for the larger trees. A combination of allometric estimates of understory tree biomass, including carefully measured ring widths on the harvested trees, and periodic remeasurement of densities by size class in nested subplots, normally will provide a sufficient basis for estimating understory tree ANPP.

### Allometric Equations for $\Delta B$

### General

Diameter increments, obtained either from repeated surveys of individual trees or from wood increment cores, are scaled to woody biomass increments, using allometric equations. These equations are used to relate the easily measured dimension of tree diameter (and sometimes height) to the biomass of various tissues. The difference in biomass between the 2 measurements is divided by the measurement intervals (years) to obtain the annual $\Delta B$.

When possible, site-specific allometric equations should be used because site abiotic and biotic conditions may generate unique tree characteristics that are not captured in general allometric equations from the literature (e.g., Swank and Schreuder 1974; Schreuder and Swank 1974). Nonetheless, many studies involving ANPP will not have the resources or time to develop site-specific allometric equations; in this case, equations developed at sites with similar growth forms can be used, although the degree of similarity is difficult to determine. A survey of the literature may generate useful species-specific or growth-form-specific equations. One informative approach in such cases would be to present the range of estimates obtained from a series of different equations. Currently, the usual approach for studies of lowland tropical forest is to apply the generalized equations of Brown (1997) across all but the most divergent tree growth forms (e.g., ferns, palms). Two extensive summaries of tree allometric coefficients have been compiled that are good first sources for allometric coefficients in boreal and temperate forests. Ter-Mikaelian and Korzukhin (1997) list biomass equations for 65 North American tree species. In addition, they have compiled multiple equations for most species, noting the state or province where the data were collected and the diameter range of the original data, thereby encouraging the user to be cautious regarding the validity of the results when predicting tree biomass. Jenkins et al. (2003) performed a similar review of data in the literature, then produced summaries and used the coefficients to develop general equations rather than to maintain site-specific allometric equations. In this case, their primary objective was to develop national-scale biomass estimators for United States tree species.

The allometric estimates of $\Delta B$ are typically developed by harvest of representative trees (see "Recommended Approaches," below). A source of error that is related to, but distinct from, mortality is wood rot and pruning of dead branches in living trees. If the former component is changing significantly in the forest under study, then even site-specific allometric equations that include truly representative trees (as opposed to only healthy ones) will provide biased ANPP estimates. Moreover, because loss of woody material (especially branches) during the sampling interval must be added into the $\Delta B$, ignoring branchfall will result in an underestimate of ANPP (Clark et al. 2001a). Unfortunately, branchfall is highly episodic and spatially variable (Fahey et al. 2005), so that long-term measurements (10+ years) must be obtained on relatively large branch-removal plots (e.g., 25 m$^2$) to obtain accurate corrections. Also, to avoid double counting, branches from standing dead trees must be excluded from these collections.

## Approaches for Developing Allometric Equations

Because site-specific allometric equations are so important, we provide a summary of the method (for additional details, see Whittaker et al. 1974; Martin et al. 1998; or Hanson and Wullschleger 2003). We encourage readers to survey the literature and to refine the methods specific to the wood, bark, branch, and foliar characteristics for the species under study. Typically, at least 10 trees of each species are selected, using a stratified random sampling design (Avery and Burkhart 1983) to ensure that the entire diameter range of trees of each species is sampled. As noted earlier, these trees should be representative of the forest, but because of limitations on sample size and the desirability of obtaining strong regression relationships, some selectivity is needed. Above all, any criteria in sample selection need to be carefully recorded and reported with the equations. Measure the diameter of each tree at breast height (DBH = 1.37 m), fell the tree, and measure the height from the base of the tree to the base of the live crown (BLC), crown length, and total height to the nearest centimeter. For smaller trees mark the live canopy into thirds, remove all branches from each section, and weigh them, using a high-capacity balance. Select a subsample branch from each crown position to determine the ratio of foliage to branch mass. For larger trees, branches may have to be removed while the tree is standing (Brown 1997), and it may be necessary to develop a separate branch mass regression based on branch diameter (and length). To estimate total branch mass, the diameter (and possibly the length) of each branch is measured.

For small trees, cut the stem into sections and weigh each section. Cut a 2-cm-thick disk from the base of each stem section to determine water content. Store tissue subsamples in plastic bags and place them in a cooler (3° C) at the end of the day to minimize moisture loss. For larger trees, it may be necessary to measure the volume of trees and remove disks to determine density. Stem volume can be determined by measuring the diameter at several points along the stem for excurrent forms and at more points for decurrent stem forms. Wood and bark density can be determined from disks by measuring the diameter and thickness to determine disk volume, and weighing the entire disk and taking subsamples to determine the moisture content. Pie-shaped subsamples are ideal because they proportionally weight tissues according to their volume.

For laboratory processing, determine the fresh mass of each subsample and separate the branch subsample into new foliage (present year), old foliage (if multiple age classes are present), new twigs, and wood components. Then dry the tissues to a constant mass at 65°C and weigh them to determine moisture content. In some cases, the drying process may take several weeks.

Calculate the total dry mass of the foliage or branches for a given canopy section by multiplying the ratio of dried foliage or dried branches to the total dry mass of the crown section subsample by the total crown section dry mass. Then sum the total foliage mass and branch mass for the tree across all three canopy positions.

For larger woody parts, weigh the fresh mass of each stem disk and then dry the disk at 65°C to a constant mass and weigh it. For large disks one may need to subsample in order to determine moisture content. Determine stem section dry mass by multiplying (1 − moisture content) by the field wet weight, and calculate total wood mass and bark mass for the section from the ratios of bark or wood to the total disk dry weight. Linear regressions are typically used to compute allometric relationships, using $\log_{10}$ transformed data (to linearize) and the following equation:

$$\log_{10} Y = a + b * \log_{10} X,$$

where X is the stem diameter in cm at breast height, Y is the dependent variable (e.g., stem wood mass, stem bark mass, foliage mass, etc.), and $a$ and $b$ are the intercept and the slope, respectively. While other forms of equations can be developed by always presenting the recommended equation, a large set of similar equations can be developed rather than a unique form for each study. Another common equation uses diameter and height as independent variables, although when site conditions are very similar, adding height may not explain significantly more variation in the population. In the case of foliage biomass, DBH is not always the best independent variable, especially for larger trees. In this case sapwood area is a better predictor of foliage mass and leaf area; and sapwood area for individual stems can be obtained using increment cores.

The optimal timing for tree harvest varies, depending upon the species and tissue components desired. If foliage mass and/or leaf area relationships are desired, we recommend harvesting trees for developing allometric relationships in the later part of the growing season, after leaf tissues have matured; for evergreen trees the mass of first-year foliage (and twigs) can be obtained, and the median longevity of leaves also should be estimated, if possible.

### Special Considerations: Losses to Herbivory, Leaching, and Volatilization

As noted earlier, Clark et al. (2001a) thoroughly reviewed the magnitude and approaches for estimating the losses of ANPP to consumers and via leaching and volatilization of organic matter. Losses to herbivory of leaves and reproductive tissues varies markedly among forest types and between years. In those situations where it is expected to be significant, substantial efforts to obtain accurate estimates are warranted. For foliage herbivory, a combination of measurements of leaf area losses from litter samples and tracking of individual leaves for entire leaf consumption is ideal (Clark et al. 2001a). Predispersal consumption of fruits also may be significant in some forests, but few estimates of this loss are available (Janzen and Vazquez-Yanes 1991; Lugo and Franzi 1993). Finally, losses to volatilization were regarded as a minor proportion of ANPP by Clark et al. (2001a), and Fahey et al. (2005) estimated canopy leaching of organic matter to be 0.9% of NPP in a temperate broadleaf forest.

### Plots and Scaling Considerations

The optimal size, number, and placement of sample plots for quantifying forest ANPP vary with the structure and dimensions of the forest and with the aims for scaling the plot measurements to the larger forest. Obviously, larger plots will be needed in lower-density forest composed of large trees than in higher-density stands, but choice of optimal plot size and number may be complex. For example, a few large, individual trees can comprise a high proportion of the biomass in some mature forests, but they could be seriously misrepresented (either over- or under-) in relatively small plots. There will always be some trade-off between plot size and replication. In general, larger plots are more desirable for NPP measurements than those often employed for vegetation composition studies in low-diversity temperate and boreal forests because edge effects are particularly serious sources of error in biomass and NPP estimates.

A general rule of thumb based on the experience of the authors is that the plot size should be chosen to encompass at least 75–100 trees larger than the lower diameter cutoff (often 10 cm DBH). In general, as the plot size increases, the variation between plots decreases (fig. 5.2). Variation should decrease as the number of replicate plots increases, although in some cases the standard deviation can increase as plot number increases (fig. 5.3A). This may be due to the fact that as more plots are added, the actual spatial variation becomes more apparent. If the plot size is adequate, the primary influence of sample size will be on the standard error of the mean, and a large sample size may be required to have standard errors within 5%–10% of the sample mean (fig. 5.3B). The placement of sample plots depends upon the problems and approaches for scaling the plot-based estimates of NPP to the larger forest under study. This consideration can also influence optimal plot size; for example, in complex terrain, if a stratified approach to sample placement is applied (e.g., stratified by slope position), then plot size may be constrained by the scale of terrain units. Researchers should be aware of the likely existence of high spatial variation of NPP within many forested landscapes (Fahey et al. 2005); hence, extrapolation of NPP values from a few plots established on convenient or uniform sites may yield erroneous large-scale estimates. Because the need for relatively large plots often precludes higher replication, the choice of sample placement may be a serious challenge. In sum, because forest structure as well as the approaches for scaling the plot-based measurements will differ markedly among NPP studies, it is not possible to specify uniform recommendations for plot size, number, and sample placement. Nevertheless, these three aspects of the NPP sampling program are critical to its success, and researchers must give them careful consideration and adequate documentation.

Both circular and rectangular plots are commonly employed. The advantage of circular plots is that surveying and marking plot boundaries is not necessary. Rather, by sighting from the plot center with a range finder, any trees near the outer boundary of the plot are checked for possible inclusion in the sample. In hilly terrain it is important to make a slope correction for the sighting distances (or boundary lengths for rectangular plots) because plot areas should reflect projected rather than ground

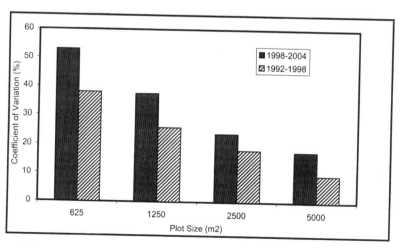

Figure 5.2. Influence of plot size on the coefficient of variation of bole-related NPP estimates on a *Pseudotsuga–Tsuga*-dominated forest in Oregon.

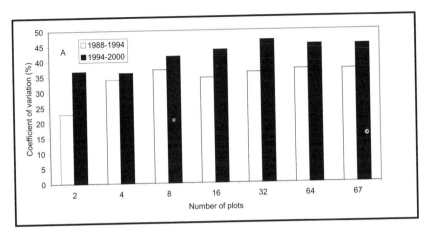

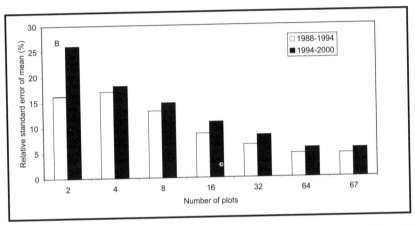

Figure 5.3. Effect of sample size on estimates of the coefficient of variation (A) and the relative standard error of the mean (B) of bole-related NPP. Each 0.1-ha plot was located in a *Pseudotsuga–Tsuga* forest, and the total number of plots was 67. Subsamples of plots were drawn from the data set 20 times to compute the average and standard deviation of NPP.

area. The plot center should be marked with a sturdy stake, and the GPS location recorded. To facilitate relocation of trees and to minimize the chances of "losing" trees, it is helpful to number the trees consecutively in a clockwise fashion.

## Future Research Needs

The need for more accurate and precise NPP data from a wider range of forest sites is likely to increase as complex ecosystem models and forest carbon seques-tration programs demand high-quality carbon (C) flux estimates for model devel-opment and project verification. Both methodological improvements and greater standardization of approaches are needed. As detailed in chapter 12 of this volume, in the future all studies of forest NPP should include explicit consideration of potential biases, sources of error, and quantitative evaluation of uncertainty; these considerations have been sorely missing from most publications reporting forest NPP. Another component that often has been inadequately addressed, but should be part of all standard measurements of forest NPP, is tree mortality (Acker et al. 2002). For this purpose tagged-tree inventories should be a central element of all NPP measurement programs. Finally, the problem of branch litterfall needs to be addressed at all sites. Periodic resurveys of branch-removal plots over several years are needed to improve estimates of this rarely measured component of ANPP.

Most of the other improvements in NPP estimates require that detailed studies, many of which are beyond the capability of routine measurement, be conducted at selected benchmark sites to provide the information needed to correct NPP estimates over a wider range of sites. Based on the analysis of Clark et al. (2001a) and our foregoing discussion, 3 key subjects of detailed, benchmark studies are most critical: (1) losses to herbivory, (2) losses to heart rot, and (3) the allometry of exceptionally large trees. It is clear that in many forest biomes, losses to herbivory are significant, and both temporally and spatially variable. Because of the difficulty of canopy access, remote-sensing approaches to the problem of herbivory losses may be most practical, and interactions between ecosystems scientists and forest pest and pathogen programs are needed. To our knowledge, no estimates are available of the magnitude of bias resulting from not accounting for rot of tree boles. Because most allometric equations are developed using healthy trees, as forests age and an increasing proportion of wood volume is subject to rot, there is the potential for systematic, large-scale overestimation of NPP unless correction factors are applied. Initially, these should be developed for benchmark sites for each of the major forest biomes, in order to evaluate the likely magnitude of error and to direct further efforts.

Finally, and perhaps most importantly, there is an urgent need for the dimensional analysis of very large trees in several forest biomes in order to improve the allometric basis for quantifying $\Delta B$. Trees beyond the diameter range over which allometric equations were developed often comprise a high proportion of forest biomass (Brown and Lugo 1992), and biomass estimates for these out-of-range trees may be highly inaccurate (Brown et al. 1995). Processing of such large trees should be conducted in conjunction with professional forest management operations. Additional measurements of tree form (e.g., diameter at midheight) may also aid in improving the accuracy of allometric estimates of the biomass of such large trees.

In conclusion, forest biomes require additional research, summary, and synthesis on ANPP to better understand annual dynamics in relation to water and nutrient availability and changes in temperature and pollution of the environment. These dynamics will continue to play a major role as global weather patterns change and affect all ecosystems, especially those such as boreal and tropical ecosystems, where marginal changes can have a marked influence on growing season length, water availability, productivity, and C allocation (Gower et al. 2001; Schuur 2003; Beedlow et al. 2004).

*Acknowledgments*    Howard Bruner and Gody Sycher provided assistance on the analysis of diameter measurement errors and effects of plot and sample sizes on NPP estimates. This synthesis and the research it was based on were funded in part by the Coweeta LTER (NSF DEB-0218001) and Andrews LTER (NSF DEB-0218088) Programs, Pacific Northwest Research Station, the Kay and Ward Richardson Endowment, and the Bullard Fellowship Program of Harvard University.

## References

Acker, S. A., C. B. Halpern, M. E. Harmon, and C. T. Dyrness. 2002. Trends in bole biomass accumulation, net primary production, and tree mortality in *Pseudotsuga menziesii* forests of contrasting age. Tree Physiology 22:213–217.

Arthur, M. A., and T. J. Fahey. 1992. Biomass and nutrients in an Engelmann spruce—subalpine fir forest in north central Colorado: Pools, annual production, and internal cycling. Canadian Journal of Forest Research 22:315–325.

Avery, T. E., and H. E. Burkhart. 1983. Forest Measurements, 3rd ed. McGraw-Hill, New York.

Barnes, B. V., D. R. Zak, S. R. Denton, and S. H. Spurr. 1998. Forest Ecology, 4th ed. Wiley, New York.

Beedlow, P. A., D. T. Tingey, D. L. Phillips, W. E. Hogsett, and D. M. Olszyk. 2004. Rising atmospheric $CO_2$ and carbon sequestration in forests. Frontiers in Ecology and the Environment 2:315–322.

Brown, I. F., L. A. Martinelli, W. W. Thomas, M. Z. Maveira, C. A. Cid Ferreira, and R. A. Victoria. 1995. Uncertainty in the biomass of Amazonian forests: An example from Rondonia, Brazil. Forest Ecology and Management 75:175–189.

Brown, S. 1997. Estimating Biomass and Biomass Change of Tropical Forests: A Primer. Forestry Paper 134. Food and Agriculture Organization, Rome.

Brown, S., and A. E. Lugo. 1992. Aboveground biomass estimates for moist forests of the Brazilian Amazon. Interciencia 17:8–18.

Clark, D. A., S. Brown, D. W. Kicklighter, J. Q. Chambers, J. R. Thomlinson, and Jian Ni. 2001a. Measuring net primary production in forests: Concepts and field methods. Ecological Applications 11:356–370.

Clark, D. A., S. Brown, D. W. Kicklighter, J. Q. Chambers, J. R. Thomlinson, Jian Ni, and E. A. Holland. 2001b. Net primary production in tropical forests: An evaluation and synthesis of existing field data. Ecological Applications 11:371–384.

Crews, T. E., K. Kitayama, J. H. Fownes, R. H. Riley, D. A. Herbert, D. Mueller-Dombois, and P. M. Vitousek. 1995. Changes in soil phosphorus fractions and ecosystem dynamics across a long chronosequence in Hawaii. Ecology 76:1407–1424.

Elliott, K. J., L. R. Boring, and W. T. Swank. 2002. Aboveground biomass and nutrient accumulation 20 years after clear-cutting a southern Appalachian watershed. Canadian Journal of Forest Research 32:667–683.

Fahey, T. J., T. G. Siccama, C. T. Driscoll, G. E. Likens, J. Campbell, C. E. Johnson, J. D. Aber, J. J. Cole, M. C. Fisk, P. M. Groffman, S. P. Hamburg, R. T. Holmes, P. A. Schwarz, and R. D. Yanai. 2005. The biogeochemistry of carbon at Hubbard Brook. Biogeochemistry 75:109–176.

Fearnside, P. M. 1995. Potential impacts of climatic change on natural forests and forestry in Brazilian Amazonia. Forest Ecology and Management 78:51–70.

Finer, L. 1989. Biomass and nutrient cycle in fertilized and unfertilized pine, mixed birch and pine and spruce stands on a drained mire. Acta Forestalia Fennica 208:654.

Flower-Ellis, J. G. K., and H. Persson. 1980. Investigation of structural properties and dynamics of Scots pine stands. Ecological Bulletin (Stockholm) 32:125–138.

Gabeev, V. N. 1990. Ecology and Productivity of Pine Forests. Nauka, Novosibirsk, Russia. (In Russian).

Gholz, H. L., and R. F. Fisher 1982. Organic matter production and distribution in slash pine (*Pinus elliottii*) plantations. Ecology 63:1827–1839.

Gower, S. T., O. Krankina, R. J. Olson, M. Apps, S. Linder, and C. Wang. 2001. Net primary production and carbon allocation patterns of boreal forest ecosystems. Ecological Applications 11:1395–1411.

Gower, S.T., S. Pongracic, and J. J. Landsberg. 1996. A global trend in belowground carbon allocation: Can we use the relationship at smaller scales? Ecology 77:1750–1755.

Gower, S. T., J. G. Vogel, J. M. Norman, C. J. Kucharik, S. J. Steele, and T. K. Stow. 1997. Carbon distribution and aboveground net primary production in aspen, jack pine, and black spruce stands in Saskatchewan and Manitoba, Canada. Journal of Geophysical Research 102(D24): 29029–29041.

Grier, C. C., and R. S. Logan. 1977. Old-growth *Pseudotsuga menziesii* communities of a western Oregon watershed: Biomass distribution and production budgets. Ecological Monographs 47:373–400.

Han, M. 1994. A study on biomass and net primary production in a Dahurian larch-birch forest ecosystem. Pages 451–458 in X. Zhou (ed.), Long-Term Research on China's Forest Ecosystems. Northeast Forestry University Press, Harbin, China. (In Chinese).

Hansen, A. J., J. J. Rotella, M. P. V. Kraska, and D. Brown. 2000. Spatial patterns of primary productivity in the Greater Yellowstone Ecosystem. Landscape Ecology 15: 505–522.

Hanson, P. J., and S. D. Wullschleger (eds.). 2003. North American Temperate Deciduous Forest Responses to Changing Precipitation Regimes. Springer-Verlag, New York.

Heinrich, I., and J. C. G. Banks. 2005. Dendroclimatological potential of the Australian red cedar. Australian Journal of Botany 53:21–32.

Janzen, D. H., and C. Vazquez-Yanes. 1991. Aspects of tropical seed ecology of relevance to management of tropical forested wildlands. Pages 137–157 in A. Gomez-Pompa, T. C. Whitmore, and M. Hadley (eds.), Rain Forest Regeneration and Management. UNESCO, Paris.

Jaramillo, V. J., J. B. Kauffman, L. Renteria-Rodriguez, D. L. Cummings, and L. J. Ellingson. 2003. Biomass, carbon, and nitrogen pools in Mexican tropical dry forest landscapes. Ecosystems 6:609–629.

Jenkins, J. C., D. C. Chojnacky, L. S. Heath, and R. A. Birdsey. 2003. National-scale biomass estimators for United States tree species. Forest Science 49:12–35.

Kira, T., H. Ogawa, K. Yoda, and K. Ogino. 1967. Comparative ecological studies on three major types of forest vegetation in Thailand. IV. Dry matter production, with special reference to Khao Chong rain forest. Nature and Life in Southeast Asia 5: 149–174.

Kitayama, K., and S. Aiba. 2002. Ecosystem structure and productivity of tropical forests along altitudinal gradients with contrasting phosphorus pools on Mount Kinabalu, Borneo. Journal of Ecology 90:37–51.

Knapp, A. K., and M. D. Smith. 2001. Variation among biomes in temporal dynamics of aboveground primary production. Science 291:481–484.

Lugo, A. E., and J. L. Frangi. 1993. Fruit fall in the Luquillo Experimental Forest, Puerto Rico. Biotropica 25:73–84.

Martin, J. G., B. D. Kloeppel, T. L. Schaefer, D. L. Kimbler, and S. G. McNulty. 1998. Aboveground biomass and nitrogen allocation of ten deciduous southern Appalachian tree species. Canadian Journal of Forest Research 28:1648–1659.

Martinez-Yrizar, A., J. Sarukhan, A. Perez-Jimenez, E. Rincon, J. M. Maass, A. Solis-Magallanes, and L. Cervantes. 1992. Above-ground phytomass of a tropical deciduous forest on the coast of Jalisco, Mexico. Journal of Tropical Ecology 363:234–240.

Müller, D., and J. Nielsen. 1965. Production brute, pertes par respiration et production nette dans la foret ombrophile tropicale. Forstlige Forsoegsvaesen (Denmark) 29:69–160.

Nadelhoffer, K. J., and J. W. Raich. 1992. Fine root production estimates and belowground carbon allocation in forest ecosystems. Ecology 73:1139–1147.

Nishioka, M. 1981. Biomass and productivity of forests in Mt. Mino. Pages 149–167 in 1980 Annual Report of the Education Committee, Mimo City, Osaka, Japan.

O'Connell, K. E. B., S. T. Gower, and J. M. Norman. 2003a. Comparison of net primary production and light-use dynamics of two boreal black spruce forest communities. Ecosystems 6:236–247.

O'Connell, K. E. B., S. T. Gower, and J. M. Norman. 2003b. Net ecosystem production of two contrasting boreal black spruce forest communities. Ecosystems 6:248–260.

Olson, R. J., J. M. O. Scurlock, S. D. Prince, D. L. Zheng, and K. R. Johnson (eds.). 2001. NPP multi-biome: Global Primary Production Data Initiative products. Data set. Available online at http://www.daac.ornl.gov from the Oak Ridge National Laboratory Distributed Active Archive Center, Oak Ridge, TN.

Rai, S. N., and J. Proctor. 1986a. Ecological studies on four rain forests in Karnataka, India. I. Environment, structure, floristics, and biomass. Journal of Ecology 74:439–454.

Rai, S. N., and J. Proctor. 1986b. Ecological studies on four rain forests in Karnataka, India. II. Litterfall. Journal of Ecology 74:455–463.

Reich, P. B., D. F. Grigal, J. D. Aber, and S. T. Gower. 1997. Nitrogen mineralization and productivity in 50 hardwood and conifer stands on diverse soils. Ecology 78:335–347.

Ruess, R. W., K. Van Cleve, J. Yarie, and L. A. Viereck. 1996. Contributions of fine root production and turnover to the carbon and nitrogen cycling in taiga forests of the Alaskan interior. Canadian Journal of Forest Research 26:1326–1336.

Runyon, J., R. H. Waring, S. N. Goward, and J. M. Welles. 1994. Environmental limits on net primary production and light-use efficiency across the Oregon transect. Ecological Applications 4:226–237.

Ryan, M. G., and R. H. Waring. 1992. Maintenance respiration and stand development in a subalpine lodgepole pine forest. Ecology 73:2100–2108.

Schoettle, A. W., and T. J. Fahey. 1994. Foliage and fine root longevity in pines. In H. L. Gholz, S.A. Linder, and R. McMurtrie, (eds.), Environental constraints on the structure and productivity of pine forest ecosystems: a comparative analysis. Ecological Bulletins (Sweden) 43: 136–153.

Schreuder, H. T., and W. T. Swank. 1974. Coniferous stand characterized with the Weibull distribution. Canadian Journal of Forest Research 4:518–523.

Schulze, E.-D., W. Schulze, F. M. Kelliher, N. N. Vygodskaya, W. Ziegler, K. I. Kobak, H. Koch, A. Arneth, W. A. Kusnetsova, A. Sogatchev, A. Issajev, G. Bauer, and D. Y. Hollinger. 1995. Aboveground biomass and nitrogen nutrition in a chronosequence of pristine Dahurian *Larix* stands in eastern Siberia. Canadian Journal of Forest Research 25:943–960.

Schuur, E. A. 2003. Productivity and global climate revisited: The sensitivity of tropical forest growth to precipitation. Ecology 84:1165–1170.

Scurlock, J. M. O., W. Cramer, R. J. Olson, W. J. Parton, and S. D. Prince. 1999. Terrestrial NPP: Toward a consistent data set for global model evaluation. Ecological Applications 9:913–919.

Sherman, R. E., T. J. Fahey, and P. Martinez. 2003. Spatial patterns of biomass and aboveground net primary productivity in a mangrove ecosystem in the Dominican Republic. Ecosystems 6:384–398.

Sizer, N. 1992. The impact of edge formation on regeneration and litterfall in a tropical rain forest fragment in Amazonia. Dissertation, Cambridge University.

Sprugel, D. G. 1984. Density, biomass, productivity, and nutrient-cycling changes during stand development in wave-regenerated balsam fir forests. Ecological Monographs 54: 165–185.

Steele, S., S. T. Gower, J. G. Vogel, and J. M. Norman. 1997. Root biomass, net primary production and turnover of aspen, jack pine and black spruce stands in Saskatchewan and Manitoba, Canada. Tree Physiology 17:577–587.

Swank, W. T., and H. T. Schreuder. 1974. Comparison of three methods of estimating surface area and biomass for a forest of young eastern white pine. Forest Science 20: 91–100.

Ter-Mikaelian, M. T., and M. D. Korzukhin. 1997. Biomass equations for sixty-five North American tree species. Forest Ecology and Management 97:1–24.

Turner, J., and J. N. Long. 1975. Accumulation of organic matter in a series of Douglas-fir stands. Canadian Journal of Forest Research 5:681–690.

Weaver, P. L., E. Medina, D. Pool, K. Dugger, J. Gonzales-Liboy, and E. Cuevas. 1986. Ecological observations in the dwarf cloud forest of the Luquillo Mountains in Puerto Rico. Biotropica 18:79–85.

Whittaker, R. H., F. H. Bormann, G. E. Likens, and T. G. Siccama. 1974. The Hubbard Brook ecosystem study: Forest biomass and production. Ecological Monographs 44: 233–252.

# 6

## Estimating Moss and Lichen Ground Layer Net Primary Production in Tundra, Peatlands, and Forests

Dale H. Vitt

M any cold temperate, boreal, and polar ecosystems have large biomass components in bryophytes and lichens. These bryophyte and lichen guilds often occupy 80%–100% of the ground cover, and form the matrix in which herbaceous and woody species are rooted. Both nutrient sequestration (Weber and van Cleve 1984; Bayley et al. 1987; Bobbink et al. 1998) and water content (Price et al. 1997; Betts et al. 1999) of the ecosystem are affected by these biomass-rich ground layers. The ground layer, here defined as the vegetation layer that is directly attached to the upper soil surface, is composed of bryophytes and lichens.

Bryophytes and lichens have very different ecological strategies for water relations, photosynthetic activity, and growth potential in comparison to vascular plants (Proctor 2000). Neither of these plant groups has roots or well-developed cuticles and water-conducting systems. Both lichens and bryophytes are poikilohydric (without ability to control evaporative water loss), and both have developed physiological drought tolerance mechanisms (Phillips et al. 2002; Proctor and Tuba 2002). Bryophyte leaves are mostly one cell thick and lack stomates, while lichens have no structures that regulate transpiration. Members of both groups attach to substrates through single-cell, nonphotosynthetic structures (rhizoids in bryophytes and rhizines in lichens), and hence lack roots. Neither lichens nor bryophytes contain secondary cell walls and true lignin (Hébant 1977), although polyphenolic networks with structures similar to vascular plant lignin or tannins have been identified in mosses (Wilson et al. 1989; Verhoeven and Liefveld 1997; Williams et al. 1998). However, despite this lack of lignin, mosses, especially peat mosses (*Sphagnum*), decompose slowly over time. Biomass may accumulate as peat in bryophyte-dominated ecosystems. These unique attributes, coupled with the small size of the individual plants, make measurement of both biomass and production difficult and time-consuming.

The development of extensive and dominant ground layers over time appears to require three preexisting conditions. First, water tables must be relatively stable; second, nutrients (especially nitrogen [N] and phosphorus [P]) need to be relatively low; and third, temperatures need to be relatively cool (Vitt 1994; Zoltai and Vitt 1995). When combined, these conditions reduce vascular plant production. Bryophytes and lichens are poor competitors (Grime et al. 1990); however, once established, they may dominate habitats because of their slow decomposition and large biomass accumulation. Much of the biomass of the ground layer is tied up in structural cell wall material, and ground layer species effectively incorporate large amounts of carbon (C). Additionally, ground layers actively absorb nutrients from precipitation and may reduce N availability for vascular plants (Svenssen 1995). Mosses are able to effectively utilize a portion of these nutrients for new annual growth (Echstein and Karlsson 1999 for forests; Li and Vitt 1997; Aldous 2002a and 2000b for peatlands).

In general, ground layers can achieve considerable importance in tundra, peatlands, and forests, and are composed of plants (mosses) and fungi (lichens) that have many morphological and ecological characteristics not present (or very uncommon) in vascular plants. Since almost all of the training and information that ecologists have at hand is based not on mosses and lichens, but on vascular plants, this chapter is written to (1) provide an overview of the general ecological attributes and previous estimates of annual net primary production (NPP) of ground layers in tundra, peatlands, and forests; (2) outline the unique problems in estimating NPP and biomass of these plants; (3) give the best available methods for measuring biomass and annual growth of species dominating ground layers; and (4) recommend methods for relating these measured parameters to NPP at the site level.

## Characteristics of Ground Layers

### Ground Layers of Conifer-Dominated Forest

The boreal forests of the northern hemisphere account for 11% of the world's terrestrial vegetation (Whittaker 1970). The boreal zone is a mosaic of these conifer-dominated forests, numerous lakes, and large, complex peatlands. Although the upland conifer-dominated forests of the boreal zone may differ geographically in their dominant tree species, structurally they remain rather constant. Ground layer floras of these forests are in particular constant across the boreal zone. Four bryophyte and three lichen species dominate much of the boreal ground layer, often developing undulating mounds and depressions (fig. 6.1). On relatively nutrient-rich, calcareous soils, *Hylocomium splendens* ("stair-step moss") dominates; on nutrient-poor, acidic soils, *Pleurozium schreberi* ("big red stem") is dominant; and in more shaded, late snowmelt areas, *Ptilium crista-castrensis* ("knight's plume") is abundant. Under oceanic climates, *Rhytidiadelpus loreus* ("lanky moss") becomes dominant. Drier, more open forest types are dominated by *Cladina rangiferina*

Figure 6.1. Ground layer dominated by *Hylocomium splendens*. *Pinus contorta* forest, Alberta, Canada.

("gray reindeer lichen") and *C. mitis* ("green reindeer lichen"), while more mature and mesic sites have *C. stellaris* ("star reindeer lichen"). Additionally, species of *Peltigera*, although never dominant, have important N-fixing roles on the forest floor (Nash 1996).

### Ground Layers of Peatlands

On the basis of vegetation and chemistry, peatlands consist of three fundamental types: bogs, poor fens, and rich fens (terminology used here is adapted from Sjörs [1948, 1950] and reviewed by Horton et al. [1979]). Each of these types has strong surface relief correlated to distance above the water table, and in each peatland type, each zone of relief has characteristic bryophyte species (table 6.1). Open water occurs as isolated pools with truly aquatic vegetation; emergent carpets of bryophytes occupy shallow standing water; and expansive bryophyte lawns occupy somewhat drier areas. Finally, there are elliptic-to-circular mounds of vegetation (hummocks) that in some conditions alternate with confined, depressed areas (hollows). Bogs and poor fens are dominated by species of the genus *Sphagnum* (peat mosses).

Hydrologically isolated (ombrogenous) bogs are treeless in oceanic areas, whereas in continental areas they have a scattered open canopy of trees (*Picea mariana* in North America, *Pinus sylvestris* in Europe, and *Larix dahurica* in Asia). The oceanic bogs have a continuum of ground layer vegetation from pools through carpets and lawns to hummocks (fig. 6.2), while in continental boreal areas, bog pools are lacking. Because bogs have extensive development of hummocks and lawns, the aerobic upper peat column (acrotelm) is relatively thick compared with that of fens. Hence the decaying biomass of bog ground layers spends considerable time in the acrotelm before being submerged in the anaerobic (catotelm) peat column.

Table 6.1. Sequence of bryophyte species along the pool-hummock microtopographic gradient for the bog-rich fen vegetation–chemical gradient

| | Permafrost Bog | Continental Bog | Poor Fen | Moderate-Rich Fen | Extreme-Rich Fen |
|---|---|---|---|---|---|
| Hummock top | *S. fuscum* <br> *S. lenense* | *S. capillifolium** <br> *S. fuscum* | *S. fuscum* <br> *T. falcifolium* | *S. fuscum* <br> *S. warnstorfii* <br> *T. nitens* | *S. fuscum* <br> *S. warnstorfii* <br> *T. nitens* |
| Hummock side | *S. magellanicum* | *S. magellanicum* | *S. magellanicum* | *S. warnstorfii* <br> *T. nitens* | *S. warnstorfii* <br> *T. nitens* |
| Lawn | *S. angustifolium* <br> *S. balticum* | *S. angustifolium* <br> *S. rubellum** | *S. angustifolium* <br> *S. papillosum** | *H. vernicosus* <br> *S. teres* | *Cm. stellatum* |
| Carpet | *S. jensenii* <br> *S. majus* <br> *S. riparium* | *S. lindbergii* | *S. jensenii* <br> *S. riparium* | *Cl. cuspidata* <br> *S. subsecundum* | *L. revolvens* <br> *S. cossonii* |
| Pool | *W. fluitans* | *S. cuspidatum** | *W. exannulata* | *D. aduncus* <br> *H. lapponicus* | *Sc. scorpioides* |

Notes: Species are the dominant ones found in subcontinental* and continental Canada. Species found in oceanic peatlands of the east and west coasts are not included. Abbreviations: *Cl.* = *Calliergonella*, *Cm.* = *Campylium*, *D.* = *Drepanocladus*, *H.* = *Hamatocaulis*, *L.* = *Limprichtia*, *S.* = *Sphagnum*, *Sc.* = *Scorpidium*, *T.* = *Tomenthypnum*, and *W.* = *Warnstorfia*.

*Source*: Table modified from Vitt 1994, 2000.

Unlike ombrogenous bogs, fens receive water that has been in contact with the surrounding uplands, potentially increasing the amount of base cations and nutrients. Also in contrast to bogs, fens have less development of hummocks and are wetter, with larger expanses of carpets and lawns. Thus, the acrotelm is less developed, and the accumulating organic material enters the anaerobic zone (catotelm) more quickly. *Sphagnum*-dominated poor fens, with acidic waters and no alkalinity, contrast with rich fens, which are characterized by true mosses, neutral or basic waters, and alkaline water chemistry. Poor fens have elevational zones characterized by individual species of peat mosses, while the elevational zones in rich fens have characteristic species of "brown mosses"—species of true mosses from a number of families that are usually reddish-brown in color. All hummocks in rich fens are dominated by a suite of similar species (table 6.1; Vitt and Chee 1990; reviewed by Vitt 2000), while carpets and lawns in rich fens may differ in characteristic species, depending on water chemistry (i.e., extreme-rich fens with pH > 7.0 and moderate-rich fens with pH around 5.5–7.0).

## Ground Layers of Tundra and Forest-Tundra

Tundra ground layers are strongly influenced by groundwater chemistry and hydrology, especially fluctuation of water tables and height above water table. Shrub-dominated tundra with abundant lichens occurs across the low Arctic (CAVM 2003), whereas sedge-moss-dominated wetlands occur under suitable hydrologic conditions in the high Arctic (Bliss 1997; Yurtsev 1994). Likewise, in the peninsular Antarctic and sub-Antarctic islands, moss-rich peat-accumulating tundra occurs

Figure 6.2. Ground layer habitat diversification in a peatland: open water (right), emergent carpet (foreground right), lawn (foreground left), and hummocks (background), all dominated by *Sphagnum* species. Oceanic bog, Nova Scotia, Canada.

(Lewis-Smith and Gimingham 1976). Calcareous tundra ground layers occupy areas of permanent water flow and resemble boreal rich fens in their bryophytic floristic composition. Acidic tundra ground layers may have abundant *Sphagnum* if sufficient water is available, or may be dominated by a suite of true moss and lichen species.

## Representative Values of Ground Layer NPP

Beginning in the late 1960s with the International Biological Programme (IBP) and the British Antarctic Programme, tundra and peatland ground layer biomass and production estimates were incorporated into ecosystem productivity studies. Estimates exist for both arctic and alpine tundra ecosystems (Wielgolaski et al. 1981); for both oceanic and continental peatlands (Rochefort et al. 1990; Campbell et al. 2000; Aerts et al. 2001; Vitt et al. 2003), especially bogs; and for boreal forests (Bisbee et al. 2001; Swanson and Flanagan 2001). At a continental scale, there is little evidence that annual peatland production increases along a north/south profile, especially when compared at similar longitudes (Rochefort et al. 1990; Campbell et al. 2000; Vitt et al. 2000). However, high interannual and spatial variability has repeatedly been shown to be evident in data sets (e.g., Rochefort et al. 1990; Aerts et al. 2001; Thormann et al. 2001). As a result of interannual variability, estimates for annual production for ground layers can vary by over an order of magnitude, ranging from less than 22 to 350 g m$^{-2}$ yr$^{-1}$ in high arctic wetlands (Vitt and Pakarinen 1977) to more than 1656 g m$^{-2}$ yr$^{-1}$ for oceanic peatlands (reviewed by Rochefort et al. 1990; Campbell et al. 2000). In peatlands, values are generally in the range of 50–300 g m$^{-2}$ yr$^{-1}$, whereas in forest systems the general range is

20–300 g m$^{-2}$ yr$^{-1}$ (table 6.2). However, some trends (or lack thereof) appear to be evident. For example, in tundra systems, productivity seems to be related to site wetness. Vitt and Pakarinen (1977) demonstrated that bryophyte production varied from 60 to 350 g m$^{-2}$ yr$^{-1}$, depending on water availability. In bogs, the high hummocks have production about 75% as high as lawns (Vitt 1990). It appears that in most cases, ground layer species of peatlands are N-limited, and additions of small amounts of N can increase annual production (Rochefort and Vitt 1988; Vitt et al. 2003), but Vitt et al. (2003) suggested that growth of *Sphagnum fuscum* is inhibited above a critical loading rate of 14–34 kg N ha$^{-1}$ yr$^{-1}$.

In forests, ground layers appear to be influenced by climatic gradients more than in peatlands. Vitt (1990) showed that growth of *Hylocomium splendens* was highly correlated to precipitation patterns at the regional scale, whereas growth of the bog species *Polytrichum strictum* was not. This is due to the wide range in variation of NPP across topographic and hydrologic gradients, and also across temporal scales. Sampling strategies and methods need to be customized for specific species and gradients.

Continuous mats of feather mosses in boreal forests can achieve 100–300 g/m$^2$ NPP; however, lower values can also be expected. In *Picea abies* forests, and when randomly selected plots having relatively low, sporadic cover of *Hylocomium* are sampled, NPP values can range from 4 to 23 g m$^{-2}$ (when actual cover values are extrapolated to 100% cover [R. Øklund, personal communication]).

## Special Problems in Estimating NPP of Ground Layers

### Terminology and Definition of the Ground Layer

Critical to partitioning biomass and production values within the ecosystem in a consistent manner is determining whether the ground layer is viewed as belowground, aboveground, or a combination of the two. Viewed from a soils perspective, it can be argued that the ground layer, including the living/growing moss/lichen shoot tips, is part of the uppermost layer of soil. Thus, all ground layer biomass and production is "belowground biomass." Alternatively, and from a vegetation point of view, it can be argued that the ground layer, including its dead, lower component, is actually a plant canopy, and that just like the forest canopy and its component parts, it is a living layer positioned above the soil layer. Thus, just like trees having mostly inactive, structural xylem making up their trunks, mosses and lichens have inactive, structural lower stems. Here the convention is applied that the ground layer canopy is included in "aboveground" production estimates.

Litter can be defined as unattached aboveground vascular plant material that falls onto the surface of the ground layer and is not the ground layer itself. In peatlands, for example, organic material received into the catotelm will be one of three types: (1) litter from the aboveground vascular plants ("vascular plant litter"), (2) belowground vascular plant roots, and (3) decomposing ground layer

Table 6.2. Estimates of aboveground net primary production (NPP) for ground layers in tundra, peatland, and forest communities

| Community | NPP (gm$^{-2}$yr$^{-1}$) | | References |
|---|---|---|---|
| | Mean | Range | |
| *Tundra* | | | |
| High arctic | | 33–77 | Vitt and Pakarinen (1977) |
| High arctic-streamside | | 350 | Vitt and Pakarinen (1977) |
| Signy Island | | 342; 223–893 | Longton (1970); Collins (1973) |
| South Georgia | | 436; 1000 | Baker (1972); Clark et al. (1971) |
| *Peatlands* | | | |
| Peat plateau | | 24–70 | *Campbell et al. (2000); Rosswall et al. (1975) |
| Continental bog | 290 | 17–380 | *Campbell et al. (2000) |
| Oceanic bog | | 70–440 | *Rochefort et al. (1990) |
| Wooded fen | 81 | 47–170 | *Campbell et al. (2000) |
| Open fen | 139 | 95–195 | *Campbell et al. (2000) |
| Fen (West Virginia) | | 540 | Wieder and Lang (1983) |
| *Sphagnum fuscum* | | 7–303 | *Campbell et al. (2000) |
| *Tomenthypnum nitens* | | 55–204 | *Campbell et al. (2000) |
| Mud-bottom | | 40–170 | *Campbell et al. (2000) |
| Carpet | | 91–105 | *Campbell et al. (2000) |
| Lawn | | 36–90 | *Campbell et al. (2000) |
| Hummock | | 55–320 | *Campbell et al. (2000) |
| *Forests* | | | |
| Feather moss | | 26–110 | Bisbee et al. (2001) |
| Feather moss and *Sphagnum* | | 22–132 | Bisbee et al. (2001) |
| Well-drained *Picea mariana* | | 18–316 | Bond-Lamberty et al. (2004) |
| Poorly drained *Picea mariana* | | 53–653 | Bond-Lamberty et al. (2004) |
| | Act. Hyloc. NPP (cover%)[a] | NPP 100% Cover Extrap[b] | |
| *Hylocomium*, in *Picea abies*, Norway | 0.2 (6) | 4.0 | R. H. Økland, pers. comm. |
| *Hylocomium*, in *Picea abies*, Norway | 0.7 (10) | 7.0 | R. H. Økland, pers. comm. |
| *Hylocomium*, in *Picea abies*, Norway | 0.6 (7) | 8.6 | R. H. Økland, pers. comm. |
| *Hylocomium*, in *Picea abies*, Norway | 3.2 (14) | 22.9 | R. H. Økland, pers. comm. |
| *Hylocomium*, in *Picea abies*, Norway | 0.5 (6) | 8.3 | R. H. Økland, pers. comm. |
| *Hylocomium*, in *Picea abies*, Norway | 6.3 (28) | 22.5 | R. H. Økland, pers. comm. |
| *Hylocomium*, in *Picea abies*, Norway | 3.7 (18) | 20.6 | R. H. Økland, pers. comm. |

*Notes*: Values were mostly derived from the reviews (*) of the primary literature. NPP estimates are based on both temporal and spatial sampling and should be used only as a guide for the expected magnitude of NPP encountered in different ground layers.

*Review article.

[a]Actual NPP for randomly sampled populations of *Hylocomium* having a variety of cover levels.

[b]Values extrapolated to 100% cover of *Hylocomium splendens*.

plants from the lower ground layer canopy. (Note: This material could be termed the "ground layer litter"). Functionally, ground layer plants in peatlands should be defined as "dead" only upon their entry into the catotelm, and partitioning of biomass in peatlands would be as follows: (1) aboveground live vascular plant material, defined as plant parts above the shoot/root transition zone; (2) living ground layer, defined as acrotelmic ground layer of mosses and lichens; (3) litter, vascular plant material that has fallen onto the ground layer surface; and (4) belowground vascular plant material, defined as roots and associated mycorrhizae. In forests, the ground layer living biomass is relatively easy to define by the presence of intact moss/lichen stems, positioned just above the duff layer. This concept then treats moss plants exactly like other long-lived perennials such as trees and shrubs, in that all of these have long-lived woody parts containing both living and dead cells. Similarly, production estimates that are determined using $CO_2$ exchange must calculate the $CO_2$ exchange rates using comparable biomass parameters (Martin and Adamson 2001).

The transition from an active photosynthetic plant layer existing as an intact growing canopy to a relatively inactive, slowly decomposing layer in peatland and tundra ecosystems is nearly always vague at best (fig. 6.3). Hence, it is doubtful whether there can be a meaningful biomass estimate that will be exactly comparable between studies (see, for example, Wielgolaski et al. 1981). A more meaningful ecological method of examining biomass in peat-accumulating systems is to estimate bulk density of the upper 5 and 10 cm intervals. In *Sphagnum*, this would include the capitulum (see below). These standard intervals would give a standardized density-biomass for peatland and tundra ecosystems that would be usable for carbon pool and flux estimates.

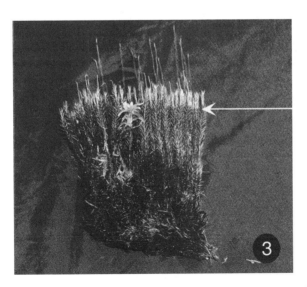

Figure 6.3. Innate variation in leaf length in *Polytrichum strictum*. Arrow indicates beginning of current year's growth. Photo was taken in the spring.

## A Primer on Bryophyte and Lichen Growth

### Bryophyte Growth Patterns

Bryophytes have several unique properties of growth that need to be considered when estimating annual production. Some bryophytes have monopodial growth; that is, they increase in length through cell division originating from two sources: the apical cell that results in increase in primary stem elongation, and the lateral buds that produce lateral branches (thus branches are developed along a support- ing axis of a lower hierarchy; La Farge-England 1996). This growth type occurs in both pleurocarpous mosses (those that produce archegonia, and hence sporophytes, without using the apical cell of the primary stem) and acrocarpous mosses (those using the apical cell for archegonial initiation). In some species, the primary stems are prostrate (plagiotropic), and in others, erect (orthotropic). It is critical (and very difficult) to incorporate both the lateral and the apical growth into growth measure- ments and annual production estimates.

Bryophytes also have sympodial growth wherein growth originates from the apical stem cell and forms a chain of connected branches of the same hierarchy. Many of these species produce archegonia terminally, using the apical cell of the main stem (acrocarpous) (fig. 6.3) or the apical cell of well-developed side branches (cladocarpous). However, the forest moss, *Hylocomium splendens*, is a pleurocarp with sympodial branching (fig. 6.4). Nearly all the acrocarpous species have erect

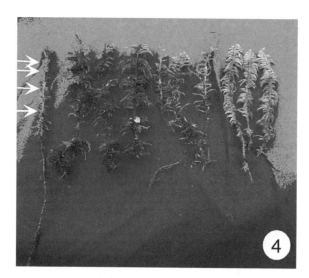

Figure 6.4. Stem lengths of the 3 feather mosses growing together in the same microhabi- tat; the photo was taken in early spring. Left: 1 stem of *Polytrichum commune* with 4 innate annual markers visible at arrows. Left center: *Hylocomium splendens* (with 6–8 annual fronds); note that the terminal frond is not fully developed. Right center: *Pleurozium schreberi*. Right: *Ptilium crista-castrensis*.

stems that do not branch or occasionally branch just below the apex (especially when injured or after production of archegonia; see La Farge-England 1996 for a review).

The genus *Sphagnum* is unique among mosses in its stem and branch organization, and deserves special attention. *Sphagnum* species grow erect or ascending, and only occasionally divide by true dichotomous branching. The stem produces, via an apical cell, a series of branches (with leaves) arranged in groups (fascicles) just below the stem apex. Between these closely spaced fascicles of branches are leaves attached directly to the stem (stem leaves). These young branches gradually elongate, and the entire group of developing branches and stem leaves remains close together at the apex of the plant (fig. 6.5). This "head," or capitulum, is found only in the genus *Sphagnum*. As the branches within the fascicles continue to grow in length, the stem elongates, with the result that the fascicles become spaced along the stem growth of the current year. Over time the capitulum remains relatively constant, continually maintaining groups of elongating young branches just below the stem apex. These branches partially mature within the capitulum and are spread along the stem as elongation just below the capitulum takes place. The branches of each fascicle are either spread outward from the stem (these spreading branches create the spacing between the *Sphagnum* stems and form the structure of the *Sphagnum* canopy), or they hang down along the stem (hanging branches) covering the stem leaves and outer stem cells.

## Lichen Growth Patterns

Foliose lichens (e.g., *Peltigera* spp.) have thalli that are bilaterally symmetrical and have differentiated upper and lower surfaces. These lichens expand laterally, with

Figure 6.5. Longitudinal section through dense canopy *of Sphagnum fuscum*. Note capitula at arrow.

growth originating near the apex of individual lobes. Colony expansion is often in all directions, with each lobe acting independently from its neighbor. Fruticose lichens (e.g., *Cladina, Cladonia, Stereocaulon, Usnea*) have thalli that are mostly radially symmetrical with an outer cortex (or algal layer) surrounding an inner set of hyphae. These lichens usually branch frequently and grow erect on the forest floor, or they are tufted or pendant as epiphytes. The genus *Cladina* has richly branched and erect secondary thalli termed podetia (and the primary thalli are lacking), whereas the closely related genus *Cladonia* possesses a foliose (squamulose) primary thallus and an erect fruticose secondary podetial thallus. *Cladina* produces annual segments, identified by major branching points, and additionally each year adds length to the previous years' segments.

The focus below is on bryophyte-dominated ground layers. Although lichens (especially reindeer lichens—*Cladina* and *Cladonia* species) may be abundant in dry tundra, forest-tundra, and pine-dominated boreal forests, and species of *Peltigera* are important components of nitrogen fixation in some ground layers, they are beyond the scope of this review. Growth dynamics of lichens have been examined by several authors (Andreev 1954; Kärenlampi 1970, 1971; Kershaw 1978; Scotter 1963). However, since many of the techniques used for bryophytes can be adapted to lichens, they are included in relevant sections.

## Methods for Measuring Annual Growth

### *Innate Markers*

#### Leaf Patterns

Longton and Greene (1967) provided a detailed study of the annual growth pattern of the polar/boreal moss *Polytrichum alpestre* (= *P. strictum*). In temperate and polar climates, bryophytes cease growth over winter. With warmer temperatures in the spring, stems produce short, widely spaced leaves that, compared with midsummer leaves, have longer limbs and are more closely spaced (figs. 6.3, 6.4). Hence, annual growth can be measured by clipping the terminal segment in late fall or, perhaps better, by measuring the previous year's growth at any time by clipping the penultimate segment (Collins 1976; Longton 1972). Several other mosses also have annual growth segments. These include *Meesia triquetra*, which is common in rich fens and arctic wetlands (Vitt and Pakarinen 1977); *Chorisodontium aciphyllum* (Baker 1972), from moss banks of the antarctic peninsula; *Catoscopium nigritum*, a rare indicator of rich fens; and, although less obvious, species of *Drepanocladus (sensu lato)* and *Scorpidium* that are characteristic species of fens.

#### Frond Patterns

*Hylocomium splendens* produces annual fronds (fig. 6.4). The bipinnately branched, arching frond has determinate annual growth and develops by sympodial branching (Økland 1995). In the autumn of the current year (year 0), a subterminal bud is initiated and begins growth very early, just at snowmelt, in the following spring (year 1). The nearly mature segment is present by the following autumn (Rydgren and Økland 2003), and in year 2 this frond matures. The mature annual fronds remain intact and visible for many years within the bryophyte canopy. Beginning with the work of Tamm (1953), *Hylocomium* has been much used for examination of forest ground layer dynamics. Using a data set from western Canada, Vitt (1990) showed that frond weight of *Hylocomium splendens* increases from about 10–20 mg frond$^{-1}$ under continental precipitation regimes (50–100 cm yr$^{-1}$) to frond masses of 50–80 mg frond$^{-1}$ under oceanic regimes (250–300 cm precipitation yr$^{-1}$). In a series of papers, R. Økland and K. Rydgren have explored various aspects of the population dynamics of this species (Økland 1995, 2000; Rydgren et al. 1998, 2001). Since frond maturation takes 2 yr, clipping the current frond in the autumn will lead to underestimation of annual growth. Under tundra conditions, *H. splendens* often does not maintain the subterminal annual growth initiation pattern, and new growth originates from the apex of the shoot. In this case, annual growth measurement is not possible.

## Stem Patterns

*Sphagnum* species form canopies that vary considerably in their stem density and shoot diameter. Compact, dense canopies are found on hummock-growing species of the genus (fig. 6.5). The dense canopy structure maintains favorable water conditions, and these species are less drought tolerant than those nearer the water surface in lawns and carpets, where canopies are loosely arranged and plants are more drought tolerant (Wagner and Titus 1984). Snowfall in the boreal zone may be considerable, and the weight of the snow causes species with loose canopies to be bent sideways and more or less parallel to the water surface. In the new season of growth, these stems remain at angles of less than 90° to the water surface; however, the new growth is completely erect (angles of 90° to the water surface). These "snow-weighted crooks" (fig. 6.6) are efficient means of measuring local spatial and temporal variation of growth in carpet and lawn habitats of poor fens and bogs. When they are combined with stem density counts and annual bulk density (see below), production can be estimated.

### Surrogate Markers

### Neighbor Comparisons

With the assumption that moss canopies of mixed species extend upward at an even rate, species with innate markers can be used to extrapolate the rate of growth for those species that do not have such growth markers. For example, in boreal peatlands, *Sphagnum fuscum*, the dominant hummock species, forms dense canopies that do not have snow-weighted markers; however, nearly always occurring within these hummocks of *S. fuscum* are gregarious plants of *Polytrichum strictum*. Since the canopy of *S. fuscum* is so dense that light becomes limiting for photosynthesis very near to the surface, *P. strictum* plants maintain a 0.3–0.8 cm vertical height above

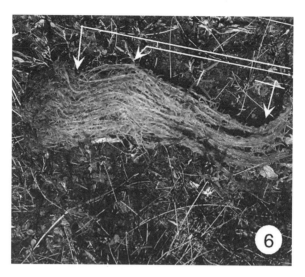

Figure 6.6. *Sphagnum jensenii* canopy showing bent stems (arrows) due to winter snow pressure.

the *Sphagnum* canopy surface. Thus the vertical increment that can be measured in *P. strictum* can be used as a surrogate (fig. 6.4) to estimate the vertical growth of *S. fuscum*, a species without innate markers.

## Cranked Wires

Species of bryophytes with erect, dense canopies commonly trap litter material in place. This trapped litter (e.g., needles of *Larix*) remains in place within the peat column. These dense canopies can also have wires implanted that will remain stationary for the growing season (Clymo 1970). These wires have been called "cranked wires" and, after some practice measuring the exposed wire ends, effectively measure the length of the annual growth of the moss shoot.

Wires commonly used are made of 1 mm stainless steel welding rod and generally are available in 1-m (or 1-yd) lengths. Each rod is cut into three 33-cm lengths (a bolt cutter works best). With a 3-inch vise and hammer, the wire is bent 90° at about 7 cm from the end. Then, leaving a space of 1 cm, the remaining wire segment is bent back straight, giving a shape shown in figure 6.7. The bends must be sharp, and the wire length must remain perfectly straight. No burrs can remain at the ends of the wires. Color-coding with several colors of spray paint helps in finding them for the recording of data at the end of the growing season.

Growth is measured with an instrument that has been termed the "sphagnometer" (fig. 6.8). This is handbuilt as follows: Cut a 1.5–2.0 cm cork transversely into a 0.75-cm slice and bore a 6 mm hole in the center. Cut a 10-cm length of 6-mm (O.D.) glass tubing, flare the ends, and glue one end of the glass tubing into the hole of the cork, making sure the end of the tubing is even with the end of the cork. Take a 15-cm

Figure 6.7. Three cranked wires made from 33-cm-long stainless steel welding rods.

plastic laboratory ruler and cut it to 9 cm in length. Cut this piece lengthwise into a strip just wide enough to show the scale and fit in the glass tube so that the strip is closer to one side, making it easy to slide the tool over the wire in the field (the wire must slide in front of the ruler scale, so that measurements can be taken). Attach the strip with glue at the end of the glass tube away from the cork. For ease of calculation, the larger numbers of the ruler should be closer to the cork. Smooth the transition of the wire through the cork by closing the portion of the cork opening that is behind the ruler with silicone caulk, forcing the wire to go in front of the scale. The upper end of the glass tube may be rounded by heating, but should not be completely closed, or there will be condensation buildup in the tube. In the field, the tube must slide easily over the wire and sit lightly on the surface of the moss canopy.

Implanting cranked wires is made easier by using a "wiseguy," or wire-setting guide (see fig. 6.8), which may be handmade as follows. Remove the barrel from a clear ballpoint pen (one that has a straight shank) and cut it to about 5.0 cm in length, so that when it is placed over the top end of the cranked wire, it ends about 1.5–2.0 cm above the crank. Cut a 1-cm diameter cork (a wine bottle cork is perfect) to 0.5-cm thickness, bore a hole in the center the size of the pen barrel, insert the pen barrel into the cork, and glue it with the end of the pen barrel even with the bottom of the cork.

In the field, place the "wiseguy" over the cranked wire and push it gently to the moss surface with the 1-cm horizontal crank just below the moss surface. This implantation tool allows all wires to be set at the same initial distance above the moss surface (fig. 6.8). Calibrate the wires by measuring and recording the set distance with a sphagnometer. The crank then will hold the wire in place, and as the moss canopy grows upward, it continues to cover the exposed 7-cm length of wire.

Wires must be implanted as soon as the ice melts from the peat mat in the spring, and measured individually at the beginning and end of each growing season. Since

Figure 6.8. Implanted cranked wires. Left: Implanting device ("wiseguy"). Center: Cranked wire in place for the season, the crank hidden just below the *Sphagnum* surface. Right: Cranked wire with device (sphagnometer) to measure length of the exposed wire.

mosses begin growth very early in the spring—before the ice cores of the hummocks are melted in continental peatlands and boreal forests—cranked wires will underestimate annual growth in continental conditions. Likewise, mosses continue to grow during moist, cool autumn conditions, and end-of-season measurements should be made as late as possible, but before ice forms in the ground layer canopy. Annual growth estimates can be checked by comparison against "snow-bend patterns." Winter ice formation, or sometimes animals, will move the wires or flatten them, and under these conditions the wires must be reset the following spring. The under-estimation of growth by using cranked wire implants needs to be carefully examined, as observations and comparisons with snow-bend markers suggest that growth may be underestimated as much as 20%–30%.

## Strings/Pins/Velcro

Measuring growth of mosses with monopodial plagiotropic branching is especially difficult. Methods that have been used can be tedious, but very accurate. For example, tying colored strings to the shoots at some distance below the growing tip can be accurate, but real problems exist in relocating the strings. Also, it may be difficult to avoid disturbing the growing apex when tying material, especially material that can be easily tied and does not decompose. Placing pins with round, colored heads just below the growing apices works well for epiphytes, as does placing small colored dots of paint on rock surfaces to form semi-permanent markers. However, these latter two methods do not work well with the soft substrates of forest floors and peatlands. Thin Velcro strips have also been utilized (Trumbore et al.

1999). Strings, pins, and Velcro measure only apical growth; however, lateral branch growth could be modeled through allometry for such monopodial species as *Tomenthypnum nitens, Pleurozium schreberi*, and *Ptilium crista-castrensis*, although this has never been done. Lateral growth added in successive years for these species may add significantly to growth derived from the apical meristem.

## Netting

Fine fiberglass netting (with 1.0- to 1.5-cm spaces) placed on the canopy surface just at the growing bryophyte tips has been used, especially in forested habitats (Gower et al. 1997; Bond-Lamberty et al. 2004). Measurement of biomass produced since the net was placed on the surface is done by clipping all material above the netting. This method severely underestimates annual production because not all second-year lateral branches are accounted for. For example, using this method in forest canopies of *Pleurozium schreberi* or on fen hummocks for *Tomenthypnum nitens* can be accurate only if bulk density is properly scaled up (see below). For these monopodial mosses that produce numerous lateral branches in both the first and the second (and perhaps the third) years, some second- (third-) year lateral growth will be missed unless the nets are clipped successively at the end of the second year. However, snow weight may adversely affect the net placement in the second year. Annual biomass production will be inaccurate if the two-year number is divided by 2. Rather, annually placed nets must be measured biennially, and successive calculations must be made in this manner, for the method to be accurate. Modelling production *of Pleurozium schreberi* has recently been done by Benscoter and Vitt (2007).

## Plugs

With a circular coring device, plugs of canopy can be extracted, measured to exact length, and reimplanted in the canopy. Annual growth can be obtained at the end of the season by remeasuring the length of the plug. Suitable coring devices are large juice cans (1.36 L, 11.4 cm diam). If one cuts the lid off with a horizontal can opener, the cut edge will be thin and sharp. If there are vascular plant roots that catch the cutting edge, it should be notched with a file or grinding wheel. The length of the plugs should be at least 15 cm, and their diameter should be at least 10 cm, in order to cause as little disturbance to the water relations as possible. *Sphagnum* may undergo some loss in stem length upon drying, so all lengths should be measured at comparable wetness.

## Surface Marker

In theory, one can use any marker that will permanently mark the surface of the canopy at the beginning of the growing season. Then, if the marker is retained in the canopy at its original location, one can relocate it at the end of the season and measure the length of growth, using the original marker as a starting point. Markers that have been considered are dyes and $^{14}C$ (Wallén 1986). Dyes are worrisome because they may interfere with photosynthesis. $^{14}$C is a perfect marker, but requires considerable forethought and time for methods and permits.

## Scaling Issues

### Relating Length Growth to Annual Net Primary Production

Growth as length of individual shoots (mm) must be converted to mass (g) per area (m$^2$) in order for annual production to be estimated.

## Shoot Density

If the length of individual shoots is known, a sampling ($n = 20–30$) of shoots from a closely associated plot can be individually weighed, and shoot density determined by counting the number of shoots in a small (100 cm$^2$) core. Shoot density varies both between and within species by habitat. Hummock species of *Sphagnum* (e.g., *S. fuscum*, S *capillifolium*) have stem densities between 13,000 and 22,000 stems/m$^2$. Robust species such as *S. magellanicum* have densities from about 10,000 to 15,000 stems/m$^2$ (unpubl. data from Line Rochefort, Laval University). Rune Øklund (personal communication) reports that normal *Hylocomium splendens* stem densities range up to 6,000 stems/m$^2$, but occasionally are higher after disturbance. These numbers clearly indicate that there is much variation both between and within habitats and species, and scaling up to the site level needs to be carried out by using densities obtained at individual plots and for individual species or habitats.

## Clipping

If surface markers or netting is used to measure growth, then clipping will yield annual biomass produced (but note the problem with branched monopodial species). Likewise, innate markers may be clipped for direct measurement of annual biomass. However, annual increment clipping will always underestimate annual production because it does not take into consideration second-year biomass additions. This will be true for orthotropic sympodial species as well, because stem outgrowths (rhizoids, paraphyllia) may be added in year 2 or year 3.

## Annual Bulk Density Determination

Cores of constant volume taken from the canopy can be weighed to give estimates of bulk density (g/cm$^3$). Bulk densities in peat and in most canopies change dramatically with depth and with species. Thus, the following method is recommended: Remove a core of about 8–10 cm diameter and 10–15 cm in depth and freeze it, preferably without removal from the coring device. A 10-cm, thin-walled, sharpened PVC pipe works well, as does a stovepipe or a large juice can, depending on the density of the canopy. Remove the core from the freezer and run warm water over it. The warm water will melt the outermost part of the plant material while allowing the majority of the material to remain frozen; this allows one to gently push the core intact from the coring device. Estimate the approximate length of the growth that occurred over the past 2–3 yr. Using a band saw, cut a slice of the top portion that represents the past 2–3 years' growth. This would be the top 1–2 cm for hummock species and the top 3–4 cm for the hollow species.

If *Sphagnum* is involved, the capitula must be removed or cut off. Dry this slice of the core at constant mass at 80°C, and cool it in a dessicator before weighing it, because mosses are especially prone to absorb water from the air. A second error that must be avoided is compaction of the volume of the moss canopy. Care should be taken in the field to see that the distance within the core occupied by the moss material is equal to the depth of the core, and that this distance remains the same when the core is extracted from the corer (the reason for freezing the core). Errors in volume are the most common problem in determining bulk densities, but can be avoided with careful techniques.

## Scaling-Up Spatially

Ground layers can form extremely variable topographic relief within the peatland, tundra, or forest community. Due to their concave and convex surface morphology, these hummocks, hollows, and lawns actually occupy a greater surface area than a flat plane. Without the surface topography taken into consideration, production will be underestimated. For example, Rochefort et al. (1990) estimated that in a boreal bog in Ontario, Canada, surface relief increased actual surface area by 38%. Estimation of ground layer relief can be done by random placement of large plots (e.g., 30 m² area) wherein all hummocks and hollows are mapped and actual surface area is calculated, or relief along a transect is determined by using a laser level. In either case, all relief points must be related to the species of moss occupying the area.

## Biomass Determinations

Ground layer biomass has been measured repeatedly since the mid-1950s. However, as noted above, it is doubtful that comparable data exist, since different researchers have used different criteria to delimit the extent of biomass. This is especially true of measuring "living ground layer biomass." There is little relationship between annual production, green vs. brown biomass color, and whether the material is living or dead (Wielgolaski 1972; Vitt and Pakarinen 1977). Green biomass changes to brown relative to hydrological conditions, so that the green portion of the mat represents anywhere from only a portion of the annual season's growth to several years' growth (Vitt and Pakarinen 1977). Likewise, both Wielgolaski (1975) and Clymo and Ducket (1986) have shown that apparently dead, brown bryophyte tissue, often from some depth in the organic column, can reactivate under suitable conditions. Thus all biomass should be considered potentially alive unless the following depths are reached: the catotelm (or practically the average annual water table) in peatlands, the duff layer in forests, or the permafrost in tundra.

## Guiding Principles and Recommendations

### *Forest Ground Layers*

#### *Hylocomium Splendens*

Annual frond increments are clipped very late in the growing season. Since bryophytes tend to produce biomass under low temperatures, it is important to wait as late as possible in the growing season in order to obtain the full season's growth. The recommended plot size is 10–20 cm², with plots placed randomly. Sampling design must account for all elevational relief. Clipping the annual fronds, weighing the fronds to obtain biomass per unit area, and accounting for variation in hummock-hollow height and variation are important variables. Sporophyte production, if present, should be included in the annual production estimate. Since mature frond development takes 2 yr, this method will underestimate growth. Clipping the penultimate fronds in late autumn provides a better estimate.

#### Other Species (Including *Pleurozium Schreberi*)

If measurements are done very carefully (without depressing the delicate shoots with the sphagnometer), cranked wires are accurate. These measurements of individual shoot lengths then must be multiplied by the 2-yr estimated bulk densities and scaled to hummock/hollow relief. Alternatively, individual shoots can be counted and the estimated 2–3-yr growth increment weighed for individuals (50 individuals per area is recommended). The number of cranked wires to be placed and the design are discussed in the peatland section below.

#### *Peltigera*

Successive photography works well for species of this genus, and the modern digital cameras make it possible to simply scan and calculate areas. Another simple method is to trace the thallus outlines onto clear plastic, transfer these outlines to a piece of paper, cut the outline out, and weigh the resulting paper. Regressions can be done to relate thallus weight to paper weight. Thallus size changes considerably with water content, so tracings and photography need to be done at comparable water-holding conditions.

### *Peatland Ground Layers*

#### *Sphagnum*

For species with high densities (*S. capillifolium, S. fuscum*) or large capitula (*S. magellanicum, S. papillosum*), cranked wires implanted in early spring are recommended. A minimum of 50 wires per plot and 250 wires per site is suggested. Sets of 50 wires per species/microhabitat nested within a site or a transect through a diversity of species/microhabitats are suitable designs. Wires should be 8–16 cm apart. In order to check if growth is underestimated due to very early or very late seasonal growth, and calibration against species with annual markers is required. *Polytrichum strictum* is a good choice for this in continental areas. Cranked wires can also be used for less dense species (*S. angustifolium, S. fallax*); however, in these cases of lawn and carpet species, the growth estimated should be calibrated against snow depth markers. Calculation of production should be done by multiplying the annual growth (in mm) by bulk density (g cm$^{-3}$) of the estimated 1- to 2-yr *Sphagnum* canopy after removal of the capitula. Density is extremely variable, and if shoot density is used to estimate production, densities must be determined at the local scale. One extrapolated value for the study site will lead to major error in production estimates. The microtopography also must be accounted for in bogs and in many fens. If experimental studies wherein *Sphagnum* plants are manipulated are carried out, the "capitulum correction" method (Clymo 1970) should be considered.

### Brown Mosses

Unfortunately, different methods need to be used for species occupying the microtopographic gradient. Growth for erect-growing hummock species (*Tomenthypnum nitens*) can be measured using cranked wires. For erect-growing lawn and carpet species (*Campylium stellatum*), surface markers (Velcro) can be used. For prostrate species of pools (*Scorpidium scorpioides*) and carpets (many *Drepanocladus* species), growth should be measured either by careful examination of annual stem markers or by Velcro strips. One- to 2-yr bulk densities can be used to estimate production for erect-growing species, but density counts should be used for the prostrate species. In some rich fens, cranked wires can be reduced in length and implanted into the carpets and lawns. The presence of *Meesia triquetra* in many rich fens in continental areas makes calibration of these methods possible. As noted earlier, nets are not reliable and are not recommended.

### *Tundra Ground Layers*

Tundra wetlands are occupied largely by a suite of species similar to that of rich fens. Often *Meesia triquetra* is an abundant species, and when it is present, growth can be measured for this species and estimated for its nearby erect-growing neighbors. With some practice, annual growth markers can be seen on species of *Drepanocladus* (*sensu lato*); these species often grow erect in tundra habitats.

In wet tundra and peatland habitats, biomass measurements are misleading and are rarely comparable among studies. Slow decomposition rates make delimiting the lower cutoff for biomass impossible. Thus, bulk density measurements are recommended for the uppermost 5 cm and 10 cm canopy increments to produce comparable estimates of active C in the system. For forest ground layers, however, the intact stems of the ground layer species do give a comparable measure of biomass above the duff layer. By using successive annual frond weights and densities for

*Hylocomium splendens,* for example, decay rates can be determined for forest stands having this species. Biomass of the ground layer will be determined by the length of time the individual *Hylocomium* shoots remain intact above the duff layer.

*Acknowledgments*  I am grateful to Line Rochefort (Laval University) for supplying density data; to Rune Økland (Haukeland University) for kindly providing unpublished production data from his long-term transects throughout Norway; to Pekka Pakarinen (University of Helsinki), who many years ago first showed me innate markers; and to Kelman Wieder (Villanova University), who provided the initial idea of a device to uniformly implant cranked wires ("wiseguy") and also located key literature. An anonymous reviewer greatly improved the text, for which I am grateful. Sandra Vitt produced the figures. Much of the background for this chapter came through research support from the Natural Sciences and Engineering Research Council of Canada and, more recently, from the National Science Foundation (United States).

References

Aerts, R., B. Wallén, N. Malmer, and H. De Caluwe. 2001. Nutritional constraints on *Sphagnum* growth and potential decay in northern peatlands. Journal of Ecology 89: 202–299.

Aldous, A. R. 2002a. Nitrogen retention by *Sphagnum* mosses: Responses to atmospheric nitrogen deposition and drought. Canadian Journal of Botany 80:721–731.

Aldous, A. R. 2002b. Nitrogen translocation in *Sphagnum* mosses: Effects of atmospheric nitrogen deposition. New Phytologist 156:241–253.

Andreev, V. N. 1954. Growth of forage lichens and methods of improving it. Geobotanika 9:11–74. (In Russian).

Baker, J. H. 1972. The rate of production and decomposition of *Chorisodontium aciphyllum* (Hook. f. and Wils.) Broth. British Antarctic Survey Bulletin 27:123–129.

Bayley, S. E., D. H. Vitt, R. W. Newbury, K. G. Beaty, R. Behr, and C. Miller. 1987. Experimental acidification of a *Sphagnum*-dominated peatland: First year results. Canadian Journal of Fisheries and Aquatic Sciences 4:194–205.

Benscoter, B. W. and D. H. Vitt. (2007). Evaluating feathermoss growth: a challenge to traditional methods and implications for the boreal carbon budget. Journal of Ecology (In press).

Betts, A. K., M. Goulden, and S. Wofsy. 1999. Controls on evaporation in a boreal spruce forest. Journal of Climate 12:1601–1618.

Bisbee, K., S. T. Gower, J. M. Norman, and E. V. Nordheim. 2001. Environmental factors as controls on the distribution and NPP of bryophytes in a southern boreal black spruce forest. Oecologia 129:261–270.

Bliss, L. C. 1997. Arctic ecosystems of North America. Pages 551–683 in F. E. Wielgolaski (ed.), Polar and Alpine Tundra. Elsevier, Amsterdam.

Bobbink, R., M. Hornung, and J. Roelofs. 1998. The effects of air-borne nitrogen pollutants on species diversity in natural and semi-natural European vegetation. Journal of Ecology 86:717–738.

Bond-Lamberty, B., C. Wang, and S. T. Gower. 2004. Net primary production of a boreal black spruce wildfire chronosequence. Global Change Biology 10:473–487.

Campbell, C., D. H. Vitt, L. A. Halsey, I. D. Campbell, M. N. Thormann, and S. E. Bayley. 2000. Net Primary Production and Standing Biomass in Northern Continental Wetlands. Northern Forestry Centre Information Report NOR-X-369. Canadian Forest Service, Ottawa.

CAVM Team. 2003. Circumpolar Arctic vegetation map, scale 1:7,500,000. Conservation of Arctic Flora and Fauna (CAFF) map no. 1. U.S. Fish and Wildlife Service, Anchorage, AK.

Clymo, R. S. 1970. The growth of *Sphagnum*: Methods of measurement. Journal of Ecology 58:13–49.

Clymo, R. S., and J. G. Duckett. 1986. Regeneration of *Sphagnum*. New Phytologist 102:589–614.

Collins, N. 1973. Productivity of selected bryophyte communities in the Antarctic. Pages 177–183 in L. C. Bliss and F. E. Wielgolaski (eds.), Primary Production and Production Processes, Tundra Biome. Tundra Biome Steering Committee, Department of Botany, University of Alberta, Edmonton, Alberta.

Collins, N. J. 1976. Growth and population dynamics of the moss *Polytrichum alpestre* in the Maritime Antarctic. Oikos 27:389–401.

Echstein, R. L., and P. S. Karlsson. 1999. Recycling of nitrogen among segments of *Hylocomium splendens* as compared with *Polytrichum commune*—implications for clonal integration in an ectohydric bryophyte. Oikos 86:87–96.

Gower, S. T., J. Vogel, J. M. Norman, C. J. Kucharik, S. J. Steele, and T. K. Stow. 1997. Carbon distribution and aboveground net primary production in aspen, jack pine and black spruce in Saskatchewan and Manitoba, Canada. Journal of Geophysical Research 102:29029–29041.

Grime, J. P., E. R. Rincón, and B. E. Wickerson. 1990. Bryophytes and plant strategy theory. Botanical Journal of the Linnean Society 104:175–186.

Hébant, C. 1977. The conducting tissues of bryophytes. Bryophytorum Bibliotheca 10.

Horton, D. G., D. H. Vitt, and N. G. Slack. 1979. Habitats of circumboreal-subarctic Sphagna. I. A quantitative analysis and review of species in the Caribou Mountains, northern Alberta. Canadian Journal of Botany 57:2283–2317.

Kärenlampi, L. 1970. Morphological analysis of the growth and productivity of the lichen *Cladonia alpestris*. Reports from the Kevo Subarctic Research Station 7:9–15.

Kershaw, K. A. 1978. The role of lichens in boreal tundra transition areas. The Bryologist 81:294–306.

La Farge-England, C. 1996. Growth form, branching pattern, and perichaetial position in mosses: Cladocarpy and pleurocarpy redefined. The Bryologist 99:170–186.

Lewis-Smith, R. I., and C. H. Gimingham. 1976. Classification of cryptogamic communities in the maritime Antarctic. British Antarctic Survey Bulletin 43:25–47.

Li, Y., and D. H. Vitt. 1997. Patterns of retention and utilization of aerially deposited nitrogen in boreal peatlands. Ecoscience 4:106–116.

Longton, R. E. 1970. Growth and productivity in the moss *Polytrichum alpestre* Hoppe in the antarctic regions. Pages 818–837 in M. W. Holdgate (ed.), Antarctic Ecology, vol. 2. Academic Press, New York.

Longton, R. E. 1972. Growth and reproduction in northern and southern hemisphere populations of the peat forming moss *Polytrichum alpestre* with reference to the estimation of productivity. Pages 259–275 in Proceedings of the 4th International Peat Congress (Helsinki), vol. 1.

Longton, R. E., and S. W. Greene. 1967. The growth and reproduction of *Polytrichum alpestre* Hoppe on South Georgia Island. Philosophical Transactions of the Royal Society of London B252:295–327.

Martin, C. E., and V. J. Adamson. 2001. Photosynthetic capacity of mosses relative to vascular plants. Journal of Bryology 23:319–324.

Nash, T. H. III. 1996. Nitrogen, its metabolism and potential contribution to ecosystems.

Pages 121–135 in T. H. Nash III (ed.), Lichen Biology. Cambridge University Press, Cambridge.

Økland, R. H. 1995. Population biology of the clonal moss *Hylocomium splendens* in Norwegian spruce forests. I. Demography. Journal of Ecology 83:697–712.

Økland, R. H. 2000. Population biology of the clonal moss *Hylocomium splendens* in Norwegian boreal spruce forests. 5. Consequences of the vertical position of individual shoot segments. Oikos 88:449–469.

Phillips, J. R., M. J. Oliver, and D. Bartels. 2002. Molecular genetics of desiccation tolerant systems. Pages 310–341 in M. Black and H. W. Pritchard (eds.), Desiccation and Survival in Plants: Drying Without Dying. CABI Publishing, New York.

Price, A. G., K. Dunham, T. Carleton, and L. Band. 1997. Variability of water fluxes through the black spruce (*Picea mariana*) canopy and feather moss (*Pleurozium schreberi*) carpet in the boreal forest of northern Manitoba. Journal of Hydrology 196: 310–323.

Proctor, M. C. F. 2000. Physiological ecology. Pages 225–247 in A. J. Shaw and B. Goffinet (eds.), Bryophyte Biology. Cambridge University Press, Cambridge.

Proctor, M. C. F., and Z. Tuba. 2002. Poikilohydry and homeohydry: Antithesis or spectrum of possibilities? New Phytologist 156:327–349.

Rochefort, L., and D. H. Vitt. 1988. Effects of simulated acid rain on *Tomenthypnum nitens* and *Scorpidium scorpioides* in a rich fen. The Bryologist 91:121–129.

Rochefort, L., D. H. Vitt, and S. E. Bayley. 1990. Growth, production, and decomposition dynamics of *Sphagnum* under natural and experimentally acidified conditions. Ecology 71:1986–2000.

Rosswall T., A. Veum, and L. Kärenlampi. 1975. Plant litter decomposition at Fennoscandian tundra sites. Pages 268–278 in F. E. Wielgolaski (ed.), Fennoscandian Tundra Ecosystems, vol. 1, Plants and Microorganisms. Springer-Verlag, Berlin.

Rydgren, K., H. de Kroon, R. H. Økland, and J. van Groenendael. 2001. Effects of fine-scale disturbances on the demography and population dynamics of the clonal moss *Hylocomium splendens*. Journal of Ecology 89:395–405.

Rydgren, K., and R. H. Økland. 2003. Short-term costs of sexual reproduction in the clonal moss *Hylocomium splendens*. The Bryologist 106:212–220.

Rydgren, K., R. H. Økland, and T. Økland. 1998. Population biology of the clonal moss *Hylocomium splendens* in Norwegian boreal spruce forests. IV. Effects of experimental fine-scale disturbance. Oikos 82:5–19.

Scotter, G. W. 1963. Growth rates of *Cladonia alpestris*, *C. mitis* and *C. rangiferina* in the Taltson River region, Northwest Territories. Canadian Journal of Botany 41:1199–1202.

Sjörs, H. 1948. Myvegetation i Bergslagen. Acta Phytogeographica Suecica 21:1–229.

Sjörs, H. 1950. Regional studies in north Swedish mire vegetation. Botaniska Notiser 1950:174–221.

Svensson, B. 1995. Competition between *Sphagnum fuscum* and *Drosera rotundifolia*: A case of ecosystem engineering. Oikos 74:205–212.

Swanson, R. V., and L. B. Flanagan. 2001. Environmental regulation of carbon dioxide exchange at the forest floor in a boreal black spruce ecosystem. Agricultural and Forest Meteorology 108:165–181.

Tamm, C. O. 1953. Growth, yield and nutrition in carpets of a forest moss (*Hylocomium splendens*). Meddelanden från Statens Skogsforskningsinstitut 43:1–40.

Thormann, M. E., S. E. Bayley, and R. S. Currah. 2001. Comparison of decomposition of belowground and aboveground plant litters in peatlands of boreal Alberta, Canada. Canadian Journal of Botany 79:9–22.

Trumbore, S. E., J. L. Bubier, J. W. Harden, and P. M. Crill. 1999. Carbon cycling in boreal wetlands: A comparison of three approaches. Journal of Geophysical Research 104:27673–27682.

Verhoeven, J. T. A., and W. M. Liefveld. 1997. The ecological significance of organochemical compounds in *Sphagnum*. Acta Botanica Neerlandica 46:117–130.

Vitt, D. H. 1990. Growth and production dynamics of boreal mosses over climatic, chemical and topographic gradients. Botanical Journal of the Linnean Society 104:35–59.

Vitt, D. H. 1994. An overview of the factors that influence the development of Canadian peatlands. Memoirs of the Entomological Society of Canada 169:7–20.

Vitt, D. H. 2000. Peatlands: Ecosystems dominated by bryophytes. Pages 312–339 in A. J. Shaw and B. Goffinet (eds.), Bryophyte Biology. Cambridge University Press, Cambridge.

Vitt, D. H., and W.-L. Chee. 1990. The relationship of vegetation to surface water chemistry and peat chemistry in fens of Alberta, Canada. Vegetatio 89:87–106.

Vitt, D. H., L. A. Halsey, I. E. Bauer, and C. Campbell. 2000. Spatial and temporal trends of carbon sequestration in peatlands of continental western Canada through the Holocene. Canadian Journal of Earth Sciences 37:683–693.

Vitt, D. H., and P. Pakarinen 1977. The bryophyte vegetation, production, and organic components of Truelove Lowland. Pages 225–244 in L. C. Bliss (ed.), Truelove Lowland, Devon Island, Canada: A High Arctic Ecosystem. University of Alberta Press, Edmonton.

Vitt, D. H., K. Wieder, L. A. Halsey, and M. Turetsky. 2003. Response of *Sphagnum fuscum* to nitrogen deposition: A case study of ombrogenous peatlands in Alberta, Canada. The Bryologist 106:235–245.

Wagner, D. J., and J. E. Titus. 1984. Comparative desiccation tolerance of two *Sphagnum* mosses. Oecologia (Berlin) 62:182–187.

Wallén, B. 1986. Above and belowground dry mass of the three main vascular plants on hummocks on a subarctic peat bog. Oikos 46:51–56.

Weber, M. G., and K. van Cleve. 1984. Nitrogen transformation in feather moss and forest floor layers of interior Alaska black spruce ecosystems. Canadian Journal of Forest Research 14:278–290.

Whittaker, R. H. 1970. Communities and Ecosystems. Macmillan, New York.

Wieder, R. K., and G. E. Lang. 1983. Net primary production of the dominant bryophytes in a *Sphagnum*-dominated wetland in West Virginia. The Bryologist 86:280–286.

Wielgolaski, F. E. 1972. Production, energy flow and nutrient cycling through a terrestrial ecosystem at a high altitude area in Norway. Pages 283–290 in F. E. Wielgolaski and T. H. Rosswall (eds.), Proceedings of the IVth International Meeting on the Biological Productivity of Tundra. Tundra Biome Steering Committee, Stockholm, Sweden.

Wielgolaski, F. E. 1975. Primary productivity of alpine meadow communities. Pages 121–128 in F. E. Wielgolaski (ed.), Tundra Biome: Fennoscandian Tundra Ecosystems, vol. 1, Plants and Microorganisms. Springer-Verlag, Berlin.

Wielgolaski, F. E., L. C. Bliss, J. Svoboda, and G. Doyle. 1981. Primary production of tundra. Pages 187–226 in L. C. Bliss, O. W. Heal, and J. J. Moore (eds.), Tundra Ecosystems: A Comparative Analysis. Cambridge University Press, Cambridge.

Williams, C. J., J. B. Yavitt, R. K. Wieder, and N. L. Cleavitt. 1998. Cupric oxidation products of northern peat and peat-forming plants. Canadian Journal of Botany 76:51–62.

Wilson, M. A., J. Sawyer, P. G. Hatcher, and H. E. Lerch. 1989. 1,3,5-hydroxybenzene structures in mosses. Phytochemistry 28:1395–1400.

Yurtsev, B. A. 1994. Floristic division of the Arctic. Journal of Vegetation Science 5: 765–776.

Zoltai, S. C., and D. H. Vitt. 1995. Canadian wetlands: Environmental gradients and classification. Vegetatio 118:131–137.

# 7

# Estimating Net Primary Production
# of Salt Marsh Macrophytes

James T. Morris

This chapter addresses methods of quantifying the net primary production (NPP) of salt marsh macrophytes. Salt marshes occur within the intertidal zone (McKee and Patrick 1988) and are typically dominated by perennial grasses of the genus *Spartina*. *S. alterniflora* is the dominant species of salt marshes of the eastern coast of the United States from New York to the middle of Florida, and along the Gulf coast from central Florida to Texas. Salt marshes are replaced in south Florida by mangroves. The National Science Foundation (NSF) supports Long-Term Ecological Research (LTER) sites with substantial salt marsh habitat in Massachusetts (Plum Island Estuary), Virginia (Virginia Coast Reserve), and Georgia (Georgia Coastal Ecosystem). NSF's Long-Term Research in Environmental Biology (LTREB) program supports a salt marsh site in South Carolina at North Inlet. Salt marsh NPP is thought by many to be dominated by macrophytes, but benthic microalgal production is significant as well (Pinckney and Zingmark 1993). Unfortunately, relatively little work has been done on the benthic algal production in salt marshes.

*Spartina alterniflora* normally grows in monospecific stands that can vary in size up to thousands of hectares (ha), depending on the geomorphology of the estuary. New England marshes are dominated by *Spartina patens* or a mixture of *S. patens-Distichlis spicata-Juncus gerardi* at higher elevations, while *S. alterniflora* is confined to the lower elevations of the marsh around the margins of tidal creeks (Chapman 1940; Bertness 1991). The occurrence of salt marsh ecosystems on the U.S. west coast is less common than in the east due to the high wave energy, but where they occur, *Spartina foliosa* is the native marsh grass. However, *S. alterniflora* has invaded the west coast and is transforming its estuaries, particularly in the northwest. European marshes are dominated by *S. anglica*, a hybrid of *S. alterniflora* that was purposely introduced into many estuaries to stabilize the sediments. *S. anglica* also was introduced, and is common, in China and elsewhere in the Pacific region.

Salt marsh NPP is most commonly defined as the net annual production of aboveground dry weight or carbon. Less commonly it is expressed in units of net $CO_2$ assimilation or energy equivalents. NPP estimates may sometimes include belowground production, but this is rare.

Annual aboveground NPP is reported to be in the range of 60 g C $m^{-2}$ $yr^{-1}$ in northern Canada to 812 g C $m^{-2}$ $yr^{-1}$ on the U.S. Gulf coast. Turner (1976) reviewed the primary production data for *S. alterniflora* and concluded that there is a latitudinal gradient that parallels solar energy and corresponds to a 0.2%–0.35% net conversion efficiency. However, there was a large variation of NPP within and among marshes. Some of this variation is a consequence of the biases inherent in the different methods that are used, and some of the variation has an environmental origin.

## Unique Aspects and Primary Drivers of NPP in Salt Marshes

Certainly the most unusual aspect of salt marsh ecosystems is their location within the intertidal zone. Stable intertidal salt marshes occupy a broad, flat expanse of landscape, often referred to as the marsh platform, at an elevation within the intertidal zone close to that of mean high tide (Krone 1985). However, the relative elevation, as well as the productivity of the salt marsh, varies with the rate of sea level rise (Morris et al. 2002). Salt marshes are also characterized by pronounced gradients in primary productivity, even within the same plant community, that are driven by factors such as inundation frequency, drainage, $O_2$ availability in the sediments, salinity, and N availability, which are interactive to varying degrees (Mendelssohn and Morris 2000). Much research has been done on these factors from the physiological (Mendelssohn et al. 1981; Pearcy and Ustin 1984; Bradley and Morris 1990; Hwang and Morris 1994). to the whole plant (Phleger 1971; De Laune et al. 1983; Giurgevich and Dunn 1979; Linthurst 1979, 1980; Mendelssohn 1979; Morris 1982; Howes et al. 1986) and landscape scales (Morris et al. 2002). Although there is no consensus on the details, a general model has emerged that *S. alterniflora* is universally limited by N availability, and that N availability on a physiological level is dictated by the presence of salts, sulfides, and $O_2$ that modify the kinetics of N uptake. However, N limitation, whether direct or indirect, is not the whole story by any means, and variables such as salinity do influence productivity independently of their effects on N availability. Salinity, and hence productivity, is sensitive to the rate of evapotranspiration, drainage, and flood frequency, which in turn are functions of the elevation of the salt marsh relative to mean sea level and tidal amplitude (Morris 1995).

On a landscape scale, relative elevation appears to be a primary driver of NPP because of its effect on flood frequency, drainage, and soil salinity, as discussed above. Experimentally, this can be demonstrated using a device termed a "marsh organ" (fig. 7.1). This simple field bioassay provides information on above- and

Figure 7.1. Relative elevation is a determinant of NPP in salt marshes, as shown experimentally here at Plum Island Sound, MA. This construction, termed a "marsh organ," consists of 6 rows of 12-cm-diameter PVC pipes stacked at different heights, 6 to a row. The pipes were filled with sediment excavated from the area and are open on the bottom to allow for drainage. Each core was planted in early spring with a few transplants taken locally.

belowground production, biogeochemical parameters, and plant physiological responses as functions of relative elevation. The marsh organ shown here was placed in a protected tidal inlet. Protection from waves and currents is an essential design element. These devices support the hypothesis that the marsh platform lies at a relative elevation that is superoptimal for NPP. This is an important feature of salt marshes that is required for marsh stability (Morris et al. 2002). Rising sea level stimulates productivity, which increases sedimentation and organic matter accumulation, and maintains the elevation of the marsh in equilibrium with mean sea level.

Belowground production in salt marsh ecosystems is significant for a variety of reasons. Where it has been measured, it is generally equal to or greater than aboveground production (Blum 1993). Root production also plays an important role in stabilizing salt marshes. As sea level rises, salt marshes maintain elevation (or not) by a combination of sediment-trapping and organic matter accretion. Turner et al. (2000, 2004) have argued that root production is vital to maintaining the relative elevation of the salt marsh in the Mississippi delta. Root production is undoubtedly the major component of the peat that characterizes many New England marshes where there is geological evidence of an accretion rate equal to the rate of sea level rise (Redfield 1965, 1972). However, root and rhizome productions are exceedingly difficult to measure (chapter 8, this volume). In one study of *S. alterniflora* in N-limited hydroponic culture, the ratio of root:shoot production was $0.93 \pm 0.16$ (Morris 1982). This study also demonstrated the trade-off that exists between root and shoot production. Because of the compounding of growth afforded by leaf production, if photosynthate is allocated to roots at the expense of leaves, then total production must decline.

## Review of Methods

Salt marsh NPP is most commonly measured by one of several harvest techniques. The Wiegert-Evans and Lomnicki methods have been discussed in chapter 4 of this volume. Milner and Hughes's (1968) method is one of the simplest of the harvest-based methods. The calculation is based on a summation of the positive changes in biomass through time:

$$NPP = \Sigma \ (B_n - B_{n-1}) \text{ for all } (B_n - B_{n-1}) > 0,$$

where $B_n$ is mean standing biomass at time $n$, obtained by harvesting replicate plots of *Spartina* grass. The plots are usually 0.0625 to 0.25 m$^2$ in area, and the number of replicates is 6 or more per site. Milner and Hughes's method is exceeded in simplicity only by estimates of NPP obtained by measuring the end-of-season standing biomass or peak standing crop, which entails only a one-time harvest of replicate plots at the time of maximum standing crop or at the end of the growing season. The end-of-season (EOS) technique can seriously underestimate annual NPP by failing to account for stem and leaf turnover, but it is a reasonable way to obtain an index of NPP, and for some questions this may be sufficient. A significant relationship has been observed between EOS standing biomass and NPP in a South Carolina salt marsh that has changed little over time (fig. 7.2), but the relationship has varied from one marsh area to another (e.g., tall vs. short *Spartina*), and it probably varies from one estuary to the next with variables such as fertility and climate.

Smalley's (1958) method compensates for changes in both living and dead components in an attempt to correct Milner and Hughes's method for turnover. Where the change in live biomass is positive, production over this interval of time is equal to this change plus any increase in the dead standing crop. If the change in live standing crop is negative, production during the interval is computed as the sum of the changes in live and dead, or as zero, whichever is larger. This method, like other harvest methods that ignore intervals of negative production, can result in a positive bias (Kirby and Gosselink 1976; Singh et al. 1984).

Kaswadji et al. (1990) measured the aboveground production of *S. alterniflora* in a Louisiana salt marsh using different methods, including harvest methods described by Milner and Hughes (1968), Smalley (1958), Wiegert and Evans (1964), and Lomnicki et al. (1968), as well as a nondestructive method based on measure-

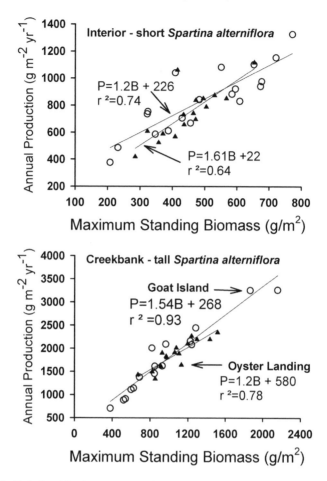

Figure 7.2. Relationships between net aboveground production and maximum standing biomass (end-of-season standing biomass) in areas of salt marsh in interior or "high" marsh locations dominated by a short morphotype of *S. alterniflora* (top panel), and areas along the margins of tidal creeks that are dominated by the tall phenotype of *S. alterniflora*. All the sites are located within the North Inlet, SC, estuary, but the specific marsh areas (Goat Island ○ and Oyster Landing ▲) are in different regions of the estuary.

ment of stem density and longevity. Annual production estimates of aboveground biomass were 831 ± 41, 831 ± 62, 1231 ± 252, 1873 ± 147, and 1437 ± 96 g/m$^2$ (dry weight), respectively. Similar discrepancies were found, also in Louisiana, by Kirby and Gosselink (1976), who reported rates of 750 to 2600 g m$^{-2}$ yr$^{-1}$ (dry weight), depending on the way the measurements were integrated. Among the five methods, Kaswadji et al. (1990) concluded that the Wiegert-Evans and Lomnicki et al. methods were more accurate than the others because they corrected for mortality losses between sampling times. The Lomnicki et al. method was preferred over the Wiegert-Evans method because of its greater simplicity. Dickerman et al.

(1986) compared several of these harvest methods with a nondestructive method and concluded that the harvest methods underestimated aboveground NPP by as much as 38% due to inadequate correction for shoot mortality and leaf turnover.

All of the methods of measuring NPP are affected by, and fail to account for in some degree, the turnover of stems and leaves between sampling dates. The importance of turnover increases with decreasing latitude, and can equal the annual changes in standing live biomass (Turner 1976). In a Louisiana marsh the average longevity of individually tagged young shoots was $5.2 \pm 0.2$ months, equivalent to an annual turnover rate of 2.3 crops per year (Kaswadji et al. 1990). Turnover rate also varies within marshes. Dai and Wiegert (1996b) found average leaf longevity was 49 and 72 days in areas supporting short and tall *Spartina*, respectively, although stem longevity was about 215 days (7 months) in both areas. Morris and Haskin (1990) reported leaf turnover in a South Carolina salt marsh of about 100% of total standing biomass for stems 12 months of age. Mean stem turnover in this marsh varied from 0.47 to 0.74 per year in areas of short *Spartina* to 1.3 to 2.2 per year along the creek bank in areas of tall *Spartina*. Stem turnover in a Plum Island, MA, salt marsh averaged 0.4 per yr, supporting Turner's (1976) observation that turnover decreases with increasing latitude.

Several nondestructive, allometric or phenometric methods of estimating NPP have been described. These techniques require dimensional analysis of the stems at intervals throughout the year, and calibration with stem height and weight. Dai and Wiegert (1996b) made a thorough study of stem densities and dimensions in a Georgia salt marsh by measuring the height and diameter of each stem in permanent plots as well as the dimensions of leaves. Living aboveground biomass was calculated as the sum of the dry mass of all the ramets (stems) in each plot, and dead biomass was the summed masses of all dead ramets and leaves appearing during each sampling interval. A variant of this technique involves parsing the stem population into size classes, using regressions to estimate the stem weights of each height class, and accounting for stem mortality (Hopkinson et al. 1980). Note that the nondestructive methods described here are conceptually equivalent to the harvest methods. The major difference lies in the technique used to estimate standing crop. NPP is still calculated as the summation of differences in total standing biomass.

Whole canopy $CO_2$ flux measurements have been used to estimate NPP and net ecosystem production (NEP) in salt marshes, but these techniques are probably not suitable for routine measurements of NPP. Typically, short-term measurements of net $CO_2$ flux are made inside a chamber under full sunlight to give net $CO_2$ uptake or NEP, and in darkness to give whole ecosystem respiration (Bartlett et al. 1989). Measurements of $CO_2$ flux from the soil, also using chambers (Morris and Whiting 1986), can then be subtracted from total respiration to derive canopy respiration. This is a reasonably good method for calculating gross primary production and then partitioning net primary production between above- and belowground components. For example, from monthly measurements of canopy and soil $CO_2$ exchange in a Danish marsh, a limit on gross belowground production of 1310 g m$^{-2}$ yr$^{-1}$ was obtained (Morris and Jensen 1998). If soil respiration is $S$, and total respiration is $R$, then canopy respiration is $C = R - S$. Gross production is GPP = NEP + $R$, where

NEP and $R$ are positive rates. If the net aboveground production (NPP) is known from harvest or census-type measurements, then the gross belowground production can be calculated as $GBP = GPP - C - NPP$. Unfortunately, the partitioning of $S$ between root respiration and decomposition is unknown. Integrated totals can be derived only with the aid of numerical models such as those proposed by Dai and Wiegert (1996a) or Morris and Jensen (1998), because the flux measures are short-term. $CO_2$ exchange methods are useful for partitioning GPP and for comparative purposes, but they are not well suited for routine monitoring of NPP.

## Recommended Methods

As a method of quantifying NPP, allometric methods that account for stem turnover are preferred over harvest methods. The allometric methods offer a number of advantages, not the least of which is the ability to measure NPP nondestructively. This advantage is greatest when NPP is to be measured over several years or in marshes of small size where destructive harvests would degrade the ecosystem. The greatest damage to a marsh can be from the trails created by field biologists on their way to visit plots. The salt marsh recovers very slowly (years) after trampling. When using allometric methods on permanent plots, it is possible and advisable to place board-walks across the marsh to access the sites. Another advantage of the allometric methods, when used on permanent plots, is the wealth of demographic data obtained. It is possible to record stem emergence and death rates, individual growth rates, age, and survivorship. Finally, when the census of plant populations is carried out on permanent plots, the resultant NPP estimates are not corrupted by spatial variability. Thus, detailed and accurate measurements of the temporal variation in growth and NPP are possible. With harvest methods, the difference in biomass between two sample dates could be due to net growth, or it could simply represent the error in estimating the biomass from random samples collected from a spatially variable landscape. This problem, and differences in the degree to which harvest methods fail to account for turnover, are the primary disadvantages of harvest methods.

### Modifications and Special Considerations

Herbivory on *Spartina* has been thought to be minor and to consist mostly of grazing by grasshoppers, which was estimated to be 7% of NPP (Teal 1962). However, the plant is also attacked by stem borers (Stilling and Strong 1983) and by a snail known as the marsh periwinkle or *Littoraria irrorata* (Silliman and Bertness 2002). Silliman and Bertness have shown that *Littoraria* can denude areas of salt marsh in the southeastern United States, where the snail is common. They employed open-topped cages constructed of galvanized hardware cloth to exclude snails from plots in order to demonstrate this effect, and there is no reason why this method could not be used in combination with one of the allometric methods to account for effects of snail herbivory. NPP measured in the absence of herbivory, in a snail exclosure, would be an overestimate due to the compounding effect of growth, while measurements made in the presence of herbivory would underestimate NPP. In addition, the potential effect of shading by cages would need to have a suitable control.

## Belowground Production

General principles of measuring belowground production are discussed in more detail in chapter 8 of this volume, but in salt marshes, belowground NPP is most commonly measured by taking sediment cores and washing the sediments from the roots in sieves (Valiela et al. 1976) or separating the roots from the sediments using a paint shaker (Hopkinson and Dunn 1984), and positive changes in root biomass are summed to calculate net belowground production. However, the spatial variability in root biomass associated with taking small-diameter cores (typically 10 cm or less) is extreme. This leads to a serious positive bias in the estimate of belowground production (Singh et al. 1984).

Another method that has been used involves some variation on the theme of directly measuring the growth of roots and rhizomes into an artificial substrate in the field. This is commonly referred to as the root ingrowth method. After removing a core of sediment, the hole is filled with sand or clean sediment, and the same spot is cored again at some later time (Valiela et al. 1976); the roots and rhizomes that have invaded the core can be measured relatively accurately. If this is done repeatedly throughout the year, with suitable replication, the summed growth of roots and rhizomes may equal net belowground production. A modification of this technique involves using litter sewn into a mesh bag as the artificial substrate (Blum 1993). However, there are several unknowns, such as the effects of a new substrate devoid of competing roots, or of severing the roots and rhizomes around the core hole. Certainly the information on seasonality of root and rhizome production obtained by this technique is very valuable, but the quality of the absolute numbers is unknown (see chapter 8, this volume).

The root ingrowth method was used in a South Carolina salt marsh site by recoring monthly over a period of several years. This method was unable to discern interannual variations in root production with only 5 replicates, due to the large spatial variation between cores. However, the accumulation of biomass within 1-month-old sand-filled core holes did provide information about the seasonality of root and rhizome production, and the cumulative total growth gave an estimated annual production (925 g/m$^2$ dry weight) that was consistent with measurements of aboveground production (780 g/m$^2$) at this site and with estimates of the ratio of root:shoot production ($0.93 \pm 0.16$) of S. alterniflora in N-limited hydroponic culture (Morris 1982).

## Specific Field and Lab Procedures

Salt marsh NPP in sites dominated by monocultures of S. alterniflora or similar vegetation should be measured by a periodic census of the heights of all stems within permanent plots. The individual stem weights should be estimated from allometric regression equations that are derived from harvests of individual stems. Stem populations censused monthly in 6 permanent 1 dm$^2$ quadrats in an area of short, dense *Spartina* gave excellent results (Morris and Haskin 1990). The site where this study was done was characterized by a dense population of *S. alterniflora* with stem heights generally less than 30 cm. Further work in other areas has shown that a plot size great enough to contain about 15 stems is sufficient, and in areas of marsh where the stem density is relatively low, the plots should be sized accordingly. The census method should be equally good for plants that present a morphology with an easily quantifiable dimension. There are marsh species, such as the New England form of *S. patens,* for which this method is impractical, and in these and all cases the choice of method really depends on the question and the need to balance labor-intensive sampling with practicality.

For illustrative purposes, an example of the recommended protocol is provided. In a South Carolina salt marsh, six permanent subplots are monitored along creek banks as well as in interior locations where different drivers are important. The subplots are grouped in pairs, one pair within each 1 m² area of marsh. Each subplot has been delineated with small dowel rods of wood or plastic. All plant stems within each plot have been tagged with plastic, number-coded bird bands. (National Band and Tag Co. manufactures these in a variety of sizes and colors.) In addition, maps of each plot have been sketched, with the locations of stems identified by a coordinate system. A combination of maps kept in field notebooks and tagging of stems is recommended.

Monitoring of the stems is required only during the growing season, which is year-round in southern marshes. The frequency of monitoring should be monthly, although it can be decreased with some sacrifice to turnover estimates during the summer months. On each census date the height of each stem, from the sediment surface to the tip of the longest, outstretched leaf, should be recorded. Newly emerged stems should be tagged or mapped, and added to the database. Stems should be monitored until they die. The plots should be located along boardwalks so as to avoid the problems associated with trampling. Following recommendations in chapter 2 of this volume, data management is expedited via the transfer of the data to an electronic database that contains the date of the census, plot identifications, stem identifications, and heights. The development of an MS Access database and data entry form that contain basic error-checking routines permits the output to an ASCII data file that can subsequently be used to convert stem heights to individual stem weights and then total biomass, total net growth since the last census, stem density, birth and death rates, and mean stem age.

The transformations of stem heights into weights are accomplished by polynomial equations that are based on regressions of individual stem weights as functions of stem height. The data used to fit these equations should be derived from harvests of individual stems representing the full range of plant sizes on the marsh. The relationship between stem weight and height varies seasonally and with morphology, and collections of specimens should accurately reflect this variability. At a minimum, collections should be made in each of the four seasons, and tall-form or creekside *Spartina* should be collected and analyzed separately from the short form of *Spartina.* Furthermore, the phenology of growth varies greatly with latitude (fig. 7.3), and this needs to be accounted for when working in disparate geographic locations.

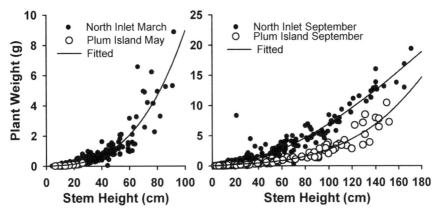

Figure 7.3. Allometric relationships between stem height and aboveground plant dry weight in populations of *S. alterniflora* from Plum Island, MA, and North Inlet, SC, at different times of year. Regression equations that describe these relationships for the Plum Island population are $W = 0.0023H - 1.09 \times 10^{-4} H^2 + 1.34 \times 10^{-5} H^3$; $r^2 = 0.96$ and $W = 0.013H - 1.55 \times 10^{-4} H^2 + 2.97 \times 10^{-6} H^3$; $r^2 = 0.94$, during May and September, respectively. For North Inlet these relationships are described by $W = 0.0038H - 1.21 \times 10^{-5} H^2 + 8.70 \times 10^{-6} H^3$; $r^2 = 0.88$ and $W = -2.7 \times 10^{-4} H + 8.45 \times 10^{-4} H^2 - 1.45 \times 10^{-6} H^3$; $r^2 = 0.95$, during March and September, respectively.

To collect specimens for calibration, all stems can be harvested from several 0.25 $m^2$ quadrats, taking care to cut the stems at the base and to preserve the leaves on each stem. In the lab, the height of each individual stem from the base to the tip of the longest leaf or inflorescence is recorded, and then stems are oven dried and weighed after they have reached a constant weight. Individual stem weights and heights are analyzed (weight is fitted to height), using a polynomial of the form

$$W_s = c_1 H + c_2 H^2 + c_3 H^3 \ldots c_n H^n, \tag{1}$$

where $W_t$ is dry stem weight (g) at time (or season) $s$, the $c_i$ are regression coefficients, and H is the stem height (cm). A third-order polynomial ($n = 3$) is generally sufficient, but some trial and error may be required in order to find the best model.

Monthly and annual rates of net aboveground production per area of delineated plot are calculated for each plot by summing the positive growth rates of each stem:

$$NPP = \Sigma_j \, \Sigma_t \, W_{it} \tag{2}$$

where $i$ represents the $i$th plant in the plot and $t$ represents the time of the sample. $W_t$ is calculated by interpolating the predicted weights calculated from equation (1):

$$W_t = W_1 + (t_t - t_1) \, (W_2 - W_1)/(t_2 - t_1), \tag{3}$$

where $t_s$ is Julian day at the time the stem height was recorded; $W_1$ is the weight, from equation (1), at time $t_1$ (Julian day) prior to $t_t$; and $W_2$ is the weight on Julian day $t_2$ after $t_t$.

Corrections for leaf turnover can be based on the age of the stem. Calibration requires tagging and monitoring leaves on a population of newly emerged stems. Calibration based on at least 20 stems on which leaves have been tagged and later censused is recommended. The weights of tagged leaves can be estimated from harvests of leaves of equal size from adjacent plants. At each leaf census, the number of leaves lost since the last census, the age of the stem, and the height of the stem are recorded. Leaf turnover was accounted for in the calculation of NPP (eq. 2) by calculating a "phantom" weight, which is a transformation of the actual weight (eq. 3) that includes the weight of leaves that have been lost by turnover:

$$W_t^* = W_t \, (1 + 0.00002 A_t^{3.91337}), \tag{4}$$

where $A_t$ is the age of the stem in months, and the coefficients are determined by a least-squares regression. Analyses have shown that leaf loss is an important component of net production for stems greater than 10 months of age. At age 10 months, stems had lost 16% of their current weight, while stems that live for 18 months, which is about the maximum life span in the North Inlet marshes, lost 160% of their current weight (Morris and Haskin 1990).

By tracking the individual histories of the stems, which will be referred to here as the census method, it is possible to compute individual growth rates. Individual growth rates are summed to arrive at NPP (eq. 2). This is an important distinction between the census method and others that do not track individual histories. If individual growth trajectories are unknown, then it is possible only to estimate total standing crop, and NPP must be calculated as the summation of positive differences in total standing crop from one time to the next. The census method advocated here gives a better account of stem turnover as well as the added benefits of demographic data.

There is significant spatial variability in NPP in salt marshes, not only along obvious linear gradients from creek edge to the uplands, but also on a complex mosaic that is the result of microtopographic features. The census method provides information on the stability of the underlying spatial variability in NPP. At North Inlet, SC, where NPP has been monitored since 1984, differences in NPP between plots have been relatively constant for many years. For example, the plot that supports the greatest average biomass had the highest peak biomass of 3 plots in 14 out of 20 years (fig 7.4). This is remarkable, considering that the plots are in the same area, and it suggests that the spatial pattern of NPP is determined by an underlying, stable geomorphological template.

## Future Directions

New technologies (see chapter 11, this volume) may allow measurements of NPP in salt marshes and elsewhere to be scaled more broadly and occur more rapidly than current methods. One promising technology is the application of light detection and ranging (LIDAR), which uses the same principle as radar. The LIDAR instrument transmits light out to a target, which scatters and reflects the signal back to the instrument, where it is analyzed. The time for the light to travel out to the target and back to the LIDAR is used to determine the range to the tar-

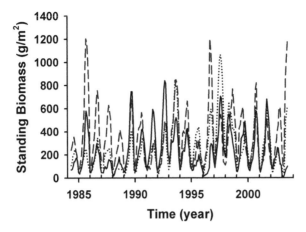

Figure 7.4. Monthly estimates of standing biomass on 3 permanent plots in a salt marsh at North Inlet, SC. Differences in biomass among plots are relatively stable. Each time series represents the mean biomass of 2 subplots contained within a 1 m² area, and is based on a monthly census of stem heights.

get. If the target is a rough surface, such as a grass canopy on the marsh surface, the pulses of light will be reflected back from the top of the canopy, the ground surface, and points in between, and the frequency distribution of distances should contain information about the geometry and biomass of the canopy. LIDAR has been used to map marsh terrain and canopy heights (Rosso et al. 2006), though this work is in its infancy. LIDAR is typically used on fixed-wing aircraft flown to map the terrain, but handheld LIDAR devices are being developed to map forest canopy structure (Parker et al. 2004), and with miniaturization such devices could be used inside a grass canopy and calibrated to biomass.

References

Bartlett, D. S., G. J. Whiting, and J. M. Hartman. 1989. Use of vegetation indices to estimate intercepted solar radiation and net carbon dioxide exchange of a grass canopy. Remote Sensing of Environment. 30:115–128.

Bertness, M. D. 1991. Interspecific interactions among high marsh perennials in a New England salt marsh. Ecology 72:125–137.

Blum, L. K. 1993. *Spartina alterniflora* root dynamics in a Virginia salt marsh. Marine Ecology Progress Series 102:169–178.

Bradley, P.M. and J.T. Morris. 1990. Influence of oxygen and sulfide concentration on nitrogen uptake kinetics in Spartina alterniflora. Ecology 71:282–287.

Bradley, P. M., and J. T. Morris. 1991. The influence of salinity on the kinetics of $NH_4+$ uptake in *Spartina alterniflora*. Oecologia 85:375–380.

Chapman, V. J. 1940. Studies in salt-marsh ecology: Sections VI and VII. Comparison with marshes on the East Coast of North America. Journal of Ecology 28:118–152.

Dai, T., and R. G. Wiegert. 1996a. Estimation of the primary productivity of *Spartina alterniflora* using a canopy model. Ecography 19:410–423.

Dai, T., and R. G. Wiegert. 1996b. Ramet population dynamics and net aerial primary productivity of *Spartina alterniflora*. Ecology 77:276–288.

DeLaune, R. D., C. J. Smith, and W. H. Patrick, Jr. 1983. Relationship of marsh elevation, redox potential and sulfide to *Spartina alterniflora* productivity. Soil Science Society of America Journal 47:930–935.

Dickerman, J. A., A. J. Stewart, and R. G. Wetzel. 1986. Estimates of net annual production: Sensitivity to sampling frequency. Ecology 67:650–659.

Giurgevich, J. R., and E. L. Dunn. 1979. Seasonal patterns of $CO_2$ and water vapor exchange of the tall and short height forms of *Spartina alterniflora* Loisel. in a Georgia salt marsh. Oecologia 43:139–156.

Hopkinson, C. S., and E. L. Dunn. 1984. Rapid sampling of organic matter in flooded soils and sediments. Estuaries 7(2):181–185.

Hopkinson, C. S., J. G. Gosselink, and R. T. Parrondo. 1980. Production of coastal Louisiana marsh plants calculated from phenometric techniques. Ecology 61(5):1091–1098.

Howes, B. L., J. W. H. Dacey, and D. D. Goehringer. 1986. Factors controlling the growth form of *Spartina alterniflora*. Feedbacks between above-ground production, sediment oxidation, nitrogen, and salinity. Journal of Ecology 74:881–898.

Hwang, Y. H., and J. T. Morris. 1994. Whole plant gas exchange responses of *Spartina alterniflora* (Poaceae) to a range of constant and transient salinities. American Journal of Botany 81:659–665.

Kaswadji, R. F., J. G. Gosselink, and R. E. Turner. 1990. Estimation of primary production using five different methods in a *Spartina alterniflora* marsh. Wetlands Ecology and Management 1:57–64.

Kirby, C. J., and J. G. Gosselink. 1976. Primary production in a Louisiana Gulf Coast *Spartina alterniflora* marsh. Ecology 57:1052–1059.

Krone, R. B. 1985. Simulation of marsh growth under rising sea levels. Pages 106–115 in W. R. Waldrop (ed.), Hydraulics and Hydrology in the Small Computer Age. Hydraulics Division, ASCE, New York.

Linthurst, R. A. 1979. The effect of aeration on the growth of *Spartina alterniflora* Loisel. American Journal of Botany 66:685–691.

Linthurst, R. A. 1980. An evaluation of aeration, nitrogen, pH and salinity as factors affecting *Spartina alterniflora* growth: A summary. Pages 235–247 in V. S. Kennedy (ed.), Estuarine Perspectives. Academic Press, New York.

Lomnicki, A., E. Bandola, and K. Jankowaska. 1968. Modification of the Wiegert-Evans method for estimation of net primary production. Ecology 59:147–149.

McKee, K. L., and W. H. Patrick, Jr. 1988. The relationship of smooth cordgrass (*Spartina alterniflora*) to tidal datums: A review. Estuaries 11:143–151.

Mendelssohn, I. A. 1979. The influence of nitrogen level, form, and application method on the growth response of *Spartina alterniflora* in North Carolina. Estuaries 2:106–112.

Mendelssohn, I. A., K. L. McKee, and W. H. Patrick, Jr. 1981. Oxygen deficiency in *Spartina alterniflora* roots: Metabolic adaptation to anoxia. Science 214:439–441.

Mendelssohn, I. A., and J. T. Morris. 2000. Ecophysiological controls on the growth of *Spartina alterniflora*. Pages 59–80 in M. P. Weinstein and D. A. Kreeger (eds.), Concepts and Controversies in Tidal Marsh Ecology. Kluwer Academic Publishers, Dordrecht, The Netherlands.

Milner, C., and R. E. Hughes. 1968. Methods for the Measurement of Primary Production of Grassland. Blackwell Scientific, Oxford.

Morris, J. T. 1982. A model of growth responses of *Spartina alterniflora* to nitrogen limitation. Journal of Ecology 70:25–42.

Morris, J. T. 1995. The mass balance of salt and water in intertidal sediments: Results from North Inlet, South Carolina. Estuaries 18:556–567.

Morris, J. T., and B. Haskin. 1990. A 5-yr record of aerial primary production and stand characteristics of *Spartina alterniflora*. Ecology 71:2209–2217.

Morris, J. T., and A. Jensen. 1998. The carbon balance of grazed and nongrazed *Spartina anglica* saltmarshes at Skallingen, Denmark. Journal of Ecology 86:229–242.

Morris, J. T., P. V. Sundareshwar, C. T. Nietch, B. Kjerfve, and D. R. Cahoon. 2002. Responses of coastal wetlands to rising sea level. Ecology 83:2869–2877.

Morris, J. T., and G. J. Whiting. 1986. Emission of gaseous carbon dioxide from salt-marsh sediments and its relation to other carbon losses. Estuaries 9:9–19.

Parker, G. G., D. J. Harding, and M. L. Berger. 2004. A portable laser altimeter for rapid determination of forest canopy structure. Journal of Applied Ecology 41:755–767.

Pearcy, R. W., and S. L. Ustin. 1984. Effects of salinity on growth and photosynthesis of three California tidal marsh species. Oecologia 62:68–73.

Phleger, C. F. 1971. Effect of salinity on growth of a salt marsh grass. Ecology 52:908–911.

Pinckney, J. L., and R. G. Zingmark. 1993. Modelling the annual production of intertidal benthic microalgae in estuarine ecosystems. Journal of Phycology 29:396–407.

Redfield, A. C. 1965. Ontogeny of a salt marsh estuary. Science 147:50–55.

Redfield, A. C. 1972. Development of a New England salt marsh. Ecological Monographs 42:201–237.

Rosso, P., S. Ustin, and A. Hastings. 2006. Use of LIDAR to study changes associated with *Spartina* invasion in San Francisco Bay marshes. Remote Sensing of the Environment 100:295–306.

Shew, D. M., R. A. Linthurst, and E. D. Seneca. 1981. Comparison of production methods in a southeastern North Carolina *Spartina alterniflora* salt marsh. Estuaries 4: 97–109.

Silliman, B. R., and M. D. Bertness. 2002. A trophic cascade regulates salt marsh primary production. Proceedings of the National Academy of Sciences 99:10500–10505.

Singh, J. S., W. K. Lauenroth, H. W. Hunt, and D. M. Swift. 1984. Bias and random errors in estimators of net root production: A simulation approach. Ecology 65:1760–1764.

Smalley, A. E. 1958. The role of two invertebrate populations, *Littorina irrorata* and *Orchelium fidicinum*, in the energy flow of a salt marsh ecosystem. Ph.D. dissertation, University of Georgia.

Stilling, P. D., and D. R. Strong. 1983. Weak competition among *Spartina* stem borers, by means of murder. Ecology 64:770–778.

Teal, J. M. 1962. Energy flow in the salt marsh ecosystem of Georgia. Ecology 43:614–624.

Turner, R. E. 1976. Geographic variations in salt marsh macrophyte production: A review. Contributions in Marine Science 20 :47–68.

Turner, R. E., E. M. Swenson, and C. S. Milan. 2000. Organic and inorganic contributions to vertical accretion in salt marsh sediments. Pages 583–596 in M. P. Weinstein and D. A. Kreeger (eds.), Concepts and Controversies in Tidal Marsh Ecology. Kluwer Academic Publishers, Boston.

Turner, R. E., E. M. Swenson, C. S. Milan, J. M. Lee, and T. A. Oswald. 2004. Belowground biomass in healthy and impaired salt marshes. Ecological Research 19:29–35.

Valiela, I., J. M.Teal, and N. Y. Persson. 1976. Production and dynamics of experimentally enriched salt marsh vegetation: Belowground biomass. Limnology and Oceanography 21:245–252.

Wiegert, R. G., and F. C. Evans. 1964. Primary production and the disappearance of dead vegetation on an old field in Southeastern Michigan. Ecology 45:49–63.

# 8

# Estimating Belowground Primary Productivity

Geraldine L. Tierney
Timothy J. Fahey

Full assessment of net primary productivity (NPP) in terrestrial ecosystems requires inclusion of production allocated belowground, but accurate measurement of NPP belowground has lagged behind aboveground measurement due to the difficulty of identifying and measuring new production in the soil. Several good reviews of methodology for measuring belowground NPP (BNPP) are Vogt et al. (1998), Fahey et al. (1999), and Lauenroth (2000); this chapter will update methodological recommendations that are based on new research in this developing field and provide guidance on method selection within various ecosystems. We also attempt to identify areas where additional research is most needed.

It is useful to begin by explicitly defining BNPP and considering its components. BNPP must be distinguished from total root allocation (TRA) of carbon (C) belowground, which includes both BNPP and C lost to root respiration ($R_r$), as shown in equation (1):

$$BNPP = TRA - R_r \tag{1}$$

BNPP cannot be measured directly, but it can be estimated from direct measurements of its components, or from estimates of TRA and $R_r$. The largest component of BNPP is allocated to the production of short-lived fine roots (FRP); however, FRP is not equivalent to BNPP, as the focus on measuring FRP in the literature suggests. A smaller fraction of BNPP is allocated to the production of relatively long-lived coarse and woody roots; woody root production can be estimated from aboveground measurements using allometric methods, as described in chapter 5 (see Whittaker et al. 1974). A substantial fraction of BNPP is conveyed to mycorrhizal symbionts or otherwise exported from vegetation via root exudation and rhizodeposition (Grayston et al. 1996). These fluxes are very difficult to quantify, but new techniques are being developed (Cheng et al. 1994; Kuzyakov and Domanski 2000; Phillips and Fahey 2005). Finally, it is difficult to quantify what fraction of BNPP is lost to herbivory. Experiments in which root herbivores are excluded suggest that this component can be substantial (Wells et al. 2002). Standard methods for measuring several of these components have yet to be developed; thus, this chapter will focus on methods which are currently available to estimate BNPP from related quantities (FRP and TRA).

Two related concepts also require definition. First, all of these approaches employ a "steady-state assumption" for soil C or fine root biomass, meaning that we assume the steady-state parameter is neither increasing nor decreasing substantially. Second, some of these approaches estimate FRP using indices of fine root turnover, a measure of the proportion of fine root biomass dying and being replaced on an annual basis. Under the steady-state assumption, the fine root turnover coefficient (TC, $yr^{-1}$) is the inverse of average fine root longevity (yr), and FRP (g dry weight $m^{-2}$ $yr^{-1}$) can be calculated as the product of TC and fine root biomass (g $m^{-2}$).

## General Considerations and Estimates of Belowground NPP

Belowground NPP (BNPP) has been estimated in a variety of ecosystems by using many different methods. Choice of methodology for a particular site will be influenced by the physiognomy and other characteristics of that ecosystem. Chief among these characteristics are the phenology of fine root production and mortality. In agricultural systems, some grasslands, and northern ecosystems with short growing seasons, fine root production is concentrated in the spring, and mortality in the fall or winter. This temporal separation of production and mortality simplifies the task of identifying new production. Conversely, in perennial vegetation, fine root production and mortality occur concurrently throughout the year, and sampling must achieve the more difficult task of distinguishing new fine root production from existing fine root biomass.

Certainly the physical characteristics of vegetation influence choice of methodology for estimating BNPP. Schenk and Jackson (2002) have shown that rooting depth varies globally across biomes, depending upon the dominant plant life-form and climate of a region. Their review of fine root data sets from around the globe shows that roots grow deepest in arid and arboreal ecosystems (Mediterranean shrublands, temperate savannas, and dry tropical savannas), while rooting depths are shallowest in tundra, boreal forest, and temperate meadows. Their analysis concluded that most studies fail to sample the deepest roots, and that this failure is particularly evident in tropical systems, where variation in rooting depth is high. Due to the difficulty of directly sampling deep roots, indirect methods for estimating BNPP will be particularly useful in deeply rooted ecosystems. Likewise, the stature of vegetation will influence choice of methodology; isotopic labeling techniques can be applied in grassland ecosystems but are difficult for mature, arboreal vegetation.

Physical characteristics of vegetation influence patterns of spatial heterogeneity in the soil, which in turn influence methodology for estimating BNPP. Heterogeneity is a defining characteristic of the soil ecosystem, and its degree and scale will influence sampling design. This is particularly true in forests and shrublands, which have greater soil heterogeneity than grasslands (Kleb and Wilson 1997). Partel and Wilson (2002) have shown that variation in fine root length is higher in an aspen forest than in a prairie, and that the scale of patchiness is substantially larger in the forest (8–12 cm) than the prairie (3–4 cm). Finally, physical characteristics of the site are also important. BNPP sampling can be hindered by rocks, water, and ice, and special methodological adaptations may be necessary (Lopez et al. 1996; Baker et al. 2001).

Taken together, these considerations indicate that BNPP is most easily measured in agro-ecosystems and grasslands, where sampling design is simplified by relatively low soil heterogeneity, temporal separation of root production and mortality, and the short stature of the vegetation. Patterns of BNPP are sufficiently understood within grasslands that models have been developed of BNPP as functions of temperature and aboveground NPP (Gill et al. 2002). In contrast, measurement of BNPP in shrublands and forests is more complex.

The fraction of NPP exported belowground varies among ecosystems (table 8.1). Not surprisingly, grasslands have the highest ratio of BNPP:NPP; typically, at least half of grassland NPP is allocated belowground. Less intuitive are differences in patterns of allocation between evergreen and deciduous boreal forests; the former appear to export a substantially greater fraction of NPP belowground. This pattern is not evident in temperate forests; belowground allocation of BNPP is similar in deciduous and evergreen temperate forests. BNPP has rarely been measured in tropical ecosystems (Clark et al. 2001) or in Mediterranean vegetation.

Understanding the processes that control patterns of BNPP across and within ecosystems is of great interest, but methodological limitations and lack of standardization have hindered comparative studies. Gill and Jackson (2000) reported a strong correlation between temperature and rate of fine root turnover across ecosystems, but found that within a site, other factors were more related to variation in fine root

Table 8.1. Estimates of belowground net primary production (BNPP) for major terrestrial biomes in relation to total NPP

| Biome | Mean BNPP $(g\ m^{-2}\ yr^{-1})$ | Mean NPP $(g\ m^{-2}\ yr^{-1})$ | BNPP:NPP | Root Turnover $(yr^{-1})$ |
|---|---|---|---|---|
| Grassland | 498 | 1032 | 0.52 | 0.47 |
| Deciduous boreal forest | 196 | 1070 | 0.18 | 0.26 |
| Evergreen boreal forest | 312 | 774 | 0.40 | 0.34 |
| Deciduous temperate forest | 380 | 1470 | 0.26 | 0.35 |
| Evergreen temperate forest | 426 | 1772 | 0.24 | — |
| Tropical forest | — | — | — | 1.0 |
| Temperate shrublands | — | — | — | 0.44 |

*Notes*: Representative root turnover coefficients $(yr^{-1})$ also are provided. Values are derived from a variety of studies using diverse methodology and are subject to many sources of error and uncertainty. The values are meant to serve as a guide to the expected magnitude of BNPP encountered in different biomes.

*Source*: Estimates derived primarily from Lauenroth and Gill (2003).

turnover. Similarly, Vogt et al. (1996) were able to develop relationships between FRP and climatic and nutrient variables within forested ecosystems across the globe by comparing data sets from the literature. Within forested sites, FRP seems to be controlled primarily by temperature and moisture at some sites, while endogenous control over FRP may be stronger at others (Tierney et al. 2003). Additional manipulative experiments are needed to better understand controls on BNPP at individual sites, and to predict how those sites may respond to global change.

## Guiding Principles and Recommendations

Currently, four approaches are recommended for estimating BNPP from measurable parameters: (1) estimating FRP from differences in fine root biomass sampled using sequential cores; (2) estimating FRP from direct observation of fine roots using minirhizotrons; (3) indirect calculation of TRA, BNPP, or FRP, using elemental budgets; and (4) estimation of FRP or other components of BNPP using carbon isotope analysis. An additional approach using ingrowth cores to estimate FRP can be useful to achieve limited objectives and is briefly discussed herein. No single approach is best overall; each has merits as well as biases and other drawbacks. The best choice for a particular study will depend on a suite of factors including specific research objectives, the ecosystem under study, time frame, and budgetary limitations.

### Overview of Methods

The sequential coring method is straightforward and has been widely used in a variety of ecosystems. In this method, FRP is estimated from changes in live and dead fine root biomass measured with soil cores at periodic intervals (Vogt and Persson 1991), often incorporating a model which estimates fine root turnover between sampling intervals (Santantonio and Grace 1987). This method is used most appropriately within ecosystems which experience a distinct and brief seasonal pulse of fine root production, such as in grasslands or agro-ecosystems; it is less accurate in forests and shrublands, in which fine root production and mortality occur concurrently throughout the growing season. In those ecosystems, sequential cores will underestimate FRP, and attempts to improve estimates by modeling turnover between intervals have been hindered by a poor understanding of fine root decomposition (Publicover and Vogt 1993). This method can require substantial labor to separate fine roots from cores, as well as assessment of fine root vitality, which is difficult.

The minirhizotron (MR) method offers potential for improvement over sequential cores in perennial ecosystems because it allows detection of concurrent fine root production and mortality. This method allows direct, repeated observation using digital or video imagery of fine roots growing on the outer surface of tubes placed in the soil (Taylor 1987). Unfortunately, this method probably has tended to overestimate FRP due to stimulation of production by MR tube installation (Joslin and Wolfe 1999), estimation of FRP from median fine root survivorship (Tierney and Fahey 2002), and tube effects on fine root survivorship (Withington et al. 2003).

Careful methodology as described herein should minimize these biases. However, this method is very labor intensive, and requires a substantial financial investment for equipment.

Elemental budget methods offer an indirect approach for estimating TRA or FRP. Assuming that soil C pools are near steady state, measurement of the major C fluxes can be used to estimate TRA and/or BNPP by closing the ecosystem C budget (Raich and Nadelhoffer 1989). This method can be used without the steady-state assumption if additional fluxes are measured (Nadelhoffer et al. 1998). One substantial drawback of this method is that root respiration ($R_r$) is needed to deduce BNPP from TRA, and measurement of root respiration is difficult (Rakonczay et al. 1997; Fahey and Yavitt 2005). An analogous method estimates FRP using the nitrogen (N) budget (Nadelhoffer et al. 1985). While the N budget method does not require measurements of root respiration, accurate estimation of in situ N mineralization is required, and errors in accuracy of that critical measurement have substantial impacts on the outcome. While the utility of these budget approaches for understanding patterns of carbon allocation across the landscape is clear, the applicability of these methods to quantifying BNPP at specific sites remains controversial (Gower et al. 1996; Nadelhoffer et al. 1998; Davidson et al. 2002a).

Isotopic methods are rapidly being developed, and offer great promise for improving quantification of BNPP and specific fluxes of C belowground in some ecosystems (Milchunas and Lauenroth 1992; Kuzyakov and Domanski 2000). Isotopic methods typically use stable ($^{13}C$) or radioactive ($^{14}C$) isotopes of carbon as tracers to quantify fluxes of carbon belowground after pulse or continuous labeling. These methods are most useful in grassland and agricultural ecosystems, where the short stature of the vegetation allows application of isotopic tracers in the field. However, one variation of this method is useful for estimating FRP in any ecosystem. Gaudinski et al. (2001) have developed a method for estimating FRP from one-time measurements of $^{14}C$ in fine root samples, based on the changing $^{14}C$ signature of the atmosphere since enrichment by atmospheric nuclear weapons testing in the 1950s and 1960s. While this method has probably underestimated FRP, its use in conjunction with additional knowledge of fine root survivorship should yield accurate estimates of FRP (Tierney and Fahey 2002; Trumbore and Gaudinski 2003). Unfortunately, the atmospheric $^{14}C$ signal is degrading over time, and this method may not be available over the long term. It is also worth noting that isotopic labeling often occurs as a by-product in enhanced $CO_2$ experiments because the isotopic signature of experimentally added $CO_2$ usually differs from that in the atmosphere; thus the network of free-air $CO_2$ experiments (FACE) offer opportunities for studying BNPP using isotopes (Hobbie et al. 2002; Matamala et al. 2003).

An additional method which uses ingrowth cores can be useful to achieve some objectives. This method estimates FRP from production into root-free soil cores placed into the soil profile (Vogt and Persson 1991). This method is easy and inexpensive, making it desirable for uses such as assessing the seasonality of FRP or comparing relative differences in FRP between experimental treatments. However, this method may not accurately quantify FRP for two reasons: first, ingrowth cores measure colonization of root-free soil which differs chemically and physically from the soil profile; and second, placement of the core into the soil severs all roots at the interface between the core and the soil profile, causing further potential bias.

## Choosing a Method

Choice of an approach will depend foremost on the specific research objectives. Preliminary or coarse assessment of BNPP, including evaluation of seasonal patterns, can be accomplished easily and quickly by using ingrowth cores. These preliminary data may be useful for establishing appropriate intervals for more intensive sampling using sequential cores or MR. However, in order to quantify BNPP with higher accuracy, it is recommended that multiple methods be used concurrently within a given site whenever possible. It is clear from comparative studies in the literature that the major approaches have yielded differing estimates of BNPP (Nadelhoffer and Raich 1992; Gill and Jackson 2000; Gaudinski et al. 2001; Hertel and Leuschner 2002); BNPP estimates from MR usually exceed estimates from cores and budgets, and all of these methods have exceeded estimates from the bomb $^{14}$C method. While methodological improvements will hopefully minimize bias in these methods, use of multiple methods will allow cross-checking between methods and increase comparability across studies.

Many site factors will determine which method or methods are best for a particular study. In remote sites, regular visits with MR equipment become overly burdensome. Likewise, MR tubes may not be practical in wetlands, where wet soils may limit visibility and preventing movement of tubes is difficult. Sequential cores or $^{14}$C may be better choices in both these cases. Additional study factors will limit choices in some cases. MR will be precluded in short-term studies or studies without substantial lead time to allow colonization of MR tubes. Studies with limited budgets may not be able to afford MR equipment or isotopic analysis. Large, comprehensive studies which are measuring many N or C fluxes for other purposes should seek to include budgetary estimates of BNPP.

## Further Methodological Considerations

(1) All of these approaches require an assumption that soil C or fine root biomass is near steady state (i.e., neither increasing nor decreasing). While this is often a good assumption, it is clearly violated in aggrading systems and is questionable in short-term studies or where shifting environmental conditions may be altering fine root dynamics or soil C storage.

(2) In the selection of sampling locations, researchers must account for site and ecosystem characteristics. While random sampling could provide unbiased estimates of BNPP, high levels of heterogeneity belowground in forested and shrubland ecosystems may make a stratified sampling design more appropriate. For example, the efficiency gained by stratification of the site according to microtopography (pits and mounds) or vegetation physiognomy (e.g., in savanna) may save considerable labor.

(3) The appropriate cutoff values for distinguishing fine vs. coarse root fractions will differ among ecosystems. These values should be based on knowledge of the fine root architecture of dominant species. For example, within the

northern hardwood forest, fine roots have commonly been defined as those less than 0.5, 1, or 2 mm in diameter. However, Pregitzer et al. (2002) have shown that more than 75% of lateral fine roots of many species of North American trees are less than 0.5 mm in diameter, leading them to conclude that defining "fine roots" in temperate forest ecosystems to be those less than 1 or 2 mm is problematic. All these classes contain so much heterogeneity in root structure, function, and survivorship that large errors in estimates of FRP could result. Thus, in order to estimate BNPP from FRP, it is important to identify the size range of the highly dynamic portion of the root system. One promising approach to this problem is to define fine roots from an architectural framework; fine roots can be considered as a branching system composed of different root orders. Recent evidence suggests that heterogeneity in root function and dynamics (including turnover) may be considerably lower within than between different branching orders (Pregitzer et al. 2002; Guo et al. 2004). Although sampling by root branching order will be impractical for coring and MR methods, this approach could be used effectively for C isotope methods.

(4) The study area must be protected from trampling during repeated measurements. This is a critical and sometimes overlooked aspect of belowground research. It is very difficult to access long-term soil sampling locations without altering the system due to soil compaction and disruption of the surface organic horizons. Careful adherence to preselected corridors for movement within plots is important; in some cases researchers have invested in boardwalks.

(5) Estimates of BNPP should be accompanied by an assessment of associated uncertainty. Lauenroth (2000) has used Monte Carlo uncertainty analysis to show that error is inflated by the calculations required to estimate BNPP from measured parameters using several methods outlined herein. His analysis shows that uncertainty in BNPP estimates can be surprisingly large, and that for some methods, threshold responses exist above which uncertainty increases quite steeply. (Readers are referred to that analysis and to chapter 12 in this volume for further considerations on estimating error and uncertainty.)

## Specific Procedures

### Sequential Cores

The sequential core method estimates FRP from changes in live and dead fine root biomass within soil cores removed at periodic intervals. Vogt and Persson (1991) provide detailed instructions for this method. In brief, soil cores are collected at regular intervals throughout the growing season. Fine roots are separated from soil by manual washing or, in the case of organic soils, by hand sorting. Some variations of this method require sorting of fine roots into live and dead categories, which is difficult and requires prompt and careful processing. Fine roots are dried and weighed to provide sequential estimates of fine root biomass.

With these data, FRP can be calculated as one of the following: (1) the difference in annual maximum and minimum fine root biomass (max-min; McClaugherty et al.

1982); (2) the sum of all positive differences in fine root biomass on successive dates (cumulative; Persson 1978); or (3) the sum of differences in fine root biomass and necromass, including estimated fine root decomposition using a "compartmental flow" model (Santantonio and Grace 1987). These approaches can yield dramatically different estimates of FRP, as discussed below.

The sampling schedule must reflect phenology of fine root growth within the ecosystem by capturing seasonal maxima and minima of fine root biomass. Ideally, sufficient samples will be collected to constrain uncertainty to reasonable levels. Lauenroth (2000) has shown that uncertainty in BNPP estimates is linearly related to uncertainty in biomass measurements (BM) using a sequential core method, such that $CV_{BNPP}$ is roughly equivalent to 5 $CV_{BM}$. Thus, uncertainty in estimates of BNPP could be controlled to 50% by sufficient sampling to obtain $CV_{BM}$ ≤10%. In practice, high spatial variability and limited labor may preclude this level of precision. The resulting high variability in biomass estimates can hinder determination of significant differences between sampling dates; thus, FRP is often calculated by including both significant and nonsignificant differences in fine root biomass within the sum (Persson 1978).

## Biases

Variations of the sequential core method can yield drastically different estimates of FRP, and comparative analyses and syntheses have concluded that the method may either over- or underestimate (Kurz and Kimmins 1987; Singh et al. 1984; Vogt et al. 1998; Nadelhoffer and Raich 1992; Hertel and Leuschner 2002). All sequential core methodology is subject to bias caused by the confounding of temporal variability with high spatial variability because destructive sampling precludes repeated measurements from permanent locations. The max-min method typically yields the lowest sequential core estimates of FRP, and most likely underestimates this parameter by missing concurrent fine root production and mortality. To a lesser degree, the cumulative method also suffers from missed production. The "compartment flow" method offers the possibility of obtaining more accurate estimates of FRP. Uncertainty in fine root decomposition rates (Silver and Miya 2001) has limited the accuracy of this approach (Publicover and Vogt 1993); however, intact core measurements of fine root decay may overcome this limitation (Dornbush et al. 2002; Cronan 2003). Finally, the inclusion of nonsignificant differences in fine root biomass within calculations summing FRP can lead to overestimation (Kurz and Kimmins 1987).

### *Minirhizotrons*

The MR method has been used to estimate FRP in a variety of ecosystems. This method uses video or digital imagery to obtain direct, repeated observation of fine root dynamics against clear tubes placed in the soil. Analysis of images follows the fate of individual fine roots through time, allowing estimation of FRP. MR is labor-intensive and requires careful attention to research protocols in order to obtain useful information. Current MR practice varies substantially between research groups, preventing meaningful comparison across studies. This method has been described in detail (Taylor 1987; Hendrick and Pregitzer 1996; Majdi 1996; Fahey et al. 1999; Johnson et al. 2001); we will summarize rather than repeat recommended methodology herein, and focus on standardization and recent developments, which have altered recommended procedures.

A standard MR system currently consists of a color microvideo camera, a camera control unit to adjust focus and lighting, a portable power source, an indexing handle to allow resampling of specific locations on tubes, and a digital image-capturing system. Direct capture of digital images in the field on a notebook computer now eliminates the need to videotape and transfer images to digital format. Data analysis is done on computers using specialized root-tracing software (Hendrick and Pregitzer 1992; Craine and Tremmel 1995).

A set of clear tubes (50 mm or 32 mm in diameter) is installed in the soil to allow observation of fine root growth. Many studies have used butyrate or acrylic access tubes, but recent evidence indicates these tubes may alter fine root life spans (Withington et al. 2003); Plexiglas is the preferred tube material. MR access tube installation is a critical step that plays a large role in determining image quality over the course of the study. An auger or soil corer is used to excavate a smooth hole of the same size as the MR access tube (usually 50 or 32 mm), often using a reverse taper bit. Care must be taken not to compact the soil against the sides of the hole during coring, which will hinder root growth against the tube. Installation must also optimize contact between tube and soil; an irregular soil surface will hinder observation or affect root growth due to air gaps between tube and soil. (See Johnson et al. 2001 for additional details on tube installation.) Tubes should be insulated to minimize differences in temperature between the tube and the bulk soil. Light must be completely excluded from the tube with an opaque cap and paint or tape covering extension of the tube above the soil.

MR tubes are imaged at regular intervals, and the fate of individual roots through time can be analyzed using specialized computer software. For estimation of FRP, imaging should begin after the effects of tube installation have subsided and tubes have been colonized by fine roots. The disturbance of tube installation can stimulate fine root production for up to one year (Joslin and Wolfe 1999) or longer (Tierney et al. 2003). Care must be taken in the field to maintain high image quality. A large swab made from a chamois cloth attached to a PVC pole can be used to wipe condensation from the tube interior prior to filming. Field personnel must adjust focus and lighting as needed during filming. Even so, differences in lighting, focus, and condensation among filming dates will occur, and must be considered during interpretation of images. Additionally, organic matter can accumulate on the upper surface of MR tubes, especially in soils rich in organic matter, blocking the view of existing roots, which seem to disappear.

## Temporal and Spatial Recommendations

MR tubes should be imaged at regular intervals to facilitate statistical analysis for determining fine root survivorship. The appropriate sampling interval must balance the need for accuracy with the effort required; longer intervals will allow increasingly larger fractions of FRP to die and decompose undetected between sampling intervals. Three studies have examined the effect of sampling interval on estimates of FRP (Dubach and Russelle 1995; Stevens et al. 2002; Tingey et al. 2003). These studies show that appropriate sampling intervals will vary, depending upon patterns of fine root survivorship of the species observed. Tingey et al. (2003) showed that measured FRP decreased by 3.1% and 4.4%, respectively, for each one-week increase in sampling interval in a study of *Pseudotsuga menziesii* and *Tilia cordata* trees. Johnson et al. (2001) suggest a simple function for determining the appropriate sampling interval to achieve a predetermined level of accuracy based upon median fine root survivorship. Assuming that fine root survivorship approximates an exponential decay function and that fine root production and mortality rates are constant, the proportion of FRP missed by a particular sampling interval will be proportional to the ratio of sampling frequency and median root lifetime.

The depth distribution of fine roots in the soil within a particular ecosystem should be considered when installing MR tubes. Most MR tubes are installed at 30° to 60° angles in order to span the soil profile. Vertical tube placement is not recommended because roots have a greater tendency to track tubes placed at this orientation (Bragg et al. 1983). Ecosystems in which fine root growth is heavily concentrated in the upper layers of the soil, such as tundra and boreal forest, may be best sampled by shallow, horizontal tubes, which maximize viewing area in the upper layers of the soil (Dubach and Russelle 1995; Tierney and Fahey 2001); these surficial tubes must be anchored to prevent movement. For sampling of deep roots, which often occur sporadically, the MR method is not recommended.

Selection of MR tube locations should take into account the considerable heterogeneity of soil that occurs at several scales. MR tubes are typically longer than the patch sizes identified by Partel and Wilson (2002) in aspen forest (8–12 cm) and prairie (3–4 cm), which should reduce variability between tubes. However, in a detailed statistical analysis, Hendrick and Pregitzer (1996) found greater variability between MR tubes within a plot than between plots, leading to the recommendation that several tubes within a treatment be pooled for subsequent statistical analysis. Additionally, pit and mound topography in forests can cause differences in fine root growth at a larger scale than the typical MR tube; thus, stratification of sampling locations is recommended.

## Calculations and Units

Two different approaches for estimating FRP from MR data have been used commonly: (1) calculation from fine root survivorship and (2) the ratio between new and preexisting fine root length. In both approaches, average fine root biomass must be measured, and is assumed to be near steady state. Both approaches have advantages and disadvantages. For the survivorship approach the fine root turnover coefficient (TC, $yr^{-1}$) is assumed to be equivalent to the inverse of average fine root survivorship (1/yr), and FRP is estimated as the product of TC and fine root biomass (Hendrick and Pregitzer 1992). We recommend that FRP ($g\ m^{-2}\ yr^{-1}$) be calculated using mean fine root survivorship (estimated from long-term MR observation of fine root length), and that fine root biomass be measured using soil cores (Hendrick and Pregitzer 1992; Fahey et al. 1999). Mean fine root survivorship can be estimated using parametric regression survivorship models. This method estimates FRP more accurately than previous methodology using median fine root survivorship because fine root survivorship in perennial ecosystems can be markedly long-tailed, causing median fine root survivorship to overestimate TC (Tierney and Fahey 2002). When substantial variation in fine root diameter exists within the ecosystem of study, FRP should be calculated independently on smaller diameter fractions of the fine root population because fine root longevity and specific root length vary substantially with fine root diameter (Wells and Eissenstat 2001; Tierney and Fahey 2002). While in theory this approach should be more accurate than the ratio approach, several practical problems exist: (1) tracking individual root life spans may be difficult because of tube movement or obscuring of images; (2) criteria for judging the death of roots may be hard to standardize; and (3) long-term observations are necessary.

Alternatively, FRP can be estimated in similar fashion using a fine root TC calculated as the ratio of new to existing fine root length or number in MR tubes (Hendrick and Pregitzer 1992; Gill and Jackson 2000). This approach avoids the difficulty of determining death of individual fine roots as well as the effort of long-term MR observation; however, MR tube installation stimulates fine root production (Joslin and Wolfe 1999) and MR tubes have been shown to alter root length density (Parker et al. 1991; Volkmar 1993). Simpler protocols in the literature have calculated annual FRP more directly as the sum of new root number or length recorded during sampling intervals over the course of a year. This approach is inadequate for two reasons in addition to those listed above: (1) it yields units (roots tube$^{-1}$ or mm fine root length tube$^{-1}$) which are difficult to convert to an areal basis; and (2) it is particularly susceptible to bias from interannual variation in FRP.

## Sample Size and Precision

As with sequential cores, labor requirements often limit the sample size in MR studies. Some studies employ subsampling schemes to maximize statistical power while controlling the effort required by data analysis. Johnson et al. (2001) assessed the implications of analyzing only every second, third, or fourth MR image. They found that analysis of every second image tended to yield only small differences in coefficient of variation of root length observed, indicating that this may be a desirable subsampling scheme.

## Biases

The MR method presents opportunity for biases from several sources, most of which can be minimized by careful adherence to established protocols. As already discussed, MR has tended to overestimate FRP for 3 reasons: (1) stimulation of production by MR tube installation (Joslin and Wolfe 1999), (2) overestimation of TC from median fine root survivorship (Tierney and Fahey 2002), and (3) effect of tube material on root life span (Withington et al. 2003), as discussed above. Careful methodology as described herein should minimize these sources of bias. Either glass or acrylic tubes may reduce tube effects in comparison with butyrate. Second, proper tube installation is crucial to successful analysis; a poor seal between tube and soil will decrease visibility, potentially elevating FRP estimates by causing fine roots to "disappear" due to condensation or accumulation of organic matter on the tube, rather than due to mortality. Careful installation, as described here and by Johnson et al. (2001), also should minimize this bias.

Finally, visual criteria for determination of fine root mortality in MR vary between researchers and can be subjective, creating the potential for bias in either direction and causing considerable variation in fine root mortality and survivorship estimates between, and even within, research groups. The simplest and most objective approach is to equate fine root disappearance with mortality by assuming that disappearance represents complete decomposition (Hendrick and Pregitzer 1992). This can be complicated by the fact that fine roots often disappear due to factors unrelated to mortality (condensation, poor image quality, masking by organic matter, subtle shifts in tube position, and even misnumbered images); careful analysis and continued observation of previous fine root locations can minimize, but not completely eliminate, this problem. Moreover, the utility of this simple approach is limited because it fails to distinguish between live and dead roots, which make vastly different contributions to biogeochemical cycling. Standardized visual criteria for root mortality in MR tubes would allow this distinction. Comas et al. (2000) found good concordance between the appearance of grape roots in MR images and root vitality assessed by staining with triphenyltetrazolium chloride (TTC). Accurate and repeatable visual mortality criteria such as this for natural ecosystems would be very informative, but will be much more difficult to develop due to differences in fine root appearance among species. We recommend that when time and funds permit, research groups develop standard visual criteria for fine root mortality within their ecosystem by careful comparison of fine root appearance with actual vitality based on TTC. However, without these criteria, we recommend the simpler and more objective approach of tracking fine roots to disappearance without distinguishing between live and dead fine roots. In all cases, researchers must carefully distinguish between true fine root decomposition and root disappearance due to other causes, and research reports should explicitly describe and justify the criteria used to determine mortality.

## Elemental Budgets

Elemental budgets have been used to estimate TRA or FRP indirectly from measurements of C or N fluxes in forested ecosystems (Raich and Nadelhoffer 1989; Nadelhoffer et al. 1998; Davidson et al. 2002a). The C budget is advantageous because it allows estimation of total BNPP, rather than just FRP, and it relies primarily on two relatively easily and widely measured fluxes: aboveground litterfall ($L_a$) and soil respiration ($R_s$). Assuming soil C storage is near steady state, C inputs to the soil, primarily TRA and $L_a$, will roughly equal C exports, primarily $R_s$, such that

$$TRA + L_a = R_s. \tag{2}$$

This equivalence allows estimation of TRA from $R_s$ and $L_a$. However, TRA encompasses both BNPP and root respiration ($R_r$; see eq. [1]). Thus estimation of BNPP also requires measurement of $R_r$, which is more difficult. Root respiration has been measured on excised roots, on intact roots in cuvettes (Cropper and Gholz 1991; Rakonczay et al. 1997), and as the difference in total soil respiration between rooted and root-free plots (Bryla et al. 2001). All these approaches have been problematic, but the methodology is improving (Amthor 2000; Burton and Pregitzer 2002). Nadelhoffer et al. (1985) developed an N budget method which uses fluxes of the often limiting element N to estimate FRP. Assuming that N is conserved by vegetation, annual N uptake ($N_u$) should be equivalent to the N content of annual production, primarily in aboveground litter ($N_{al}$), wood and bark ($N_{wb}$), and fine roots ($N_{fr}$), such that:

$$N_u = N_{al} + N_{wb} + N_{fr}. \qquad (3)$$

Thus, $N_{fr}$ can be estimated from easily and widely measured fluxes of $N_{al}$ and $N_{wb}$, together with the more complex measurement $N_u$. Measurement of fine root N content then allows estimation of FRP from $N_{fr}$. To estimate $N_u$, Nadelhoffer et al. (1985) assumed that all available N was taken up by vegetation, so that $N_u$ is the sum of N inputs from mineralization ($N_m$) and precipitation ($N_p$), less N lost to leaching ($N_l$):

$$N_u = N_m + N_p - N_l. \qquad (4)$$

$N_m$ is a large component of this equation, and estimates of FRP using this method will depend upon its accurate measurements, typically using field incubations. This N budget method has been used only in forested sites to date.

The fluxes of C and N in terrestrial ecosystems vary seasonally and annually, depending on environmental conditions. Proper use of these methods requires successive measurements throughout the growing season, and it is recommended that data be collected over more than a single year. Annual soil respiration is typically estimated from shorter-term measurements, using relationships between soil respiration and temperature. (See Davidson et al. 2002b for a useful analysis of sampling protocols for soil respiration.) Soil heterogeneity must be taken into account when sampling soil respiration and N mineralization and leaching. Ideally, these measurements should be taken repeatedly from a large number of permanent sampling locations.

## Biases and Limitations

Like the direct methods, budget methods have important biases. The C budget method has been limited by the accuracy of soil C efflux estimates (Davidson et al. 2002b) and by the difficulty of measuring root respiration. Continued methodological development will be necessary to improve the accuracy of this method. Similarly, the N budget method can be substantially biased by inaccuracy of N uptake estimates. All of the components of $N_u$ (eq. 4) are prone to measurement errors. Most significantly, $N_m$ is typically measured by incubation within in situ cores, which may alter rates of $N_m$ due to altered soil moisture availability, soil compaction, or the exclusion of live fine roots (Robertson et al. 1999). Moreover, $N_m$ can exhibit high spatial variability within a small area. Atmospheric deposition and soil efflux measurements are complicated by gaseous forms of N (Groffman et al. 1999). Also, Lauenroth (2000) has shown that the calculations needed to estimate BNPP by using budget methods can dramatically inflate uncertainty; the N budget method is particularly sensitive to error in the N concentration of fine roots. Finally, discrepancies in the relationship between litterfall and soil respiration at some temperate forest sites indicate that this method may be more appropriate for broad-scale analyses than for quantification of BNPP at individual sites (Gower et al. 1996; Davidson et al. 2002a). Despite all these considerations, the budget methods offer useful checks on direct methods by constraining estimates of BNPP.

## Isotope Decay

The isotope decay method of measuring BNPP (Dahlman and Kucera 1965) estimates the turnover coefficient (TC) for roots on the basis of disappearance of a carbon isotope label from the root system. Production is calculated by dividing the time for complete turnover by the average root biomass (Milchunas and Lauenroth 1992). Because of difficulties in applying the label, this method is feasible only in low-stature vegetation. To date, only the radioactive C isotope ($^{14}$C) has been used successfully, so that regulations and dangers of using this isotope limit applications. Although this method appears to provide a simple and cost-effective approach for certain situations, Milchunas and Lauenroth (2001) identified some causes of uncertainty.

In the isotope decay method an intact area of vegetation is pulse-labeled with $^{14}$CO$_2$ under a clear tent. After an appropriate time for stabilization of the label in structural tissue (e.g., a single growing season in Colorado shortgrass steppe; Milchunas and Laurenroth 1992), disappearance of the isotope from the root system is quantified by washing live roots from soil cores collected sequentially and estimating $^{14}$C activity by liquid scintillation. The timing and depth of sampling should be adapted to the particular ecosystem characteristics. In shortgrass steppe, Milchunas and Lauenroth (1992) collected separate samples at 0–20 cm and 0–40 cm depths twice annually for 3 years after labeling. In principle, the same approach could work using a $^{13}$C label and analysis of $^{13}$C:$^{12}$C ratio by mass spectrometry; however, initial trials have reported insufficient differences between background and labeled roots (Laurenroth and Gill 2003).

A somewhat analogous approach can take advantage of free-air CO$_2$ enrichment experiments (FACE) in which added CO$_2$ in FACE treatment plots is highly depleted in $^{13}$C. For example, Matamala et al. (2003) measured the $^{13}$C signature of roots from intact soil cores of pine and sweetgum plantations with that for roots collected from ingrowth cores. They reasoned that the $^{13}$C in the original standing crop of roots in the plantation would be progressively altered by root turnover and replacement by new roots bearing the signature represented in the ingrowth roots. They calculated the mean residence time of carbon in different root size classes from the exponential decay coefficient of "old" carbon remaining in the root system through time. As noted below (under "Bomb $^{14}$C"), this approach likely underestimates TC because of heterogeneity in survivorship within root size classes and because of skewed survivorship curves of fine roots; nevertheless, this approach appears highly promising for FACE sites.

## Calculations

The TC of roots can be estimated by extrapolating regressions of isotope mass remaining through time to the time of complete disappearance of label (Dahlman and Kucera 1965). Root production is estimated as the ratio of TC to average fine root biomass, best measured on adjacent unlabeled plots. As for the other approaches, this method assumes that root biomass is near steady state. It also assumes that root growth potential in the soil profile is relatively uniform and that the proportion of label entering structural tissues is not influenced by the season of label application (Milchunas and Laurenroth 2001).

## Biases

One source of potential bias in the isotope decay method appears to be from the extrapolation to zero isotope remaining. Milchunas and Laurenroth (2001) collected root samples from isotope-labeled plots through 10 years after labeling and observed a marked reduction in the rate of isotope disappearance after 3–4 yr. They tentatively attributed the 2-phase pattern of isotope disappearance to 2 likely causes: (1) contamination of roots with soil organic matter and (2) long life span of a small proportion of the root system. Both of these effects would result in biased estimates of TC from short-term extrapolations. The former cause could apply at all stages, and would be obviated by applying a correction for soil contamination (Milchunas and Lauenroth 2001). The latter cause would result in overestimation of the TC by linear extrapolation in the same way that median survivorship can result in overestimation in the MR method (Tierney and Fahey 2002). Ideally, knowledge of the long-term survivorship curve is needed to correct for this bias when extrapolating from short-term isotope decay measurements.

### Bomb $^{14}$C

A new, developing method estimates FRP from fine root turnover as indicated by fine root radiocarbon ($^{14}$C) content (Gaudinski et al. 2001). This is possible due to labeling of the atmosphere with $^{14}$C by atmospheric nuclear weapons testing in the 1950s and 1960s (Levin and Hesshaimer 2000). Thus far, this method has been used within only a few temperate forest sites. While it offers the ability to estimate fine root turnover from one-time measurements, it tends to underestimate FRP unless used in conjunction with knowledge of fine root survivorship (Tierney and Fahey 2002) or with careful subsampling of root systems. This method may be available only in the short term because continued depletion of the $^{14}$C content of the atmosphere will degrade the $^{14}$C signal.

A representative sample of fine roots must be collected from the study site; this can be accomplished by separating fine roots from soil cores. The estimate of TC using this method is particularly sensitive to the manner in which the roots are sorted and chosen for analysis. First, it is critical to thoroughly remove dead roots (Gaudinski et al. 2001). Moreover, because of high heterogeneity in fine root survivorship, it is important to sort fine roots by diameter or branch order class. For example, in the northern hardwood forest, fine root survivorship increases 43% with every 0.1 mm increase in fine root diameter for roots < 0.5 mm diameter (Tierney and Fahey 2002). It was recently demonstrated (Guo et al. in press) that most of the underestimation bias for TC by this and other isotope methods results from this heterogeneity, and Guo et al. (in press) recommend sampling by root branching order. Once sorted, fine root samples should be cleaned of soil and dead organic matter by sonication.

In preparation for radiocarbon analysis, root samples are treated using an acid-alkali-acid procedure to remove secondary carbon components that may be unrepresentative of fine root age. This treatment reduces the sample to structural carbon components, such as cellulose and lignin. It is performed by applying a series of hot, HCl washes to eliminate carbonates, followed by NaOH washes to remove secondary organic acids, followed by a final acid rinse to neutralize the solution. Samples are then converted to graphite and analyzed for radiocarbon content using accelerator mass spectroscopy (AMS). AMS allows analysis of much smaller samples (e.g., 1 mg) than traditional radiocarbon analysis using beta-counting; AMS analysis is performed at a small number of laboratories around the world, most of which are listed at the Radiocarbon Web site (www.radiocarbon.org). AMS analysis is expensive, costing on the order of $500 per sample. Conversion to graphite can be performed by the AMS lab or by the researcher, as desired.

Study design must consider spatial and temporal variation in both atmospheric $^{14}C$ content and fine root biomass. The atmosphere is considered a relatively well-mixed reservoir of $^{14}C$. However, local and regional atmospheric $^{14}C$ content varies due to dilution by fossil fuel combustion, and this dilution is more pronounced in winter than summer (Levin and Hesshaimer 2000). This variation in atmospheric $^{14}C$ content may be substantial enough to affect estimates of FRP. Thus, it is recommended that turnover models be fine-tuned using atmospheric $^{14}C$ data collected from the study site during the growing season. Seasonal variation in fine root biomass will also influence turnover estimates in some ecosystems, and fine root samples should be collected during the season of peak fine root biomass (Gaudinski et al. 2001). As with the other approaches described herein, fine root sampling must consider the rooting depth and patterns of spatial heterogeneity of the ecosystem under study.

## Calculations and Units

Radiocarbon values detected by AMS must be corrected for mass-dependent isotope fractionation that occurs during photosynthesis, as well as during laboratory analysis. This is done by measuring sample $^{13}C$:$^{12}C$ ratio, and correcting $^{14}C$ values to a common $^{13}C$:$^{12}C$ ratio. $^{14}C$ can then be expressed as $\Delta^{14}C$, the difference in parts per thousand between the $^{14}C$:$^{12}C$ ratios of the sample and a universal oxalic acid standard, corrected to a common $^{13}C$:$^{12}C$ ratio of $-25‰$ (Stuiver and Polach 1977).

Fine root turnover must be estimated from fine root $^{14}C$ content by modeling uptake and loss of $^{14}C$ from the atmosphere into fine root biomass over time, assuming fine root biomass is near steady state. FRP (g m$^{-2}$ yr$^{-1}$) can then be calculated as the product of fine root TC (yr$^{-1}$) and fine root biomass (g m$^{-2}$); assuming that TC is estimated for multiple root diameter classes or root orders, biomass values for each class are needed. Long-term measurements of the $^{14}$C content of the atmosphere are available from several sites to calibrate $^{14}$C models (Burchuladze et al. 1989; Nydal and Lovseth 1996; Levin and Kromer 1997; Levin and Hesshaimer 2000).

### Biases

Gaudinski et al. (2001) have developed a simple, time-dependent model to calculate fine root turnover from $^{14}$C content. However, their model assumes that the probability of fine root death does not vary as fine roots age, and it does not account for heterogeneity of survivorship within the fine root system. These limitations may substantially affect estimates of fine root turnover and associated FRP (Tierney and Fahey 2002). Thus, more complex models incorporating more realistic fine root dynamics will need to be applied (Trumbore and Gaudinski 2003). This bias can be exacerbated by the high radiocarbon content of the largest and longest-lived fine roots, and of dead roots, unless fine root samples are divided into relatively homogeneous size and vitality classes. Additionally, seasonal and local variation in atmospheric $^{14}$C content can bias this method unless atmospheric $^{14}$C is sampled on-site during the growing season.

### Comparisons of Methods: Case Study

Comparisons of methods show that these different approaches can yield substantially different estimates of BNPP (Singh et al. 1984; Nadelhoffer and Raich 1992; Gaudinski et al. 2001). For example, Hertel and Leuschner (2002) examined BNPP within an old-growth, temperate beech-oak forest, using two sequential core methods (max-min and compartment flow), ingrowth cores, an ingrowth chamber method, and a C budget approach. Their C budget approach was based upon estimates of C captured by photosynthesis, rather than the respiration and litterfall-based budget described herein. They found that estimates of FRP using these five methods varied by a factor of 12. At the low end, their estimates from ingrowth cores and their ingrowth chamber method (147 g and 108 g dry biomass m$^{-2}$ yr$^{-1}$, respectively) were low compared with their C budget estimate ((890 g m$^{-2}$ yr$^{-1}$) and the literature describing patterns of FRP production in global forests (Nadelhoffer and Raich 1992). In contrast, their max-min core and compartment-flow estimates (915 and 1360 g m$^{-2}$ yr$^{-1}$) were exceptionally high. Based on the similarity between the C budget method and the max-min cores, Hertel and Leuschner (2002) concluded that the max-min method was preferable to the other direct methods they used, but that it may have overestimated FRP. However, they recognized that their C budget method had the opportunity for large bias, and suggested that future comparisons include a method allowing direct observation of fine roots (i.e., MR). Other studies have found that max-min methods have underestimated FRP due to the failure of those methods to account for concurrent production and mortality. Large differences in fine root necromass among sampling dates led to the high FRP estimates by the two sequential coring methods. While these differences may truly have reflected high FRP at this site, they may also have been due in part to high spatial variability in necromass at this site.

References

Amthor, J. S. 2000. Direct effect of elevated $CO_2$ on nocturnal in situ leaf respiration in nine temperate deciduous tree species is small. Tree Physiology 20:139–144.

Baker, T. T., W. H. Conner, B. G. Lockaby, J. A. Stanturf, and M. K. Burke. 2001. Fine root productivity and dynamics on a forested floodplain in South Carolina. Soil Science Society of America Journal 65(2):545–556.

Bragg, P. L., G. Govi, and R. Q. Cannell. 1983. A comparison of methods, including angled and vertical minirhizotrons, for studying root growth and distribution in a spring oat crop. Plant and Soil 73:435–440.

Bryla, D. R., T. J. Bouma, U. Hartmond, and D. M. Eissenstat. 2001. Influence of temperature and soil drying on respiration of individual roots in citrus: Integrating greenhouse observations into a predictive model for the field. Plant, Cell and Environment 24:781–790.

Burchuladze, A. A., M. Chudy, I. V. Eristavi, S. V. Pagava, P. Povinec, A. Sivo, and G. I. Togonidze. 1989. Anthropogenic $^{14}C$ variations in atmospheric $CO_2$ and wines. Radiocarbon 31:771–776.

Burton, A. J., and K. S. Pregitzer. 2002. Measurement of carbon dioxide concentration does not affect root respiration of nine tree species in the field. Tree Physiology 22(1): 67–72.

Cheng, W. X., D. C. Coleman, C. R. Carroll, and C. A. Hoffman. 1994. Investigating short-term carbon flows in the rhizospheres of different plant species, using isotopic trapping. Agronomy Journal 86(5):782–788.

Clark, D. A., S. Brown, D. W. Kicklighter, J. Q. Chambers, J. R. Thomlinson, J. Ni, and E. A. Holland. 2001. Net primary production in tropical forests: An evaluation and synthesis of existing field data. Ecological Applications 11(2):371–384.

Comas, L. H., D. M. Eissenstat, and A. N. Lakso. 2000. Assessing root death and root system dynamics in a study of grape canopy pruning. New Phytologist 147:171–178.

Craine, J., and D. Tremmel. 1995. Improvements to the minirhizotron system. Bulletin of the Ecological Society of America 76:234–235.

Cronan, C. S. 2003. Belowground biomass, production and carbon cycling in mature Norway spruce, Maine, USA. Canadian Journal of Forest Research 33:339–350.

Cropper, W. P., and H. L. Gholz. 1991. In situ needle and fine root respiration in mature slash pine (*Pinus elliottii*) trees. Canadian Journal of Forest Research 21: 1589–1595.

Dahlman, R. C., and C. L. Kucera. 1965. Root productivity and turnover in native prairie. Ecology 46:84–89.

Davidson, E. A., K. Savage, P. Bolstad, D. A. Clark, P. S. Curtis, D. S. Ellsworth, P. J. Hanson, B. E. Law, Y. Luo, K. S. Pregitzer, J. C. Randolph, and D. Zak. 2002a. Belowground carbon allocation in forests estimated from litterfall and IRGA-based soil respiration measurements. Agricultural and Forest Meteorology 113(1–4):39–51.

Davidson, E. A., K. Savage, L. V. Verchot, and R. Navarro. 2002b. Minimizing artifacts and biases in chamber-based measurements of soil respiration. Agricultural and Forest Meteorology 113(1–4):21–37.

Dornbush, M. E., T. M. Isenhart, and J. W. Raich. 2002. Quantifying fine-root decomposition: An alternative to buried litterbags. Ecology 83(11):2985–2990.

Dubach, M., and M. P. Russelle. 1995. Reducing the cost of estimating root turnover with horizontally installed minirhizotrons. Agronomy Journal 87(2):258–263.

Fahey, T. J., C. S. Bledsoe, F. P. Day, R. Ruess, and A. J. M. Smucker. 1999. Fine root production and demography. Pages 437–455 in G. P. Robertson, D. C. Coleman, C. D. Bledsoe, and P. Sollins (eds.), Standard Soil Methods for Long Term Ecological Research. Oxford University Press, New York.

Fahey, T. J., and J. B. Yavitt. 2005. An in situ approach for measuring root-associated respiration and nitrate uptake of forest trees. Plant and Soil 272:125–131.

Gaudinski, J. B., S. E. Trumbore, E. A. Davidson, A. C. Cook, D. Markewitz, and D. D. Richter. 2001. The age of fine-root carbon in three forests of the eastern United States measured by radiocarbon. Oecologia 129:420–429.

Gill, R. A., and R. B. Jackson. 2000. Global patterns of root turnover for terrestrial ecosystems. New Phytologist 147:13–31.

Gill, R. A., R. H. Kelly, W. J. Parton, K. A. Day, R. B. Jackson, J. A. Morgan, J. M. O. Scurlock, L. L. Tieszen, J. V. Castle, D. S. Ojima, and X. S. Zhang. 2002. Using simple environmental variables to estimate belowground productivity in grasslands. Global Ecology and Biogeography 11(1):79–86.

Gower, S. T., S. Pongracic, and J. J. Landsberg. 1996. A global trend in belowground carbon allocation: Can we use the relationship at small scales? Ecology 77:1750–1755.

Grayston, S. J., D. Vaughn, and D. Jones. 1996. Rhizosphere carbon flow in trees, in comparison with annual plants: The importance of root exudation and its impact on microbial activity and nutrient availability. Applied Soil Ecology 5:29–56.

Groffman, P. M., E. A. Holland, D. D. Myrold, G. P. Robertson, and X. Zou. 1999. Denitrification. Pages 272–288 in G. P. Robertson, D. C. Coleman, C. D. Bledsoe, and P. Sollins (eds.), Standard Soil Methods for Long Term Ecological Research. Oxford University Press, New York.

Guo, D. L., R. J. Mitchell, and J. J. Hendricks. 2004. Fine root branch orders respond differentially to carbon source-sink manipulations in a longleaf pine forest. Oecologia 140:450–457.

Guo, D.L., H. Li, R.J. Mitchell, W. Han, J.J. Hendricks, T.J. Fahey, and R.L. Hendrick. 2007. Fine root heterogeneity by branch order: resolving the discrepancy in root longevity and turnover estimates between minirhizotron and C isotope methods. Global Change Biology, in press.

Hendrick, R. L., and K. S. Pregitzer. 1992. The demography of fine roots in a northern hardwood forest. Ecology 73:1094–1104.

Hendrick, R. L., and K. S. Pregitzer. 1996. Applications of minirhizotrons to understand root function in forests and other natural ecosystems. Plant and Soil 185:293–304.

Hertel, D., and C. Leuschner. 2002. A comparison of four different fine root production estimates with ecosystem carbon balance data in a *Fagus-Quercus* mixed forest. Plant and Soil 239(2):237–251.

Hobbie, E. A., D. T. Tingey, P. T. Rygiewicz, M. G. Johnson, and D. M. Olszyk. 2002. Contributions of current year photosynthate to fine roots estimated using a $^{13}$C-depleted $CO_2$ source. Plant and Soil 247(2):233–242.

Johnson, M. G., D. T. Tingey, D. L. Phillips, and M. J. Storm. 2001. Advancing fine root research with minirhizotrons. Environmental and Experimental Botany 45:263–289.

Joslin, J. D., and M. H. Wolfe. 1999. Disturbances during minirhizotron installation can affect root observation data. Soil Science Society of America Journal 63:218–221.

Kleb, H. R., and S. D. Wilson. 1997. Vegetation effects on soil resource heterogeneity in prairie and forest. American Naturalist 150:283–298.

Kurz, W. A., and J. P. Kimmins. 1987. Analysis of some sources of error in methods used to determine fine root production in forest ecosystems: A simulation approach. Canadian Journal of Forest Research 17:909–912.

Kuzyakov, Y., and G. Domanski. 2000. Carbon input by plants into the soil. Review. Journal of Plant Nutrition and Soil Science 163(4):421–431.

Lauenroth, W. K. 2000. Methods of estimating belowground net primary production. Pages 58–71 in O. E. Sala et al. (eds.), Methods in Ecosystem Science. Springer-Verlag, New York.

Lauenroth, W. K., and R. Gill. 2003. Turnover of root systems. Pages 61–90 in H. deKroon and E. J. W. Visser (eds.), Root Ecology. Springer, New York.

Levin, I., and V. Hesshaimer. 2000. Radiocarbon—A unique tracer of global carbon cycle dynamics. Radiocarbon 42:69–80.

Levin, I., and B. Kromer. 1997. Twenty years of atmospheric $^{14}CO_2$ observations at Schauinsland Station, Germany. Radiocarbon 39:205–218.

Lopez, B., S. Sabate, and C. Gracia. 1996. An inflatable minirhizotron system for stony soils. Plant and Soil 179(2):255–260.

Majdi, H. 1996. Root sampling methods—Applications and limitations of the minirhizotron technique. Plant and Soil 185(2): 255–258.

Matamala, R., M. A. Gonzalez-Meler, J. D. Jastrow, R. J. Norby, and W. H. Schlesinger. 2003. Impacts of fine root turnover on forest NPP and soil C sequestration potential. Science 302:1385–1387.

McClaugherty, C. A., J. D. Aber, and J. M. Melillo. 1982. The role of fine roots in the organic matter and nitrogen budgets of two forested ecosystems. Ecology 63:1481–1490.

Milchunas, D. G., and W. K. Lauenroth. 1992. Carbon dynamics and estimates of primary production by harvest, C-14 dilution and C-14 turnover. Ecology 73:1593–1607.

Milchunas, D.G., and W. K. Lauenroth. 2001. Belowground primary production by carbon isotope decay and long-term root biomass dynamics. Ecosystems 4(2):139–150.

Nadelhoffer, K. J., J. Aber, and J. M. Melillo. 1985. Fine roots, net primary production, and soil nitrogen availability: A new hypothesis. Ecology 66:1377–1390.

Nadelhoffer, K. J., and J. W. Raich. 1992. Fine root production estimates and belowground carbon allocation in forest ecosystems. Ecology 73:1139–1147.

Nadelhoffer, K. J., J. W. Raich, and J. D. Aber. 1998. A global trend in belowground carbon allocation: Comment. Ecology 79:1822–1825.

Nydal, R., and K. Lovseth. 1996. Carbon-14 Measurements in Atmospheric $CO_2$ from Northern and Southern Hemisphere Sites 1962–1993. ORNL/CDIAC-93 NDP-057, Oak Ridge National Laboratory, Oak Ridge, TN.

Parker, C. J., M. K. V. Carr, N. J. Jarvis, B. O. Puplampu, and V. H. Lee. 1991. An evaluation of the minirhizotron technique for estimating root distribution in potatoes. Journal of Agricultural Science 116:341–350.

Partel, M., and S. D. Wilson. 2002. Root dynamics and spatial pattern in prairie and forest. Ecology 83:1199–1203.

Persson, H. 1978. Root dynamics in a young Scots pine stand in central Sweden. Oikos 30:508–519.

Phillips, R.P. and T.J. Fahey. 2005. Patterns of rhizosphere carbon flux in sugar maple (Acer saccharum) and yellow birch (Betula allegheniensis) saplings. Global Change Biology 11:983–995.

Pregitzer, K. S., J. L. DeForest, A. J. Burton, et al. 2002. Fine root architecture of nine North American trees. Ecological Monographs 72(2):293–309.

Publicover, D. A., and K. A. Vogt. 1993. A comparison of methods for estimating forest fine root production with respect to sources of error. Canadian Journal of Forest Research 23:1179–1186.

Raich, J. W., and K. J. Nadelhoffer. 1989. Belowground carbon allocation in forest ecosystems: Global trends. Ecology 70:1346–1354.

Rakonczay, Z., J. R. Seiler, and L. J. Samuelson. 1997. A method for the in situ measurement of fine root gas exchange of forest trees. Environmental and Experimental Botany 37:107–113.

Robertson, G. P., D. Wedin, P. M. Groffman, J. M. Blair, E. A. Holland, K. J. Nadelhoffer, and D. Harris. 1999. Soil carbon and nitrogen availability. Pages 258–271 in G. P. Robertson, D. C. Coleman, C. D. Bledsoe, and P. Sollins (eds.), Standard Soil Methods for Long Term Ecological Research. Oxford University Press, New York.

Santantonio, D., and J. C. Grace. 1987. Estimating fine root production and turnover from biomass and decomposition data: A compartment flow model. Canadian Journal of Forest Research 17:900–908.

Schenk, H. J. and R. B. Jackson. 2002. The global biogeography of roots. Ecological Monographs 72:311–328.

Silver, W. L., and R. K. Miya. 2001. Global patterns in root decomposition: Comparisons of climate and litter quality effects. Oecologia 129(3):407–419.

Singh, J. S., W. K. Lauenroth, and R. K. Steinhorst. 1984. Bias and random errors in estimators of net root production: A simulation approach. Ecology 65:1760–1764.

Stevens, G. N., R. H. Jones, and R. J. Mitchell. 2002. Rapid fine root disappearance in a pine woodland: A substantial carbon flux. Canadian Journal of Forest Research 32(12):2225–2230.

Stuiver, M., and H. Polach. 1977. Discussion: Reporting of $^{14}$C data. Radiocarbon 19: 355–363.

Taylor, H. M. (ed.). 1987. Minirhizotron Observation Tubes: Methods and Applications for Measuring Rhizosphere Dynamics. American Society of Agronomy Special Publication no. 50. American Society of Agronomy, Madison, WI.

Tierney, G. L., and T. J. Fahey. 2001. Evaluating minirhizotron estimates of fine root longevity and production in the forest floor of a temperate broadleaf forest. Plant and Soil 229:167–176.

Tierney, G. L., and T. J. Fahey. 2002. Fine root turnover in a northern hardwood forest: A direct comparison of the radiocarbon and minirhizotron methods. Canadian Journal of Forest Research 32(9):1692–1697.

Tierney, G. L., T. J. Fahey, P. M. Groffman, J. P. Hardy, R. D. Fitzhugh, C. T. Driscoll, and J. B. Yavitt. 2003. Environmental control of fine root dynamics in a northern hardwood forest. Global Change Biology 9:670–679.

Tingey, D. T., D. L. Phillips, and M. G. Johnson. 2003. Optimizing minirhizotron sample frequency for an evergreen and deciduous tree species. New Phytologist 157(1):155–161.

Treseder, K. K., and M. F. Allen. 2000. Mycorrhizal fungi have a potential role in soil carbon storage under elevated $CO_2$ and nitrogen deposition. New Phytologist 147(1): 189–200.

Trumbore, S. E., and J. B. Gaudinski. 2003. The secret lives of roots. Science 302: 1344–1345.

Vogt, K. A., and H. Persson. 1991. Measuring growth and development of roots. Pages 477–501 in J. P. Lassoie and T. M. Hinckley (eds.), Techniques and Approaches in Forest Tree Ecophysiology. CRC Press, Boca Raton, FL.

Vogt, K. A., D. J. Vogt, and J. Bloomfield. 1998. Analysis of some direct and indirect methods for estimating root biomass and production of forests at an ecosystem level. Plant and Soil 200(1):71–89.

Vogt, K. A., D. J. Vogt, P. A. Palmiotto, P. Boon, J. O'Hara, and H. Asbjornsen. 1996. Review of root dynamics in forest ecosystems grouped by climate, climatic forest type and species. Plant and Soil 187:159–219.

Volkmar, K. M. 1993. A comparison of minirhizotron techniques for estimating root length density in soil of different bulk density. Plant and Soil 157:239–245.

Wells, C. E., and D. M. Eissenstat. 2001. Marked difference in survivorship among apple roots of different diameters. Ecology 82:882–892.

Wells, C. E., D .M. Glenn, and D. M. Eissenstat. 2002. Soil insects alter fine root demography in peach (*Prunus persica*). Plant, Cell and Environment 25(3):431–439.

Whittaker, R. H., F. H. Bormann, G. E. Likens, and T. G. Siccama. 1974. The Hubbard Brook Ecosystem Study: Forest biomass and production. Ecological Monographs 44: 233–254.

Withington, J. M., A. D. Elkin, B. Bulaj, J. Olesinski, K. N. Tracey, T. J. Bouma, J. Oleksyn, L. J. Anderson, J. Modrzynski, P. B. Reich, and D. M. Eissenstat. 2003. The impact of material used for minirhizotron tubes for root research. New Phytologist 160: 533–544.

# 9

# Measuring and Modeling Primary Production in Marine Pelagic Ecosystems

Maria Vernet
Raymond C. Smith

The measurement of primary production in the ocean is key to our estimates of ecosystem function and the role of the ocean and its biota in the planetary carbon (C) cycle. Accurate estimates are critical to a broad suite of biological. questions across a wide range of space and time scales. The methods developed to measure primary production reflect the diversity of our research interests and encompass a range of approaches: from in situ to airborne and spaceborne observations, from intracellular to global systems, and from experimental to modeling.

A variable of interest in quantifying primary productivity is the rate of population increase within a pelagic community (McCormick et al. 1996). Growth rate ($\mu$ [$t^{-1}$]) can be expressed as the rate of change in the number of individuals (n) per unit time (t) or as a chemical constituent (C, nitrogen [N], etc.) within the community,

$$\mu = dn/dt \ (1/n). \tag{1}$$

In the field, the estimate of growth rates is limited by a number of factors because the terms dn and n, or alternatively dC and C, from phytoplankton are not readily measured. Several characteristics of plankton challenge the precision and accuracy of our present methods. One problem is that the target autotrophic algal population is suspended in seawater and has a spatial and size distribution which overlaps that of heterotrophic organisms. The plants in the plankton are microscopic (usually 2–200 µm) and multiply very quickly (from 0.1 to 2.0 divisions per day or a doubling rate of 0.5 to 10 days); the herbivores usually ingest whole cells and not parts; and the herbivores themselves are microscopic, with body size and division rate similar to the plants. In the absence of loss terms, these factors make measurements of primary production difficult (Waterhouse and Welschmeyer 1995). Also, the plant eaters are mostly omnivores, so catabolic and anabolic reactions in both groups of organisms are difficult to differentiate, and often interact with each other.

Methods that are specific to autotrophic organisms are thus necessary to overcome the challenges in working with planktonic systems. There are numerous methods, such as the use of radioactive tracers, the determination of biophysical processes in photosynthetic pigments, and models aimed at the mechanistic estimation of photosynthesis. In this chapter the most common techniques used to estimate primary production in marine pelagic ecosystems are discussed, their strengths and limitations are described, and the comparability of the results from the different methods are considered. An important source of discrepancy among techniques originates from the different temporal and spatial scales that each method addresses (Li and Maestrini 1993). Our focus is a coastal marine ecosystem in the western Antarctic peninsula, the site of the Palmer Long-Term Ecological Research Program since 1990.

## General Considerations and Concepts

### Gross and Net Primary Production

Photosynthesis is often expressed in units of moles (or its mass equivalent) of carbon per unit cell (or volume of water containing cells) per unit time. This is an instantaneous rate (measured in milliseconds) which is integrated over time in order to be operational for estimations made in the field (Platt and Sathyendranath 1993). Over ecologically relevant periods (daily, annual, etc.), primary production is the organic C produced within that period that is made available to other trophic levels (Lindeman 1942). Methods of estimating primary production at the molecular and single-cell scale need to be scaled up in order to obtain a daily rate within a volume of seawater. When interest is aimed at primary production rates of a certain taxon, primary production rates are combined with cell size determinations or photosynthetic pigment complements (Gieskes et al. 1993).

Gross primary production (GPP) is the total number of electron equivalents originating from the photolysis of water (Fogg 1980; Falkowski and Raven 1997). Photosynthesis is defined as the conversion of light into metabolic energy (Fogg 1980); it is identical to gross photosynthesis, $P_g$. Net photosynthesis, $P_n$, is the difference $P_g - R_l$, where $R_l$ is the respiratory loss in the light. Respiration is the conversion of metabolic energy into heat. These photosynthetic parameters are all rates; that is, time-dependent processes with dimensions of mass/time. Within planktonic communities, GPP is defined as photosynthesis not affected by respiration or the metabolism of heterotrophic organisms in the same body of water. Net primary production (NPP) is estimated as GPP corrected for algal respiration. Net ecosystem production (NEP) is GPP corrected for the metabolism of the entire autotrophic and heterotrophic community (community respiration, CR) and is defined as GPP minus CR (Williams 1993a). While the previous variables are based on C units, there exist parallel terms to express phytoplankton production in units of N (Dugdale and Goering 1967; Minas and Codispoti 1993).

## An Overview of Methods

Methods and instrumentation for estimating primary production in the field are constantly evolving. Although the $^{14}C$ incubation remains the standard method against which most other methods are compared or calibrated (Williams 1993b), a new suite of methods has been introduced since the mid-1980s. The traditional method of cell enumeration with microscopy (Hewes et al. 1990) has been extended to include flow cytometry (Li 1993) based on cell fluorescent and size-related properties of single cells, molecular techniques with emphasis on understanding gene expression and controlling mechanisms in photosynthetic processes (La Roche et al. 1993), and isotope tracers including not only $^{14}C$ or $^{13}C$ (Goes and Handa 2002), but also $^{15}N$ (Le Bouteiller 1993) and $^{18}O$ (Bender et al. 1987). For fieldwork these techniques require sampling of a parcel of water which is isolated from the environment and is considered representative of the target population. Other methods involve direct, noninvasive measurements in the water column, such as the use of cellular fluorescence, both solar-induced (Doerffer 1993) and active fluorescence (Falkowski and Kolber 1993); diel variability in optical properties in the water column affected by particle dynamics (Siegel et al. 1989); and remote sensing of ocean color based on water-leaving surface reflectance (Hovis et al. 1980; Gordon et al. 1980; Gordon and Morel 1983). These latter techniques scale from seconds to months and from cm to hundreds of km. Specific application of these methods is dependent upon their suitability to address a particular research question. Finally, the use of nonconservative tracers such as $O_2$, $CO_2$, and $NO^-_3$ on ocean mesoscales (Emerson et al. 1993; Robertson and Watson 1993; Minas and Codispoti 1993) is designed to integrate whole community processes over time scales of days to months. This chapter will focus on radioactive tracers, fluorescence, and remote sensing techniques which are widely used in biological oceanography.

## Space and Time Scales in Marine Ecosystems

The oceans cover nearly three-fourths of the earth's surface and exhibit physical and biological variability over a wide range of space and time scales (Steele 1978). The space/time scales of marine and terrestrial systems can be significantly different (Steele 1991), and these differences often influence both our approach to studying the system and our way of understanding how various components of the system are interconnected. One important difference includes sampling strategies; that is, the way we obtain data from the field. Phytoplankton are embedded in a continually changing environment that regulates factors controlling cell growth rates (temperature, light, and nutrients) as well as factors that control the accumulation rate of cells in the euphotic zone and, hence, population growth (grazing, water column stability, and sinking). A second important difference is the trophic structure of, as well as the related size and growth structure within, the system. Third, although physical processes of the ocean and atmosphere follow the same basic laws of fluid dynamics (Pedlosky 1987), they have very different temporal and spatial scales of their underlying processes. In marine systems the space/time scales of the physics and biology are close, and their interactions are tightly coupled (Steele 1985). Thus there can be a significant difference of emphasis, with focus on internal mechanisms in terrestrial studies and on external physical forcing in marine studies.

Multiplatform sampling strategies (Steele 1978; Esaias 1981; Smith et al. 1987; Dickey 2003) utilizing buoys, ships, aircraft, and satellites have been developed to meet the need to measure distributions of physical and biological properties of the ocean over large areas synoptically and over long time periods. Figure 9.1 compares the space/time domains of several physical and biological oceanic processes with space/time sampling regimes of various measurement platforms. Due to the wide range of space/time scales encompassed by marine organisms and the corresponding physical, chemical, and biological mechanisms that regulate their distributions, no single platform of data sensors is adequate to provide a comprehensive synoptic picture. With respect to estimates of primary production, ships can provide relatively accurate point data plus a wide variety of complementary physical, optical, chemical, and biological data, including water samples, from a range of depths in the water column. Ships, however, are disadvantaged by their limited spatial coverage. Moored buoys yield even less spatial coverage but have been utilized to provide long time series data at selected locations and to provide information as a function of depth. Aircraft and satellites permit regional and global coverage, and a wealth of horizontal detail impossible to obtain from ships and buoys alone, but these data are restricted to the upper few attenuation lengths in the water column.

Autonomous underwater vehicles (AUVs), drifters, and floats (Dickey 2003) are sampling platforms that have been developed to cover intermediate space/time scales. Optical sensors, providing proxy measures of various biological parameters, are typically deployed on in-water platforms such as buoys, AUVs, drifters, and floats, as well as aircraft and satellites. Indirect methods (discussed below) are used to estimate phytoplankton biomass and productivity from optical sensors deployed on these various platforms. The accuracy of NPP estimates, particularly in eutrophic coastal and upwelling areas, is hindered by the dynamic variability of the processes affecting production and the inability of a single-platform sampling strategy to provide the required synoptic data. Multiplatform sampling strategies and progress in more accurate quantification of remotely sensed observations have been used to lower the variances in estimates of NPP and have helped to identify the physical and biological factors responsible for these variances.

## Experimental Approaches to Primary Production: The Radiocarbon ($^{14}$C) Method

The most widespread experimental approaches to estimating primary production in marine systems are based on incubation of a water sample, spiked with a radioactive isotope, for a known period of time. Typically, samples are obtained over a range of depths within the water column, where solar radiation stimulates photosynthesis. There are various experimental approaches with respect to the number of depths sampled and how these depths are selected. Also, the design and physical setup of incubators varies, as does the timing of the incubation start and end point. For some field experiments, in situ (IS) incubations can be used, in which samples

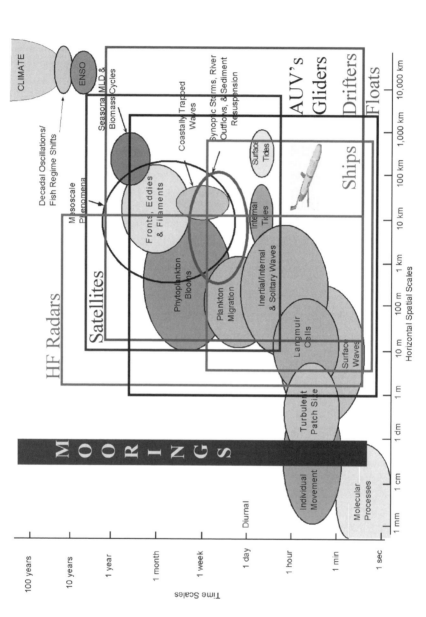

Figure 9.1. Time-horizontal space scale diagram. Physical and biological processes representative of ocean processes = ovals. Approximate sampling domain capabilities for sensor platforms = rectangles. Reproduced from Dickey (2003) with permission from Elsevier.

are returned to the depth and light conditions from which they were obtained. Alternatively, incubations can be carried out elsewhere (such as on the deck of a ship), under diverse conditions, usually with the attempt to simulate in situ conditions, especially light and temperature. Under such simulated in situ (SIS) conditions, factors are needed to convert estimated production rates to in situ estimates. Finally, there are several approaches to data analysis and presentation.

## In Situ and Simulated In Situ Experiments

The $^{14}C$ method was introduced by Steeman Nielsen (1952) and measures the $CO_2$ incorporation by addition of trace amounts of $^{14}C$ bicarbonate in seawater (Vollenweider 1965; Parsons et al.1984; Rai 2002; Scott 2002). This method is specific for autotrophic photosynthesis and can be used in mixed populations. Large amounts of $^{14}C$ data exist, and it has become the standard method in marine research against which other methods are compared.

Samples for the $^{14}C$ method are obtained from the euphotic zone, defined as the layer where there is sufficient irradiance to support net primary production (NPP > 0). The compensation depth, where photosynthetic fixation balances respiratory losses over a day, is the base of the euphotic zone (see Platt et al. 1989 for a review). Since the euphotic depth is seldom measured directly, it is often estimated to be equal to the 1% (or sometimes the 0.1%) depth of the incident photosynthetically available radiation (PAR), although it is recognized that the compensation depth is probably variable (Falkowski and Owens 1978; Platt et al. 1990). It is assumed that phytoplankton is freely mixed within the upper mixed layer and that the mixed layer is shallower than the euphotic zone, permitting cells to remain exposed to light and production to exceed respiratory losses. The term "critical depth" was introduced to characterize the depth in the water column where net carbon production (NCP) > 0 (Nelson and Smith 1991).

For primary production determinations at a given oceanographic station, samples are typically taken throughout the upper water column with Niskin or Go-Flo bottles attached to a conductivity-temperature-depth (CTD) rosette. A water sample is placed in an incubation bottle for a known period of time. A major limitation of this method is that it requires incubation of a sample in a confined volume that can introduce "bottle effects" (Gieskes et al. 1979). In the early 1980s "clean methods" (principally taking extreme care to exclude minute concentrations of toxic trace metals) were introduced (Fitzwater et al.1982). Data prior to the introduction of these clean methods are generally considered to underestimate true photosynthetic rates (Martin 1992).

For shipboard observations, ideally and whenever possible, samples are taken before dawn for incubations to start at sunrise. Samples for productivity measurements need to be processed quickly after collection to avoid contamination and to minimize phytoplankton changes. These processes include filtering out larger zooplankton, transferring the sample to light and dark incubation bottles, spiking the incubation bottles with $^{14}C$, and incubating the spiked samples. For in situ incubations the incubation bottles are replaced at the depth from which they were sampled for the duration of incubation. Alternatively, samples are incubated on deck in a setup simulating in situ conditions for light and temperature (Lohrenz et al. 1992; Lohrenz 1993).

$^{14}$C incorporation into the sample is measured in units of disintegration per minute (DPM). The intensity of the signal is proportional to the beta particle emission from the $^{14}$C incorporated into the cells. Primary production is calculated as

$P_z$ = (DPM in the light bottle – DPM in the dark bottle)/volume
sample filtered * 24,000 * 1.05 * hrs of incubation/(specific
activity in the sample/volume specific activity)                    (2)

in units of [mg C m$^{-3}$ h$^{-1}$], where $P_z$ is production at depth z, total HCO$_3^-$ in the water is ~24,000 [mg C kg$^{-1}$] (Carrillo and Karl 1999), and 1.05 is the discrimina-

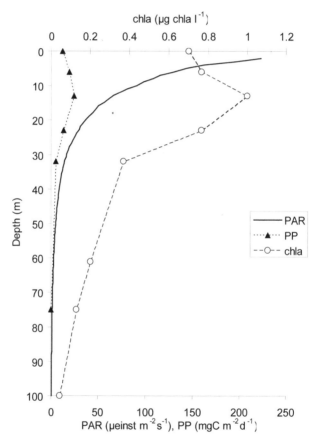

Figure 9.2. Profiles of phytoplankton biomass as chlorophyll $a$ [mg chl$a$ m$^{-3}$], photosynthetically active radiation (PAR) measured with a $4\pi$ collector model QSP-200L4S from Biospherical Instruments, Inc. [μEinst m$^{-2}$ s$^{-1}$], and primary production [mg C m$^{-3}$ d$^{-1}$] measured with simulated in situ incubations on board ship for a coastal station in the western Antarctic peninsula (64.893S, 64.173W) in January 2003. Triangles and circles denote the depth of sampling at 100%, 50%, 25%, 10%, 5%, and 0.5% of incident radiation (E$_o$). Euphotic zone was calculated as 1% of incident radiation at 61 m with corresponding integrated primary production of 602 mg C m$^{-2}$ d$^{-1}$ and integrated chlorophyll $a$ of 33 mg m$^{-2}$.

tion factor between incorporation of $^{14}C$ and $^{12}C$. $P_z$ is the primary production expressed as C incorporated per unit volume of water per unit time (fig. 9.2). Daily $P_z$ is calculated by converting the hours into a 24–hr day, and it is considered as NPP where the balance of photosynthesis – respiration is $> 0$. Furthermore, $P^B$ [mg C mg chl$a^{-1}$ h$^{-1}$] is defined as the assimilation number and is calculated as $P_z$ per unit of biomass in the sample, usually chlorophyll $a$, in units [mg chl$a$ m$^{-3}$]. The latter is used to standardize NPP when comparing different regions, and it is a measure of photosynthetic efficiency.

The $^{14}C$ incorporation in the light bottle is considered to account both for biotic (i.e., photosynthesis and $CaCO_3$ incorporation) and for abiotic (i.e., adsorption) processes (Banse 1993). Thus, the $^{14}C$ incorporated is corrected by the dark bottle to account for biological $^{14}C$ uptake that can occur outside photosynthesis. The incorporation of $^{14}C$ into $CaCO_3$ is corrected by sublimation with acid. Finally, a time-zero determination corrects for abiotic processes. In general, time-zero values should remain low (i.e., $<5\%$) to indicate quality of the incubation.

Marra (1995) argues that the relevant time interval for estimation of ocean primary production is 24 hr. This time scale includes a whole photoperiod with maximum irradiance as well as night catabolism. In many instances, metabolic processes balance within a day. For experiments starting before dawn, production is positive during daylight and negative at night, and balancing daily primary production to initial values before dawn is recommended (Marra 2002). Cell division rates vary from hours to days; thus a 24–hr estimate fits with the ecology of most phytoplankton groups. Experimental approaches that last longer need to take into account biomass changes within the population and the efficiency of carbon transfer to other trophic levels. Shorter time scales will be more dependent on physiological properties of phytoplankton and will necessitate knowledge of physiological responses and how they vary within dominant groups in the sample.

### Determination of Light Field

In order to estimate water column productivity, it is necessary to sample as a function of depth. Typically, sampling depths are selected on the basis of the distribution of solar radiation within the water column. Light decreases exponentially, and sampling depths are defined as percentages of incident irradiance at the water surface, using the Lambert Beer law for predetermined light percentages:

$$E = E_o \exp - (K_{PAR} * z), \qquad (3)$$

where E is PAR [$\mu$Einst m$^{-2}$ sec$^{-1}$] at depth z, $E_o$ is incident PAR just below the air-water interface, $K_{PAR}$ is the attenuation coefficient in [m$^{-1}$], and z [m] is depth. $K_{PAR}$ is estimated from measurement of PAR versus depth, where

$$K_{PAR} = - \ln (E_{z2}/E_{z1})/(z_2 - z_1). \qquad (4)$$

$E_{z1}$ and $E_{z2}$ are irradiances at two different depths, and $(z_2 - z_1)$ is the depth interval of the irradiance readings $(z_2 > z_1)$. To determine a sampling depth, for example, 50% of $E_o$,

$$z = -\ln (E_z/E_o)/K_{PAR} = -\ln (0.5)/K_{PAR}. \tag{5}$$

Ideally, incubations should replicate the light field from which a sample has been obtained. If incubations are done in situ, the light bottle is exposed to the irradiance and light quality at the depth sampled within the water column, and the production value is thus representative of environmental conditions at that depth. In simulated in situ experiments, the different irradiance levels are simulated by the use of neutral density filters placed over the incubation bottles that screen surface irradiance to simulate the percent PAR from the depth sampled. Simulating the change in light quality (i.e., spectral characteristics) with depth is not achieved by neutral density filters. In general there is no consistent and accepted method to simulate spectral characteristics with depth, but the difference in light exposure can be corrected by modeling (Barber et al. 1997). Depending on the body of water under study, blue or green filters have been recommended for the deep samples: green filters in coastal productive waters and blue filters for more oceanic or oligotrophic environments. The addition of color filters increases primary production estimates at depth by decreasing potential photoinhibition of cells suddenly exposed to white light (Tilzer et al. 1993).

[14]C estimates of primary production usually lie somewhere between "true" GPP and NPP. The degree to which the [14]C incorporation approximates GPP or NPP is dependent on incubation time and photosynthetic rate (Williams 1993b). Models of the [14]C incorporation at varying photosynthetic and respiratory rates and time of incubation show that at low respiration rates and short incubation times, [14]C-derived production is a reasonable approximation to GPP. In phytoplankton cultures under controlled conditions, when comparing [14]C production and particulate organic carbon (POC) accumulation (an index of NPP), experiments show that at low growth rates ($<0.1$ d$^{-1}$), [14]C production is about 5 times higher than POC accumulation (Peterson 1978), and thus more closely approximates GPP. Under conditions of high respiration (rates similar to production), [14]C production also approximates GPP (Calvario-Martinez 1989). On the other hand, at high growth rates ($>0.5$ d$^{-1}$), [14]C production and POC accumulation agree, indicating that under these conditions the [14]C method more closely approximates NPP.

Scaling up daily primary production estimates measured on a per volume basis includes interpolation of data points. First, to estimate integral water column photosynthesis in units of [mg C m$^{-2}$ d$^{-1}$], individual sample depths are integrated over depth by polynomial interpolation. It is assumed that the production between two consecutive depths changes linearly and that any incubation less than 24 hr can be prorated to a full day. Second, time integration is carried out by interpolating between sampling dates, as is done when calculating seasonal primary production. This provides a seasonal or annual estimate in units of [mg C m$^{-2}$ mo$^{-1}$] or [mg C m$^{-2}$ y$^{-1}$]. Finally, when estimating primary production in a region such as an embayment or a continental shelf, sampling stations are interpolated spatially and divided by the time interval under analysis, providing a measure in [mg C d$^{-1}$].

The frequency of sampling is determined by the question to be addressed and the dominant process controlling primary production in the biome of interest. In long-term ecological studies in coastal Antarctica, where a key question is to de-

termine the seasonal evolution and the interannual variability, sampling is carried out twice weekly (Vernet, unpubl. data). Thus, determining factors in scales
less than 1 week is not possible. Mixing events that control the phytoplankton
accumulation within surface waters are driven by large storms that on average
pass through the region every few weeks. Each bloom is then characterized by an
average of 5–7 data points, which provides detail on productivity increase, peak
value, and decrease within each cycle. Within one growth season, defined by sun
angle and ice cover to last between October and April, tens of sampling points
provide definition of the bloom events within a season. Similar sampling carried
out during the next season provides the additional data to compare annual NPP
among seasons as well as the difference in frequency, intensity, and timing of the
bloom events within each season.

## Laboratory Incubations

Photosynthesis versus irradiance curves (P vs. E curves) have been recommended
over in situ or simulated in situ experiments as the best method to estimate NPP in
predictive models of photosynthesis in ocean waters (Coté and Platt 1984). Estimates of productivity are based on determination of the response of phytoplankton
incubated over a range of irradiances at in situ temperature. Two parameters are
necessary to describe the P vs. E relationships: alpha ($\alpha$), the initial slope of the
light-limited portion photosynthesis, and $P_{max}$, the light-saturated rate of photosynthesis. The photosynthetic response is modeled by curve-fitting. By transferring the
modeled curve into the vertical gradient of the underwater light field, the vertical
distribution of photosynthesis can be estimated.

Three models of curve-fitting have been the most commonly used in the literature, but care must be taken to recognize their intrinsic differences (Frenette et al.
1993). When no photoinhibition is present, production can be modeled as suggested
by Webb et al. (1974):

$$P^B = P_M^B * (1 - \exp(-\alpha * E / P_M^B)), \tag{6}$$

or, as given by Platt and Jasby (1976):

$$P^B = P_M^B \tanh(\alpha * E / P_M^B), \tag{7}$$

where $P^B$ is photosynthesis per unit biomass (or chlorophyll $a$) in units of [mg C
(mg chl$a$)$^{-1}$ h$^{-1}$], $P_M^B$ is maximum rate of photosynthesis per unit chl a, $\alpha$ is the
initial slope in units of [mg C (mg chl$a$)$^{-1}$ h$^{-1}$ ($\mu$Einst m$^{-2}$ sec$^{-1}$)$^{-1}$], and E is irradiance in units of [$\mu$Einst m$^{-2}$ sec$^{-1}$].

When photoinhibition is present, photosynthesis can be modeled as an extension of equation (6) (Platt et al.1980):

$$P^B = P_S^B * (1 - \exp - (\alpha * E / P_S^B)) * \exp - (\beta * E/P_S^B), \tag{8}$$

where all variables are defined as before and beta (ß) is the photoinhibition parameter with the same units as $\alpha$, and where

$$P_{max} = P_s [\alpha/(\alpha + \beta)] [\beta/(\alpha + \beta)]^{\beta/\alpha}. \tag{9}$$

Unlike IS and SIS, where one determination is taken at each depth, in P vs. E curves a suite of light and dark bottles are incubated at different irradiances for each depth sampled. All incubations are thus done in vitro, and neutral filters are used to simulate varying irradiances (but see Lohrenz et al. 1992). Incubations are usually short, from 1 to 4 hr, because the response curve is determined before photoacclimation. Most recently, P vs. E curves have been determined with increased numbers of light treatments (e.g., 25) and decreased volume of incubation (e.g., 2 ml; fig. 9.3). Sensitivity of the determination is preserved by increasing the specific activity of the sample. Irradiance levels usually range from 0 to

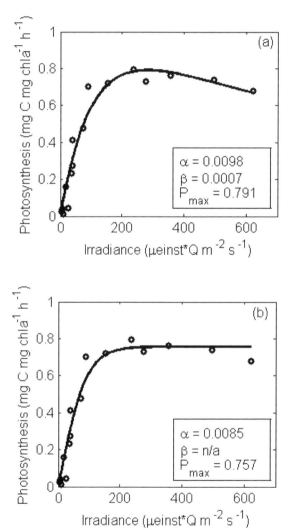

Figure 9.3. Photosynthesis [mg C chl$a^{-1}$ h$^{-1}$] versus irradiance [µEinst m$^{-2}$ s$^{-1}$] determined with $^{14}$C incubations in the western Antarctic peninsula. Curvefitting with (a) equations (8) and (9) (Platt et al. 1980) and (b) equation (7) (Platt and Jasby 1976).

1000 $\mu$Einst m$^{-2}$ sec$^{-1}$, although adjustments in the light range are necessary in different geographic locations and depend on the time of the year. For example, Antarctic samples are usually exposed to 0–600 $\mu$Einst m$^{-2}$ s$^{-1}$, while open ocean samples in the North Pacific are exposed to 0–2000 $\mu$Einst m$^{-2}$ s$^{-1}$. Relatively high irradiances are needed to determine the photoinhibition factor ($\beta$).

Value of the modeled productivity is dependent on the analytical quality of the P vs. E curves and their accuracy with respect to in situ production rates. It is recommended that incubators also simulate in situ light quantity with the addition of neutral filters. Similar to SIS incubations, determination of the light quality can be achieved through the use of colored filters. More accurate determination can be achieved with the use of solar simulators. Furthermore, the value of $P_{max}$ is a function of ambient temperature (Geider and MacIntyre 2002). Acclimation is sufficiently fast that $P_{max}$ may differ for the same sample incubated at several temperatures (fig. 9.4). It is recommended that incubations be carried out at in situ temperatures. Alternatively, if the temperature in both the water and the incubator is known, a predetermined $Q_{10}$ can be applied for correction (Tilzer et al. 1993). Finally, the accuracy of the P vs. E determination is compromised if the natural variability of $\alpha$ or $P_{max}$ is not included in the primary production estimate. Changes in irradiance with depth determine the value of $\alpha$ in situ. The value of $\alpha$ is proportional to the light acclimation of the cells in the field. Thus, for each water column, several P vs. E determinations are needed. Furthermore, if temperature changes with depth (e.g., the euphotic zone is deeper than the mixed layer and the bottom of the euphotic zone is at different temperature), then $P_{max}$ will change with depth in the water column, thus influencing the estimated productivity and requiring a further correction for temperature (see above).

## Experimental Approaches to Primary Production: Oxygen Methods

### Oxygen Production

Oxygen evolution is a primary by-product of the splitting of the $H_2O$ molecule during photon absorption (Falkowski and Raven 1997). Increased $O_2$ concentration in a water sample is proportional to photosynthesis, and thus to production. Under light, $O_2$ production is measured as the difference between initial and final $O_2$ concentration in a light bottle. In the water column, bottles are incubated at different irradiances, as explained for the $^{14}$C method, to estimate water column production. Oxygen evolution from photosynthesis can be masked by $O_2$ consumption by respiration, since both happen simultaneously in the cells. To account for this process, dark bottles are incubated concurrently with light bottles. Assuming respiration is the same under dark and light conditions, gross production is calculated from $O_2$ increase in the light bottle + consumption in the dark bottle. The proportion of $O_2$ produced to C uptake or $O_2$ evolved to $CO_2$ assimilated is the photosynthetic quotient (PQ). For healthy, nutrient-replete cultures, PQ is 1.2 to 1.8, consistent with protein and lipids as the major products of photosynthesis (Laws 1991). (For further discussion of this method, see chapter 10 of this volume.)

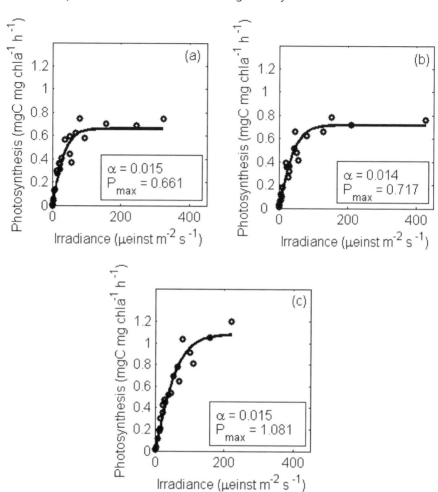

Figure 9.4. Photosynthesis [mg C chl$a^{-1}$ h$^{-1}$] versus irradiance [µEinst m$^{-2}$ s$^{-1}$] determined with $^{14}$C incubations in the western Antarctic peninsula at different temperatures to show dependence of $P_{max}$ on temperature: (a) −1.5° C lower than ambient temperature, (b) incubation at ambient temperature, and (c) +1.5° C higher than ambient temperature.

Oxygen consumption by respiration in plankton samples has both autotrophic and heterotrophic components. Heterotrophic respiration by microzooplankton and bacteria can be higher than autotrophic $O_2$ production, so that net community production is negative.

### The $^{18}$O Method

Similar to the $^{14}$C method, the $^{18}$O tracer method was developed to measure gross production in vitro with light and dark bottles (Bender et al. 1987). This is an extension of $O_2$ production, but in this case $O_2$ is measured not by concentration but by using a radioactive tracer. $^{18}O$ is an oxygen isotope with natural abundance of 0.204 atom%, while the major isotope $^{16}O$ has an abundance of 99.758%. The $^{18}O$ method involves spiking a water sample with $H_2^{18}O$, incubating in the light, and measuring the amount of $^{18}O_2$ produced during photosynthesis. All $O_2$ is in a dissolved phase and the ambient $O_2$ is so large (150 µM) that only a negligible amount of $O_2$ will be recycled by respiration during the incubation. Consumption has a very small effect on the $^{18}O{:}^{16}O$ ratio, such that the ratio can be considered constant throughout the incubation. The only unknown source of error would be intracellular recycled $O_2$.

## Further Considerations on Experimental Methods

### What Is Estimated Using the $^{14}C$ Method?

By comparing the method of $^{14}C$ incorporation with the $O_2$-based methods, we can evaluate what is estimated by using the $^{14}C$ method in field measurements. As mentioned above, $^{14}C$ estimates approximate gross or net primary production or something in between, depending upon conditions. In the North Atlantic, Marra (2002) observed that $^{14}C$ underestimated gross primary production (as measured by the $^{18}O$ method) (fig. 9.5). $^{14}C$ agreed only with net primary production measured with $O_2$ production bottles over a 24 hr period. According to these comparisons, the $^{14}C$ method seemed to best approximate net community production. This result might be due to the fact that gross C uptake and gross $O_2$ production cannot be equated because they are associated with different biochemical pathways within the cell. Ryther (1956) encountered similar discrepancies in culture experiments. He concluded that respired $CO_2$ is reassimilated in photosynthesis, whereas $O_2$ released in photosynthesis is not reassimilated by respiration. This conclusion agrees with modeled data where most, if not all, respired $CO_2$ is refixed as photosynthesis (Williams 1993b). Thus, as measured by $^{18}O$, there is an imbalance between $CO_2$ and $O_2$ dynamics. If that is so, the cells use proportionally more $H_2O$ (and $H_2^{18}O$) than external $CO_2$, because internal $CO_2$ from respiration is a source. This would mean higher $^{18}O$ uptake than $^{14}C$ uptake for the same production rate. Thus, it would appear that $^{18}O$ more closely approximates gross photosynthesis and $^{14}C$ more closely approximates net photosynthesis. If respiration is low (i.e., low $CO_2$ from respiration is available for photosynthesis), the $^{14}C$ method would approximate gross production. Under all other conditions the $^{14}C$ uptake approximates net production (Marra 2002; Williams 1993a).

### Errors and Limitations

Accurate estimation of daily water-column primary production is challenging by its very definition: the extrapolation of results from short incubations to daily rates; from results obtained in small containers scaled to ecologically relevant spatial scales; and the influence of respiration and heterotrophic activity on gross vs. net estimations.

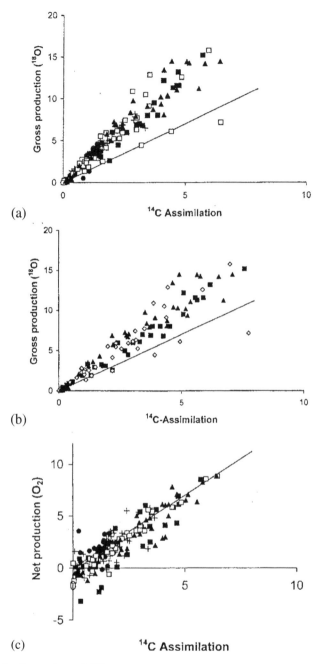

Figure 9.5. Comparison of different approaches to measuring primary production in marine phytoplankton. (a) Gross primary production measured with $^{18}O$ (Bender et al. 1987) and daily primary production measured with 24-hr in situ $^{14}C$ incubations. (b) Gross primary production measured with $^{18}O$ and primary production estimated with daytime in situ $^{14}C$ incubations. (c) Daily net primary production measured with light-dark $O_2$ production compared to daily $^{14}C$ assimilation. Data obtained during several cruises in the North Atlantic and equatorial Pacific. Reproduced from Marra (2002) with permission from Blackwell Publishing.

The action spectrum of photosynthesis, the solar spectrum, and the underwater light field all vary as a function of wavelength. The spectral characteristics of underwater irradiance change as the irradiance is transmitted downward through the water column. Maximum penetration occurs in the green (530 nm) in coastal waters and in the blue (485 nm) in open ocean (Tyler and Smith 1970). The differential absorption through the water column is due to absorption by water per se, to phytoplankton particles (via their photosynthetic pigments), to dissolved organic matter (DOC), and to any suspended inorganic material. If the measurements are done in situ, this potential problem is minimized (Dandonneau 1993). If the profile of primary production with depth within the euphotic zone is measured with SIS incubations on ship deck, then the matching of the vertical variability in the water column requires a more rigorous treatment. The addition of either blue or green filters to better simulate natural light conditions at low irradiances has been found necessary for accurate estimates of both $\alpha$ and $P_{max}$ (Tilzer et al. 1993).

It has been calculated that if the water column is uniformly mixed, ignoring spectral effects can lead to an error as high as 30% of the integrated primary production (Platt and Sathyendranath 1991). When biomass distribution is nonuniform with depth, error can reach 60%. The key factor to consider is the depth dependence of the attenuation coefficient of light. These errors may be further minimized with information and modeling of the spectral attenuation coefficient (Tilzer et al. 1993).

Heterotrophic activity and phytoplankton physiological state can adversely affect estimates using the $^{14}C$ method. The onset of nutrient limitation or the production of $NH_4$ by microzooplankton during the incubation period can either depress or stimulate production estimates, but experimental evidence to date indicates this influence is typically insignificant (Harrison 1993). DOC released by the cells during incubation can lead to underestimation of the amount of $^{14}C$ fixed if the $DO^{14}C$ returns to the dissolved pool (Jackson 1993). If the $DO^{14}C$ is taken up by heterotrophs, thus returning $^{14}C$ to the particulate pool, the analytical technique used to concentrate phytoplankton (i.e., the pore size of the filters used) will determine if this fraction is or is not accounted as primary production.

## Indirect, Noninvasive Methods of Measuring Primary Production

A new generation of instruments and methods, based on fluorescent properties of photosynthesis, has emerged in oceanography during the past few decades. These methods are noninvasive, and do not depend on incubation of small samples captured from the water column. An important advantage of these measurements is that they permit higher temporal sampling rates that are more closely matched to sampling rates for physical variables (e.g., temperature, salinity, oxygen, etc.), which allows for a better coupling between environmental and production measurements.

Fluorescence is the production of visible light emitted by specific molecules at longer (or less energetic) wavelengths than the wavelengths absorbed. In the case of photosynthesis, chlorophyll $a$ absorbs energy in the blue region of the spectrum (430–440 nm) and emits in the red region (680–685 nm), corresponding to the Sorel maxima of absorption for chlorophyll $a$ (Jeffrey et al. 1997). After photon absorption by chlorophyll $a$, the energy can be used for photochemistry, lost as heat, or emitted as light through fluorescence. As a first approximation, it would seem that fluorescence would be inversely proportional to photosynthesis. The relationship is not strong, however, because fluorescence is highly dependent on intensity and quality of the incident light. Low irradiance levels of incident light induce fluorescence that has a positive correlation with chlorophyll $a$ concentration in the cell. High irradiance levels of incident light quench chlorophyll fluorescence in a nonphotochemical process. Furthermore, the dynamics of chlorophyll fluorescence shows a time-dependent response which can be used to infer several biophysical variables related to photosynthesis (Falkowski and Kolber 1993).

### Passive Fluorescence Methods

Fluorescence can be induced by both solar radiation and artificial illumination. In vivo solar-induced fluorescence can be measured passively and detected at 683 nm in near-surface waters (Kiefer et al. 1989; Chamberlin et al. 1990). Measurement of solar-induced fluorescence is accomplished by lowering a photometer with appropriate band-pass filter into the water to obtain a continuous vertical profile of fluorescence. Photometers to detect in vivo fluorescence can be deployed on buoys to obtain data over diel cycles. In vivo fluorescence can also be measured as a component of water-leaving radiance at 683 nm by new satellite sensors with multispectral resolution (Topliss and Platt 1986; Doerffer 1993).

Natural fluorescence emitted mostly from Photosystem II by the cells ($F_f$ in [Einst $m^{-3}$ $s^{-1}$]) is a product of the flux of absorbed light ($F_a$ in [Einst $m^{-3}$ $s^{-1}$]) and the quantum yield of fluorescence ($\Phi_f$ in [Einst emitted/Einst absorbed]).

$$F_f = \Phi_f * F_a \qquad (10)$$

and

$$F_a = a_c * E_o, \qquad (11)$$

where $a_c$ is the absorption coefficient for phytoplankton [$m^{-1}$] and $E_o$ is irradiance in [Einst $m^{-2}$ $s^{-1}$]. Similarly,

$$F_c = \Phi_c * F_a, \qquad (12)$$

where $F_c$ is the rate of carbon incorporation in [g-at C $m^{-3}$ $s^{-1}$] and $\Phi_c$ is the quantum yield of photosynthesis in [g-at C fixed/Einst absorbed]. Combining the last three equations, primary production can be estimated (where the relevant parameters are determined at each depth z in the water column) from the model,

$$F_c = (\Phi_c / \Phi_f) * \Phi_f * a_c * E_o, \qquad (13)$$

for 24 hr and at depth z. Field tests using $^{14}C$ incubations have shown that $F_c$ can be modeled from natural fluorescence over a range of three orders of magnitude in production. This method approximates GPP as it relates to photon absorption. Fluorescence measured in the field can be overestimated because of fluorescence from detrital chlorophyll or phaeopigments, and can be underestimated by the presence of planktonic cyanobacteria because most of the chl a is associated with the Photosystem I. Modeling primary production based on fluorescence measurements is also subject to variability because $\Phi_f$ is a function not only of light but also of nutrient status (Chamberlin et al. 1990).

## Active Fluorescence Methods

Lamp-induced fluorescence measurements are based on the dynamics of fluorescence decay in the first few milliseconds after a light flash (Rabinowitch and Govindjee 1969). Under weak flashes, pigment reaction centers remain open (i.e., they continuously receive photons because some molecules within the antenna pigment always remain in the ground state). Under strong flashes, all the chlorophyll molecules in the antenna pigment saturate, the reaction centers close (i.e., no more photons are absorbed). By using an appropriate combination of weak and strong flashes, several parameters of the fluorescence decay can be determined ($F_o$, or baseline fluorescence, and $F_m$, maximum fluorescence). A third term, variable fluorescence ($F_v$), is defined as the difference between maximum and baseline fluorescence ($F_m - F_o$). The quantum yield of photochemistry ($\Phi_f$), related to photosynthesis and thus to productivity, is defined as $F_v/F_m$ or $(F_m - F_o)/F_m$. The pulse-amplitude-modulated (PAM) fluorometer uses repeating strong flashes of light against a continuous background of weak light in order to determine $F_m$ and $F_o$. This technique may be used to model productivity (Neale and Priscu 1998; Hartig et al. 1998; see also chapter 10, this volume).

A second-generation fluorometer was designed to address some of the limitations of PAM fluorometry (Kolber et al. 1998). The fast repetition rate fluorometry (FRRF) was developed to obtain specific parameters needed to model production (e.g., the cross section for absorption of irradiance ($\sigma_{PSII}$) and the parameter for photochemical quenching ($q_p$), $1/\tau$, which gives the rate of electron transport from initial donor ($H_2O$) to final acceptor ($CO_2$). In FRRF, plankton cells are exposed to a series of flashes at subsaturating intensities. The rapid series of flashes produces an increase in fluorescence as the antenna pigment reaches saturation. The rate of fluorescence increase is related to the functional cross section of Photosystem II, while the subsequent rate of fluorescence decay at subsaturating light is a measure of the time constant of reoxidation of Qa–, which can be related to the turnover time of photosynthesis at irradiance levels that completely reduce the PQ pool. Turnover time of photosynthesis is $1/I_k \sigma_{PSII}$. Quantum yield for fluorescence ($\Phi_{max}$) is calculated from these variables.

By measuring $\sigma_{PSII}$, $\Delta\Phi_{max}$ of fluorescence, $q_p$, and incident PAR ($E_o$), we can calculate the noncyclic electron transport rate of each PSII reaction center as

$$P_f = [\Delta\Phi F_{max}/0.65] \, q_p * E_o * \sigma_{PSII}. \tag{14}$$

It is assumed that there is a constant ratio of PSII reaction centers to chl$a$ ($\sim$1500, in moles). Furthermore, to derive photosynthetic rates it is assumed that 4 electrons are required to reduce a molecule of $CO_2$ to the level of carbohydrate, and that the only terminal electron acceptor is $CO_2$—this is the upper limit approximation.

Then

$$P_C{}^B = P_f * b/4,  \qquad (15)$$

where $P_C{}^B$ is the chlorophyll-specific rate of C fixation [moles of $CO_2$ mole$^{-1}$ chl$a$ t$^{-1}$], $P_f$ is the fluorescence-based rate of photosynthetic electron flow [e- reaction center$^{-1}$ t$^{-1}$], and $\Delta\Phi_{max}$ is scaled to the maximal value of 0.65. Short-term photosynthetic rates calculated from $F_v/F_m$, as measured with an FRRF in the field (fig. 9.6) correlate positively with hourly $^{14}$C incubation in field samples (fig. 9.7), suggesting this is a viable method for fast, incubation-free, and noninvasive determination of photosynthetic electron transport (Kolber and Falkowski 1993).

### Errors and Limitations

Estimating primary production from passive solar-induced fluorescence requires the assumption of a constant quantum yield of fluorescence. The FRRF technique has shown that this is not a valid assumption for fieldwork, since nutrient conditions as well as irradiance levels affect this yield (Falkowski and Kolber 1993). In addition, there is no estimate of fluorescence quenching at high irradiance. This effect cannot be corrected without active measurements of fluorescence. Finally, the

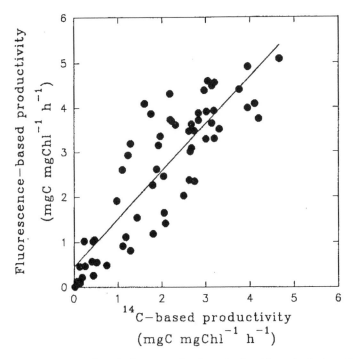

Figure 9.6. Fluorescence-based primary production calculated from the fast repetition rate fluorometer (FRRF), compared to short-term incubations with $^{14}$C in [mg C mg chl$a^{-1}$ h$^{-1}$]. Reproduced from Falkowski and Kolber (1993) with permission from International Council for the Exploration of the Sea.

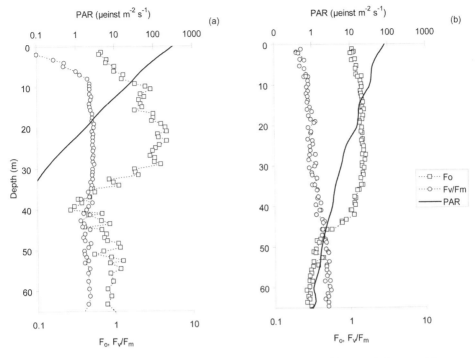

Figure 9.7. Profiles of baseline fluorescence ($F_o$) (squares) and $F_v/F_m$ (rounded squares) from the fast repetition rate fluorometer (FRRF), collected on the western Antarctic peninsula. PAR (solid line) at each station is also shown. (a) Coastal station in Marguerite Bay, western Antarctic peninsula. (b) Slope waters of the continental shelf at similar latitude as (a).

method assumes that the quantum yield of fluorescence changes similarly to the quantum yield of carbon (Kiefer et al. 1989). Laboratory studies show that the two quantum yields vary as passive fluorescence signal increases almost linearly over the whole range of irradiances, while C fixation saturates at irradiance levels above $E_k$. The consensus is that fluorescence methods are very promising and that we need more studies to interpret the fluorescent signal in the field as it relates to NPP estimates (Laney 1997). As the method becomes more widely used, a better characterization of its results and limitations is becoming available (Laney 2003).

All fluorescence methods make use of short time intervals, from milliseconds to minutes, and necessitate a knowledge of their response to environmental variability and an estimate of that variability in order to scale up to daily rates. The challenge is to integrate biophysics with ecological scales of interest.

## Remote Sensing

The most effective (and perhaps the only practical) way to adequately sample the space/time variability of the 75% of the earth's surface covered by oceans is by means of remote sensing. Phytoplankton is mechanistically linked with optical properties of the ocean, so the determination of in-water optical properties offers the possibility of both synoptic (e.g., via satellite) and continuous (e.g., via moorings) estimation of pigment biomass parameters over a range of space and time scales. As a consequence, there has been considerable progress in the development and use of optical proxy measures of pigment biomass and phytoplankton production, and in the use of bio-optical models that can accommodate data from satellites and aircraft as well as a range of in-water platforms such as ships, moorings, autonomous underwater vehicles, drifters, and gliders (Dickey 2003). These approaches are in many ways analogous to those for terrestrial ecosystems described in chapter 11 of this volume.

## Biomass Estimates by Remote Sensing

It has long been recognized (Kalle 1938; Jerlov 1951; Yentsch 1960; Morel and Smith 1974; Morel and Prieur 1977) that the color of ocean waters varies with the concentration of dissolved and suspended material (i.e., that the spectrum of backscattered sunlight shifts from deep blue to green as the concentration of phytoplankton increases). That ocean color could be detected by remote optical sensors led to the desire to relate ocean optical properties, in particular upwelled spectral radiance from the sea surface, to the various constituents of the medium (Duntley et al. 1974). These early studies led to the development and launch of the coastal zone color scanner on the Nimbus-7 satellite in October 1978 (Hovis et al. 1980; Gordon et al. 1980) and to subsequent advances in ocean color satellite systems. A more recent (May 2004) accounting by the International Ocean-Color Coordinating Group (IOCCG) lists ocean color satellite missions deployed by various international space agencies: eight historical sensors, nine current sensors, and five scheduled sensors (http://www.ioccg.org/semsprs_ioccg.html). This advancement in satellite technology has been accompanied by significant advances in bio-optical field instruments and methods, and improved theoretical analyses, both of which are enhancing our understanding of marine ecosystems.

Early workers using ocean color satellite observations focused on the retrieval of regional and global near-surface chlorophyll a ($Chl_{sat}$, [mgChl m$^{-2}$]) concentrations and the quantitative comparison with ship-based observations (Gordon et al. 1980; Smith and Baker 1982; Gordon and Morel 1983). Early algorithms for estimating $Chl_{sat}$ were empirically derived by statistical regression of radiance ratios at different wavelengths versus chlorophyll a. In spite of their simplicity, these algorithms captured roughly two-thirds of the variation in radiance band ratios and the three orders of magnitude variation in $Chl_{sat}$. When limited to waters in which phytoplankton and their derivative products play a dominant role in determining their optical properties (so-called Case 1 waters; Morel and Prieur 1977), these pigment algorithms enabled the retrieval of chlorophyll a from satellite observations with an accuracy of roughly ±35% (Smith and Baker 1982; Gordon and Morel 1983). This estimated accuracy is a baseline against which more recent and improved algorithms can be compared.

Ocean color pigment algorithm development is an ongoing process. O'Reilly and a host of coauthors (1998, 2000) evaluated numerous pigment algorithms suitable for operational use by the SeaWiFS (Sea-viewing Wide Field-of-view Sensor) Project Office (Firestone and Hooker 1998; McClain et al. 2004). Their goal was to permit estimation of in situ $Chl_{sat}$ concentrations with the highest possible accuracy and precision over a wide range of bio-optical conditions and with due consideration to the atmospheric correction algorithms necessary for accurate retrievals. There has also been advancement in so-called semi-analytic algorithms that seek improvements in understanding the theoretical linkages between biological constituents and their corresponding optical properties (Gordon et al. 1988; Morel and Berthon 1989; Morel 1991; Platt et al. 1992; Garver and Siegel 1997; Carder et al. 1991). To date, empirical algorithms generally perform better than semi-analytic algorithms when considering both statistical and graphical criteria (O'Reilly et al. 1998, 2000). Also, it is recognized that algorithms designed for global scales may be less accurate than algorithms tuned for local and regional scales, and considerable current research is devoted to improving both regional and global algorithms. Because algorithm development progresses rapidly, interested readers should consult Web sites for specific satellite sensors to obtain the most recent developments (e.g., http://www.ioccg.org).

### Modeling Primary Production

Prior to the advent of satellite ocean color sensors, estimations of regional and global ocean production were biased by the errors associated with the inability to sample on the appropriate time and space scales (Harris 1986). Bidigare et al. (1992) discuss the scaling of discrete measurements to remote observations and note that this linkage requires mathematical models relating measurable optical properties to desired biological parameters. They also review the evolution of bio-optical production models which can accommodate ship, mooring, and satellite data. Early workers (Talling 1957; Rodhe 1966; Ryther and Yentsch 1957) related NPP to the product of chlorophyll biomass, daily integrated surface solar radiation, a parameter to estimate attenuation of photosynthetically available radiation (PAR) within the water column, and a variety of variables associated with the photophysiological, quantum or assimilation efficiencies of phytoplankton. Behrenfeld and Falkowski (1997b) reviewed the development of phytoplankton primary productivity models and showed a "fundamental synonymy" between nearly two dozen models developed since the 1960s. These authors noted that "all of these models can be related to a single formulation equating depth-integrated primary production ($PP_{eu}$ [$mgC\ m^{-2}\ d^{-1}$]) to surface phytoplankton biomass ($Chl_{sat}$ [$mgChl\ m^{-3}$]), a photoadaptive variable ($P^b_{opt}$ [$mgC(mgChl)^{-1}\ h^{-1}$]), euphotic depth ($Z_{eu}$ [m]), an irradiance-dependent function ($f(E_{par})$), and day length (DL [$h\ d^{-1}$])":

$$PP_{eu} = Chl_{sat}\ Z_{eu}\ f(E_{par})\ DL\ P^b_{opt}, \qquad (16)$$

where $PP_{eu}$ is the daily C fixation integrated from the surface to the euphotic depth ($Z_{eu}$) and $P^b_{opt}$ is the maximum chlorophyll-specific C fixation rate observed within a water column measured under conditions of variable irradiance during incubations typically spanning several hours (equivalent to $P_{max}$ as determined by P vs. E curves). PP$_{eu}$ may be considered a measure of net primary production (NPP) because this equation is based on $^{14}$C incubations.

Behrenfeld and Falkowski (1997a) assembled a data set of 11,283 $^{14}$C-based measurements of daily C fixation from 1698 oceanographic stations in both ocean and coastal waters. When they partitioned the variability in PP$_{eu}$ into the variability associated with each of the variables in equation (16), they found that nearly all (~85%) could be attributed to changes in depth-integrated biomass (i.e., Chl$_{sat}$ Z$_{eu}$) and the horizontal variability in the photoadaptive variable P$^b_{opt}$. Making use of their large database, they developed a vertically generalized production model (VGPM; eq. [16]), discussed the limitations of productivity models, estimated total global annual productivity, and compared their results with those of earlier ship-based global estimates (Eppley and Peterson 1979; Longhurst et al. 1995; Antoine et al. 1996).

Integrated estimates of primary production based on satellite measurements for both oceanic and terrestrial ecosystems have been presented by Field et al. (1998) and Behrenfeld et al. (2001). For both land (Monteith 1972) and oceans (Morel 1991), NPP can be computed as the product of the absorbed photosynthetically active (400–700 nm) solar radiation (APAR) and an average light utilization efficiency ($\varepsilon$):

$$NPP = APAR \, \varepsilon. \qquad (17)$$

These authors note that while models based on this approach are "diverse in terms of mechanistic detail, they are all strongly connected to global-scale observations." The uncertainty in $\varepsilon$ is a primary source of error for both land and ocean NPP estimates. Both the Carnegie-Ames-Stanford for land (CASA; Potter et al. 1993) and the VGPM (eq. [16]) models are conceptually similar (eq. [17]), and can be used to estimate primary production for the whole biosphere. The SeaWiFS sensor was the first satellite instrument with both the spectral coverage and the dynamic range necessary to derive both Chl$_{sat}$ and NDVI (the Normalized Difference Vegetation Index used in the CASA and other terrestrial models) (Behrenfeld et al. 2001). Their observations allowed the comparison of simultaneous ocean and land NPP responses to a major El Niño-to-La Niña transition, and were the first single-sensor global observations of the photosynthetic biosphere. The CASA-VGPM model gave an NPP estimate for the total biosphere of 104.9 [Pg C yr$^{-1}$] (annual mean for the period September 1997 to August 2000) with a contribution of 56.4 [Pg C yr$^{-1}$] for the terrestrial component and 48.5 [Pg C yr$^{-1}$] for the oceanic component (P = 10$^{15}$).

Ocean color satellite data now routinely provide estimates of chlorophyll biomass (Chl$_{sat}$) and incident PAR. The conversion to C in these chlorophyll-based ocean NPP models is then made via a chlorophyll-specific physiological variable (e.g., P$^b_{opt}$ [mgC (mgChl)$^{-1} \cdot$ h$^{-1}$]). For example, in the VGPM model all parameters, save for the physiological variable, could be estimated from ocean color-related satellite data. P$^b_{opt}$ was then assumed to be known from laboratory data or estimated from satellite sea surface temperature (SST) via empirical models previously determined (Antoine et al. 1996; Behrenfeld and Falkowski 1997b; Balch et al. 1992). In contrast to these chlorophyll-based models, Behrenfeld et al. (2005) have proposed a C-based model. They show that derived Chl:C ratios are consistent with expected physiological dependencies on light, nutrients, and temperature. With this information, they make global estimates of phytoplankton growth rates ($\mu$ [divisions d$^{-1}$]) and carbon-based NPP, using

$$NPP \text{ [mgC m}^{-2}\text{ d}^{-1}] = C_{sat} \text{ [mgC m}^{-3}\text{ ]} \cdot \mu \text{ [divisions d}^{-1}] \, Z_{eu} \text{ [m] } \cdot h \text{ (Eo)}, \quad (18)$$

where $C_{sat}$ is the estimate of surface C and h (Eo) describes how changes in surface light influence the depth-dependent profile of carbon fixation. Equation 18 is the same form as eq. [16] except that Chl is replaced by C and the empirical estimate of $P^b_{opt}$ is replaced by the phytoplankton growth rate $\mu$ (where C and $\mu$ are now directly estimated from remotely sensed data). Global estimates of $\mu$ and C-based NPP are comparable with earlier chlorophyll-based NPP estimates. Notably, the C-based estimates, when compared with the chlorophyll-based estimates, provide a different perspective on how ocean productivity is distributed over space and time. In particular, one expects the physiological differences between C and chlorophyll biomass models to differ in response to changing light, nutrient, and temperature conditions.

Remote sensing provides the most consistent method of estimating NPP at regional and global scales. An example is given in table 9.1. Annual NPP is estimated for different marine biomes, such as the polar, west wind drift, trade winds, and coastal biomes in the Pacific, Atlantic, Indian, and Antarctic oceans.

## Errors and Limitations

Quantitative estimates of the accuracy of variables retrieved from satellite data are an ongoing process. Some disagreement between modeled and in situ $^{14}C$ measured production is due to methodological differences and errors in the in situ data.

Table 9.1. Annual primary production estimated from ocean color remote sensing of chlorophyll *a* (1978–1986) on a 1° grid

| Ocean | Biome | ANPP(gC m$^{-2}$ yr$^{-1}$) | | Province (n) |
|---|---|---|---|---|
| | | *Average* | *Std. Dev.* | |
| Atlantic Ocean | Atlantic Polar | 350.83 | 48.61 | 3 |
| | Atlantic Westerly Winds | 183.30 | 64.03 | 4 |
| | Atlantic Trade Winds | 130.66 | 44.05 | 5 |
| | Atlantic Coastal | 525.38 | 161.91 | 8 |
| Indian Ocean | Indian Ocean Trade Winds | 88.40 | 24.32 | 2 |
| | Indian Ocean Coastal | 360.72 | 157.76 | 6 |
| Pacific Ocean | Pacific Polar | 359.00 | — | 1 |
| | Pacific Westerly Winds | 177.00 | 25.78 | 4 |
| | Pacific Trade Winds | 89.33 | 19.78 | 6 |
| | Pacific Coastal | 382.31 | 141.75 | 8 |
| Southern Ocean | Antarctic Westerly Winds | 126.50 | 9.90 | 2 |
| | Antarctic Polar | 170.75 | 66.11 | 2 |

*Notes*: A total of 51 provinces were identified within 12 biomes, based on monthly composites of surface chlorophyll *a* measured by the coastal zone color scanner. Primary production was modeled on the basis of monthly averages of surface chlorophyll *a*, 21,872 sets of oceanographic profiles determining vertical distribution of chlorophyll *a* ($Z_{eu}$), a photosynthesis-irradiance relationship (similar to $P^b_{opt}$ by Behrenfeld and Falkowski (1997b), and climatologies on surface solar irradiance.

*Sources*: Longhurst (1998); Longhurst et al. (1995); Platt and Sathyendranath (1988); Sathyendranath et al. (1995).

However, Behrenfeld and Falkowski (1997a) suggest that much of the discrepancy must also result from limitations of the models. For example, the differences observed between the C-based and Chl-based models depend upon differences in the conceptual framing and parameterization of physiological variables. How the models handle the physiological complexity of phytoplankton productivity remains a continuing research effort. Maritorena and Siegel (2005) have addressed the issue of retrieval accuracy within the context of how data from different and/or sequential ocean color satellites can be used together. They use the normalized water-leaving radiances ($L_{wN}(\lambda)$) from SeaWiFS and MODIS in a semi-analytical merging model to produce global retrievals of chlorophyll $a$, dissolved plus detrital absorption coefficient, and particulate backscattering coefficient. These authors show that, compared with the individual data sources, the merged products provide enhanced global daily coverage and lower uncertainties in the retrieved variables. Ultimately, the overall accuracy of multiplatform sampling strategies will hinge on the space/time integration of diverse data sets by means of increasingly robust mathematical models. Success will also be measured by a more complete view of the space/time abundance and distribution of ocean primary producers and by increased understanding of fundamental processes governing marine ecosystems.

## Summary and Recommended Methods

Methods to improve our ability to estimate primary production are constantly being developed. Molecular methods (LaRoche et al. 1993), single-cell determinations as in the use of the flow cytometer (Li 1993), or biophysical approaches such as the fast repetition rate fluorometer have appeared since the 1990s. The use of new radioactive isotopes as tracers, the introduction of stable isotopes, and the extensive development of analytical as well as experimental models, spurred in part by remote sensing of ocean color, indicate that a careful evaluation of the method of estimating primary production is important to better answer the question posed. We have presented in this chapter the more widespread approaches to the estimate of NPP in marine ecology, but we recommend that the reader consider other methods as well.

[14]C is still the most commonly used method to estimate photosynthesis and primary production in marine pelagic systems. Long studied in detail, it is the standard against which most other methods are compared and/or are calibrated. Thus, most marine ecological projects include [14]C in one or more of their approaches (IS, SIS, P vs. E, etc.), and a careful evaluation of its performance for any particular study is of the utmost importance. The limitations of the method, if not always corrected, are usually well understood, and that broadens its usefulness for comparative studies.

### The Nature of the Biome

The diversity of phytoplankton communities in the ocean makes it difficult to recommend any single method for the measurement of primary production. As discussed above, different methods provide approximations to GPP or NPP, depending on circumstances. Thus, the various methods offer different tools to better understand the system, the cycling of C within phytoplankton, and the transfer of C among trophic levels. Furthermore, the range of response of any given method under different environmental conditions argues that the method of choice should be based on the scientific question at hand and the space/time scales under investigation.

For example, in the tropical gyres of the oceans, the system is highly heterotrophic, and on average R > P, so that the NCP is negative. These systems are dominated by small phytoplankton cells and the microbial loop. Thus, $^{14}C$ incorporation into particulate C is significantly affected by recycled intracellular $CO_2$, a large proportion of the new organic C can be exuded as dissolved organic C, and active microzooplankton grazing is occurring during the length of the incubation, changing phytoplankton biomass and possibly composition. Experimentally, in these heterotrophic areas the $^{14}C$ uptake in dark bottles can be as high as the uptake in light bottles, and it is not usual to subtract one from the other. All these characteristics make the $^{14}C$ method less than ideal in heterotrophic dominated areas of the oceans, and complementary approaches are sometimes needed to better understand the results obtained with the $^{14}C$ method (Laws et al. 1984; Grande et al. 1989). In contrast, ice-edge blooms in polar areas can be highly autotrophic, dominated by large cells, with low microzooplankton grazing and low DOC and bacterial activity. Under these conditions the $^{14}C$ method is ideal, and the estimates provide relatively accurate data for the estimation of NPP. In general this is true for most of the eutrophic areas of the world's ocean, where relatively high levels of primary productivity lead to high levels of upper trophic level biomass.

## Scale Considerations

With observations covering spatial scales from molecular to global, the consideration of scale is critical when selecting a method. As noted above, multiplatform sampling strategies are necessary in order to effectively sample the wide range of space/time variability in the oceans. High accuracy in estimating productivity in a single incubation bottle can provide valuable physiological insight for the system. However, the value of a point measurement for scaling to larger scales and longer times is dependent upon how representative the sample is within the context of greater scale. This context can be provided by various sensors on multiple platforms, and the overall accuracy of the combined data can largely be a function of the robustness of the integrative models used to merge disparate data.

Small-scale (shipboard) methods provide a level of detail that various remote sensing methods do not currently offer. For example, remote sensing (both in-water and satellite sensors) of pigment biomass currently focuses on chlorophyll $a$, whereas shipboard observations permit detailed analysis of pigment composition. Consequently, studies aimed at a greater understanding of community composition must currently rely on shipboard methods of analysis. Possible future advancement beyond this stage would require more complete models of phytoplankton growth that include community (and pigment) composition, and another generation of sensors aimed at more detailed physiological information. In short, the accurate estimation of phytoplankton production requires observations across a range of space/time scales and robust integrative phytoplankton models.

*Acknowledgments* We would like to thank C. Johnson and W. Kozlowski for technical help in manuscript preparation, W. Kozlowski, K. Sines, and many volunteers to the PAL LTER program for data collection and analysis. J. Grzymski provided helpful comments and reviewed an early version of the manuscript. This project was funded by NSF award OPP-0217282. It is Palmer LTER Contribution no. 257.

## References

Antoine, D., J. M. Andre, and A. Morel. 1996. Oceanic primary production 2. Estimation at global scale from satellite (CZCS) chlorophyll. Global Biogeochemical Cycles 10:57–69.

Balch, W., R. Evans, J. Brown, G. Feldman, C. McClain, and W. Esaias. 1992. The remote-sensing of ocean primary productivity—Use of a new data compilation to test satellite algorithms. Journal of Geophysical Research—Oceans 97:2279–2293.

Banse, K. 1993. On the dark bottle in the $^{14}$C method for measuring marine phytoplankton production. ICES Marine Science Symposia 197:132–140.

Barber, R. T., L. Borden, Z. Johnson, J. Marra, C. Knudson, and C. C. Trees. 1997. Ground truthing modeled $k_{PAR}$ and on deck primary productivity incubations with in situ observations. Ocean Optics XIII Society of Photo-Optical Instrumentation Engineers 2963:834–849.

Behrenfeld, M. J., E. Boss, D. A. Siegel, and D. M. Shea. 2005. Carbon-based ocean productivity and phytoplankton physiology from space. Global Biogeochemical Cycles 19:GB1006.

Behrenfeld, M., and P. Falkowski. 1997a. Consumers guide to phytoplankton primary productivity models. Limnology and Oceanography 42:1479–1491.

Behrenfeld, M., and P. Falkowski. 1997b. Photosynthetic rates derived from satellite-based chlorophyll concentration. Limnology and Oceanography 42:1–20.

Behrenfeld, M. J., J. T. Randerson, C. R. McClain, G. C. Feldman, S. O. Los, C. J. Tucker, P. G. Falkowski, C. B. Field, R. Frouin, W. E. Esaias, D. D. Kolber, and N. H. Pollack. 2001. Biospheric primary production during an ENSO transition. Science 291:2594–2597.

Bender, M., K. Grande, K. Johnson, J. Marra, P. J. L. Williams, J. Sieburth, M. Pilson, C. Langdon, G. Hitchcock, J. Orchardo, C. Hunt, P. Donaghay, and K. Heinemann. 1987. A comparison of four methods for determining planktonic community production. Limnology and Oceanography 32:1085–1098.

Bidigare, R. R., B. B. Prezelin, and R. C. Smith. 1992. Bio-optical models and the problems in scaling. Pages 175–212 in P. G. Falkowski and A. Woodhead (eds.), Primary Productivity and Biogeochemical Cycles in the Sea, vol. 43. Plenum Press, New York.

Calvario-Martinez, O. 1989. Microalgal photosynthesis: Aspects of overall carbon and oxygen metabolism. Ph.D. dissertation, University of Wales.

Carder, K. L., S. K. Hawes, K. A. Baker, R. C. Smith, R. G. Steward, and B. G. Mitchell. 1991. Reflectance model for quantifying chlorophyll *a* in the presence of productivity degradation products. Journal of Geophysical Research 96:20599–20611.

Carrillo, C. J., and D. M. Karl. 1999. Dissolved inorganic carbon pool dynamics in northern Gerlache Strait, Antarctica. Journal of Geophysical Research 104:15873–15884.

Chamberlin, W. S., C. R. Booth, D. A. Kiefer, J. H. Morrow, and R. C. Murphy. 1990. Evidence for a simple relationship between natural fluorescence, photosynthesis and chlorophyll in the sea. Deep-Sea Research A37:951–973.

Coté, B., and T. Platt. 1984. Utility of the light-saturation curve as an operational model for quantifying the effects of environmental conditions on phytoplankton photosynthesis. Marine Ecology Progress Series 18:57–66.

Dandonneau, Y. 1993. Measurement of in situ profiles of primary production using an automated sampling and incubation device. ICES Marine Science Symposia 197:172–180.

Dickey, T. 2003. Emerging ocean observations for interdisciplinary data assimilation systems. Journal of Marine Systems 40–41:5–48.

Doerffer, R. 1993. Estimation of primary production by observation of solar-stimulated fluorescence. ICES Marine Science Symposia 197:104–113.

Dugdale, R. C., and J. J. Goering. 1967. Uptake of new and regenerated forms of nitrogen in primary productivity. Limnology and Oceanography 12:196–206.

Duntley, S. Q., R. W. Austin, R. L. Ensminger, T. J. Petzold, and R. C. Smith. 1974. Experimental Time Varying Intensity (TVI) System, Part I July 1974, Part II October 1974. Report no. SIO Ref. 74–1. University of California, San Diego, Scripps Institution of Oceanography, Visibility Laboratory, La Jolla, CA.

Emerson, S., P. Quay, C. Stump, D. Wilbur, and R. Schudlich. 1993. Determining primary production from the mesoscale oxygen field. ICES Marine Science Symposia 197:196–206.

Eppley, R. W., and B. J. Peterson. 1979. Particulate organic matter flux and planktonic new production in the deep ocean. Nature 282:677–680.

Esaias, W. E. 1981. Remote sensing in biological oceanography. Oceanus 24:32–38.

Falkowski, P. G., and K. Kolber. 1993. Estimation of phytoplankton photosynthesis by active fluoresence. ICES Marine Science Symposia 197:92–103.

Falkowski, P. G., and T. G. Owens. 1978. Effects of light intensity on photosynthesis and dark respiration in 6 species of marine phytoplankton. Marine Biology 45:289–295.

Falkowski, P. G., and J. A. Raven (eds.). 1997. Aquatic Photosynthesis. Blackwell Science, Malden, MA.

Field, C. B., M. J. Behrenfeld, J. T. Randerson, and P. Falkowski. 1998. Primary production of the biosphere: Integrating terrestrial oceanic components. Science 281: 237–240.

Firestone, E. R., and S. B. Hooker. 1998. SeaWiFS Prelaunch Technical Report Series, Final Cumulative Index, vol. 43. NASA/GSFC, Code 970.2., Greenbelt, MD.

Fitzwater, S. E., G. A. Knauer, and J. H. Martin. 1982. Metal contamination and its effects on primary production measurements. Limnology and Oceanography 27:544–551.

Fogg, G. E. 1980. Phytoplanktonic primary production. Pages 24–45 in R. S. K. Barnes and K. H. Mann (eds.), Fundamentals of Aquatic Ecosystems. Blackwell, Oxford.

Frenette, J-J,, S. Demers, L. Legendre, and J. Dodson. 1993. Lack of agreement among models for estimating the photosynthetic parameters. Limnology and Oceanography 38:679–687.

Garver, S. A., and A. D. Siegel. 1997. Inherent optical property inversion of ocean color spectra and its biogeochemical interpretation: I. Time series from the Sargasso Sea. Journal of Geophysical Research 102:18607–18625.

Geider, R. J., and H. L. MacIntyre. 2002. Physiology and biochemistry of photosynthesis and algal carbon acquisition. Pages 44–77 in P. J. L. Williams, D. N.Thomas, and C. S. Reynolds (eds.), Phytoplankton Productivity: Carbon Assimilation in Marine and Freshwater Ecosystems. Blackwell Science, Oxford.

Gieskes, W. W. C., G. W. Kraay, and M. A. Baars. 1979. Current $^{14}$C methods for measuring primary production: Gross underestimates in oceanic waters. Netherlands Journal of Sea Research 13:58–78.

Gieskes, W. W. C., G. W. Kraay, and A. G. J. Buma. 1993. $^{14}$C labelling of algal pigments to estimate the contribution of different taxa to primary production in natural seawater samples. ICES Marine Science Symposia 197:114–120.

Goes, J. I., and N. Handa. 2002. $^{13}$C tracer-GC-MS methodology for estimating production rates of organic compounds in phytoplankton. Pages 307–316 in D. V. S. Rao (ed.), Pelagic Ecology Methodology. A. A. Balkema, Tokyo.

Gordon, H. R., O. B. Brown, R. H. Evans, J. W. Brown, R. C. Smith, K. S. Baker, and D. K. Clark. 1988. A semianalytic radiance model of ocean color. Journal of Geophysical Research 93:10909–10924.

Gordon, H. R., D. K. Clark, J. L. Mueller, and W. A. Hovis. 1980. Phytoplankton pigments from the Nimbus-7 Coastal Zone Color Scanner: Comparisons with surface measurements. Science 210:63–66.

Gordon, H. R., and A. Y. Morel. 1983. Remote Assessment of Ocean Color for Interpretation of Satellite Visible Imagery: A Review., vol. 4. Springer-Verlag, New York.

Grande, K. D., P. J. L. Williams, J. Marra, D. A. Purdie, K. Heinemann, R. W. Eppley, and M. L. Bender. 1989. Primary production in the North Pacific gyre: A comparison of rates determined by the C-14, O-2 concentration and O-18 methods. Deep-Sea Research, Part A—Oceanographic Research Papers 36:1621–1634.

Harris, G.. 1986. Phytoplankton Ecology: Structure, Function and Fluctuation. Chapman & Hill, London.

Harrison, G. 1993. Nutrient recycling in production experiments. ICES Marine Science Symposia 197:149–158.

Hartig, P.., K. Wolfstein, S. Lippemeier, and F. Colijn. 1998. Photosynthetic activity of natural microphytobenthos populations measured by fluorescence (PAM) and C-14-tracer methods: A comparison. Marine Ecology Progress Series 166:53–62.

Hewes, C. D., E. Sakshaug, F. Reid, and O. Holm-Hansen. 1990. Microbial autotrophic and heterotrophic eucaryotes in antarctic waters: Relationships between biomass and chl, adenosine triphosphate and particulate organic carbon. Marine Ecology Progress Series 63:27–35.

Hovis, W., D. Clark, F. Anderson, R. Austin, W. Wilson, E. Baker, D. Ball, H. Gordon, J. Mueller, S. El-Sayed, B. Sturm, R. Wrigley, and C. Yentsch. 1980. Nimbus-7 Coastal Zone Color Scanner: System description and initial imagery. Science 210:60–63.

Jackson, G. A. 1993. The importance of the DOC pool for primary production estimates. ICES Marine Science Symposia 197:141–148.

Jassby, A. D., and T. Platt. 1976. Mathematical formulation of the relationship between photosynthesis and light for phytoplankton. Limnology and Oceanography 21:540–547.

Jeffrey, S. W., R. F. C. Mantoura, and S. W. Wright. (eds.). 1997. Phytoplankton Pigments in Oceanography: Guidelines to Modern Methods. UNESCO, Paris.

Jerlov, N. 1951. Optical studies of ocean water. Reports of the Swedish Deep-Sea Expedition 1947–1948. vol. 3: zoology, No. 1, pp.1–59.

Kalle, K. 1938. Zum problem der meereswasserfarbe. Annalen der Hydrographie und Maritimen Meteorologie 66:1–13.

Kiefer, D. A., W. S. Chamberlin, and C. R. Booth. 1989. Natural fluorescence of chlorophyll $a$: Relationship to photosynthesis and chlorophyll concentration in the western South Pacific gyre. Limnology and Oceanography 34:868–881.

Kolber, Z., and P. G. Falkowski. 1993. Use of active fluorescence to estimate phytoplankton photosynthesis in-situ. Limnology and Oceanography 38:1646–1665.

Kolber, Z. S., O. Prasil, and P. G. Falkowski. 1998. Measurements of variable chlorophyll fluorescence using fast repetition rate techniques: Defining methodology and experimental protocols. Biochimica et Biophysica Acta—Bioenergetics 1367:88–106.

Laney, S. R. 1997. Fast repetition rate fluorometry—Exploring phytoplankton fluorescence. Sea Technology 38:99–102.

Laney, S. R. 2003. Assessing the error in photosynthetic properties determined by fast repetition rate fluorometry. Limnology and Oceanography 48:2234–2242.

LaRoche, J., R. J. Geidere, and P. G. Falkowski. 1993. Molecular biology in studies of oceanic primary production. ICES Marine Science Symposia 197:42–51.

Laws, E. A. 1991. Photosynthetic quotients, new production and net community production in the open ocean. Deep-Sea Research 38:143–167.

Laws, E. A., D. G. Redalje, L. W. Haas, P. K. Bienfang, R. W. Eppley, W. G. Harrison, D. M. Karl, and J. Marra. 1984. High phytoplankton growth and production rates in oligotrophic Hawaiian coastal waters. Limnology and Oceanography 29:1161–1169.

Le Bouteiller, A. 1993. Comparison of in-bottle measurements using $^{15}$N and $^{14}$C. ICES Marine Science Symposia 197:121–131.

Li, W. K. W. 1993. Estimation of primary production by flow cytometry. ICES Marine Science Symposia 197:79–91.

Li, W. K. W., and S. Y. Maestrini. 1993. Measurement of primary production from the molecular to the global scale. ICES Marine Science Symposia 197:1–2.

Lindeman, R. 1942. Experimental simulation of winter anaerobiosis in a senescent lake. Ecology 23:1–13.

Lohrenz, S. E. 1993. Estimation of primary production by the simulated in situ method. ICES Marine Science Symposia 197:159–171.

Lohrenz, S. E., D. A. Wiesenburg, C. R. Rein, R. A. Arnone, C. D. Taylor, G. A. Knauer, and A. H. Knap. 1992. A comparison of in situ and simulated in situ methods for estimating oceanic primary production. Journal of Plankton Research 14:201–221.

Longhurst, A. 1998. Ecological Geography of the Sea. Academic Press, San Diego

Longhurst, A., S. Sathyendranath, T. Platt, and C. Caverhill. 1995. An estimate of global primary production in the ocean from satellite radiometer data. Journal of Plankton Research 17:1245–1271.

Maritorena, S., and D. A. Siegel. 2005. Consistent merging of satellite ocean color data sets using a bio-optical model. Remote Sensing of Environment 94:429–440.

Marra, J. 1995. Primary production in the North Atlantic: Measurements, scaling, and optical determinants. Philosophical Transactions of the Royal Society of London B348: 153–160.

Marra, J. 2002. Approaches to the measurement of plankton production. Pages 78–108 in P. J. L. Williams, D. N. Thomas, and C. S. Reynolds (eds.), Phytoplankton Productivity: Carbon Assimilation in Marine and Freshwater Ecosystems, Blackwell Science, Oxford.

Martin, J. H. 1992. Iron as a limiting factor in oceanic productivity. Pages 123–137 in P. G. Falkowski and A. D. Woodhead (eds.), Primary Productivity and Biogeochemical Cycles in the Sea. Plenum Press, New York.

McClain, C., G. C. Feldman, and S. B. Hooker. 2004. An overview of the SeaWiFS project and strategies for producing a climate research quality global ocean bio-optical time series. Deep-Sea Research II 51:5–42.

McCormick, M. J., G. L. Fahnenstiel, S. E. Lohrenz, and D. G. Redalje. 1996. Calculation of cell-specific growth rates: A clarification. Limnology and Oceanography 41:182–189.

Minas, H. J., and L. A. Codispoti. 1993. Estimation of primary production by observation of changes in the mesoscale nitrate field. ICES Marine Science Symposia 197:215–235.

Monteith, J. L. 1972. Solar radiation and productivity in tropical ecosystems. Journal of Applied Ecology 9:747–766.

Morel, A. 1991. Light and marine photosynthesis: A spectral model with geochemical and climatological implications. Progress in Oceanography 26:263–306.

Morel, A., and J. F. Berthon. 1989. Surface pigments, algal biomass profiles, and potential production of the euphotic layer: Relationships reinvestigated in view of remote-sensing applications. Limnology and Oceanography 34:1545–1562.

Morel, A., and L. Prieur. 1977. Analysis of variations in ocean color. Limnology and Oceanography 22:709–722.

Morel, A., and R. C. Smith. 1974. Relation between total quanta and total energy for aquatic photosynthesis. Limnology and Oceanography 19:591–600.

Neale, P. J., and J. C. Priscu. 1998. Fluorescence quenching in phytoplankton of the McMurdo Dry Valley lakes (Antarctica): Implications for the structure and function of the photosynthetic apparatus. Pages 241–253 in J. C. Priscu (ed.), Ecosystem Dynamics in a Polar Desert, the McMurdo Dry Valleys, Antarctica. American Geophysical Union, Washington, DC.

Nelson, D. M., and W. O. Smith. 1991. Sverdrup revisited: Critical depths, maximum chlorophyll levels, and the control of Southern Ocean productivity by the irradiance-mixing regime. Limnology and Oceanography 36:1650–1661.

O'Reilly, J. E., S. Maritorena, B. G. Mitchell, D. A. Siegel, K. L. Carder, S. A. Garver, M. Kahru, and C. McClain. 1998. Ocean color chlorophyll algorithms for SeaWiFS. Journal of Geophysical Research 103:24937–24953.

O'Reilly, J. E., S. Maritorena, M. C. O'Brien, D. A. Siegel, D. Toole, D. Menzies, R. C. Smith, J. L. Mueller, B. G. Mitchell, M. Kahru, F. P. Chavez, P. Strutton, G. F. Cota, S. B. Hooker, C. R. McClain, K. L. Carder, F. Muller-Karger, L. Harding, A. Magnuson, D. Phinney, G. F. Moore, J. Aiken, K. R. Arrigo, R. Letelier, and M. Culver. 2000. SeaWiFS postlaunch calibration and validation analysis, part 3. Pages 1–49 in S. B. Hooker and E. R. Firestone (eds.), SeaWiFS Postlaunch Technical Report Series, vol 11. NASA Goddard Space Flight Center, Greenbelt, MD.

Parsons, T. R., Y. Maita, and C. M. Lalli. 1984. A Manual of Chemical and Biological Methods for Seawater Analysis. Pergamon Press, New York.

Pedlosky, J. 1987. Geophysical Fluid Dynamics. Springer-Verlag, New York.

Peterson, B. J. 1978. Radiocarbon uptake: Its relation to net particulate carbon production. Limnology and Oceanography 23:179–184.

Platt, T., K. L. Denman, and A. D. Jassby. 1975. The mathematical representation and prediction of phytoplankton productivity. Fisheries and Marine Service Technical Report #523, 110 pages, Ottowa, Canada.

Platt, T., C. Gallegos, and W. Harrison. 1980. Photoinhibition of photosynthesis in natural assemblages of marine phytoplankton. Journal of Marine Research 38:687–701.

Platt, T., W. G. Harrison, M. R. Lewis, W. K. W. Li, S. Sathyendranath, R. E. Smith, and A. F. Vezina. 1989. Biological production of the oceans: The case for consensus. Marine Ecology Progress Series 52:77–88.

Platt, T., and A. D. Jassby. 1976. The relationship between photosynthesis and light for natural assemblages of coastal marine phytoplankton. Journal of Phycology 12:421–430.

Platt, T., and S. Sathyendranath. 1988. Oceanic primary production: Estimation by remote sensing at local and regional scales. Science 241:1613–1620.

Platt, T., and S. Sathyendranath. 1991. Biological production models as elements of coupled, atmosphere-ocean models for climate research. Journal of Geophysical Research 96:2585–2592.

Platt, T., and S. Sathyendranath. 1993. Fundamental issues in measurement of primary production. ICES Marine Science Symposia 197:3–8.

Platt, T., S. Sathyendranath, and P. Ravindran. 1990. Primary production by phytoplankton: Analytic solutions for daily rates per unit area of water surface. Proceedings of the Royal Society of London B241:101–111.

Platt, T.S., P. Jauhari, and S. Sathyendranath. 1992. The importance and measurement of new production. In, Primary Productivity and Biogeochemical Cycles in the Sea. P. Falkowski and A. Woodhead. New York, Plenum Press: 273–284.

Potter, C. S., J. T. Randerson, C. B. Field, P. A. Matson, P. M. Vitousek, H. A. Mooney, and S. A. Klooster. 1993. Terrestrial ecosystem production: A process model based on global satellite and surface data. Global Biogeochemical Cycles 7:811–841.

Rabinowich, E., and Govindjee. 1969. Photosynthesis. Wiley, New York.

Rai, H. 2002. Radioactive isotope ($^{14}C$ incorporation) technique for measuring rate of primary production (photosynthesis) and photosynthetically fixed dissolved organic carbon (PDOC) of phytoplankton. Pages 155–161 in D. V. S. Rao (ed.), Pelagic Ecology Methodology. A. A. Balkema, Tokyo.

Robertson, J. E., and A. J. Watson. 1993. Estimation of primary production by observation of changes in the mesoscale carbon dioxide field. ICES Marine Science Symposia 197:207–214.

Rodhe, W. 1966. Standard correlations between pelagic photosynthesis and light. Pages 249–381 in C.R. Goldman (ed.), Primary Productivity in Aquatic Environments. University of California Press, Berkeley.

Ryther, J. H. 1956. The measurement of primary production. Limnology and Oceanography 1:72–84.

Ryther, J., and C. Yentsch. 1957. The estimation of phytoplankton production in the ocean from chlorophyll and light data. Limnology and Oceanography 2:281–286.

Sathyendranath, S., A. Longhurst, C. M. Caverhill, and T. Platt. 1995. Regionally and seasonally differentiated primary production in the North Atlantic. Deep-Sea Research Part I—Oceanographic Research Papers 42:1773–1802.

Scott, B. D. 2002. Carbon-14 uptake incubation methods. Pages 145–153 in D. V. S. Rao (ed.), Pelagic Ecology Methodology. A. A. Balkema, Tokyo.

Siegel, D. A., T. D. Dickey, L. Washburn, M. K. Hamilton, and B. G. Mitchell. 1989. Optical determination of particulate abundance and production variations in the oligotrophic ocean. Deep-Sea Research 36:211–222.

Smith, R. C., and K. S. Baker. 1982. Oceanic chlorophyll concentrations as determined by satellite (Nimbus-7 coastal zone color scanner). Marine Ecology Progress Series 66: 269–279.

Smith, R. C., O. B. Brown, F. E. Hoge, K. S. Baker, R. H. Evans, R. N. Swift, and W. E. Esaias. 1987. Multiplatform sampling (ship, aircraft, and satellite) of a Gulf Stream warm core ring. Applied Optics 26:2068–2081.

Steele, J. H. 1978. Spatial Patterns in Plankton Communities. Plenum Press, New York.

Steele, J. H. 1985. A comparison of terrestrial and marine ecological systems. Nature 313:355–358.

Steele, J. H. 1991. Marine functional diversity: Ocean and land ecosystems may have different time scales for their responses to change. BioScience 41:470–474.

Steeman-Nielsen, E. 1952. The use of radio-active carbon (C14) for measuring organic production in the sea. Journal du conseil international pour l'exploration de la mar. 18:117–140.

Talling, J. 1957. The phytoplankton population as a compound photosynthetic system. The New Phytologist 56:133–149.

Tilzer, M. M., C. Hase, and I. Conrad. 1993. Estimation of in situ primary production from parameters of the photosynthesis-light curve obtained in laboratory incubators. ICES Marine Science Symposia 197:181–195.

Topliss, B. J., and T. Platt. 1986. Passive fluorescence and photosynthesis in the ocean: Implications for remote sensing. Deep-Sea Research 33:849–864.

Tyler, J. E., and R. C. Smith. 1970. Measurements of Spectral Irradiance Underwater. Gordon & Breach, New York.

Vollenweider, R. A. 1965. Calculation models of photosynthesis-depth curves and some implications regarding day rate estimates in primary production measurement. Pages 425–457 in C. R. Goldman (ed.), Primary Productivity in Aquatic Environments. University of California Press, Berkeley.

Waterhouse, T. Y., and N. A. Welschmeyer. 1995. Taxon-specific analysis of microzooplankton grazing rates and phytoplankton growth rates. Limnology and Oceanography 40:827–834.

Webb, W. L., M. Newton, and D. Starr. 1974. Carbon dioxide exchange of *Alnus rubra*: A mathematical model. Oecologia 17:281–291.

Williams, P. J. L. 1993a. Chemical and tracer methods of measuring plankton production. ICES Marine Science Symposia 197:20–36.

Williams, P. J. L. 1993b. On the definition of plankton production terms. ICES Marine Science Symposia 197:9–19.

Yentsch, C. 1960. The influence of phytoplankton pigments on the colour of the sea water. Deep-Sea Research 7:1–9.

# 10

# Measuring Freshwater Primary Production and Respiration

Robert O. Hall, Jr.
Serge Thomas
Evelyn E. Gaiser

M easuring primary production and respiration has a long history in limnology because the transformation of carbon (C) between inorganic and organic forms is a central ecosystem process. As a result, there are many sources for procedures on how to measure primary production, ranging from manuals (Vollenweider 1969) to book chapters (Hall and Moll 1975) and many primary methods papers (Steeman Nielsen 1951; Odum 1956; Fee 1969, 1973a, 1973b; Marzolf et al. 1994; Dodds and Brock 1998). Despite a long history of metabolic assessment in limnology, it has hardly gone out of date as an important ecosystem process to measure. Recent studies have addressed the degree of heterotrophy in lakes (Bachmann et al. 2000; Cole et al. 2000, 2002; Hanson et al. 2003), related metabolism to nutrient use (Hall and Tank 2003; McCormick et al. 2001), compared production among streams in different biomes (Mulholland et al. 2001), examined ecosystem stability (Uehlinger 2000), and examined impacts of land use on streams (Young and Huryn 1999). A summary of representative rates of primary production in a variety of freshwater ecosystems is provided as a guide to the values investigators might anticipate in the systems under study (table 10.1).

Our objective in this chapter is to describe methods of measuring metabolism (i.e., gross primary production and respiration) in freshwater ecosystems and to point out recent methodological improvements. We address advantages and disadvantages of each technique, then let the reader decide the most suitable option, since the best method is one that is modified to fit the ecosystem being studied and the particular questions being addressed.

We describe a variety of methods to quantify production in microbially based food webs that can be found in lakes, wetlands, and rivers, focusing more on benthic communities than pelagic, because approaches for the latter are covered in large

Table 10.1. Representative rates of primary production from various freshwater ecosystems

| Ecosystem | Primary Production g C m$^{-2}$ d$^{-1}$ Mean | Range | References |
|---|---|---|---|
| Forested small streams | 0.2 | 0.05–0.7 | Mulholland et al. 2001; Hall and Tank 2003 |
| | 0.1 | 0–0.2 | Webster and Meyer 1997 |
| Open-canopy streams | 0.5 | 0.08–1.3 | Mulholland et al. 2001; Hall and Tank 2003 |
| | 0.5 | 0.15–1.1 | Webster and Meyer 1997 |
| Spring streams, desert streams | 4.1 | 1.7–6.1 | Mulholland et al. 2001; Hall and Tank 2003 |
| | 3.5 | 1.5–7.4 | Webster and Meyer 1997 |
| Ultraoligotrophic lakes | | <0.05 | Wetzel 2001 |
| Oligotrophic lakes | | 0.05–0.1 | Wetzel 2001 |
| Mesotrophic lakes | | 0.25–1 | Wetzel 2001 |
| Eutrophic lakes | | >1 | Wetzel 2001 |
| Everglades periphyton | | 0.8–7.4 | E. E. Gaiser unpublished data |
| Tropical wetland periphyton | | 0.8–3.04 | Rejmankova and Komarkova 2000 |
| Lake Okeechobee epiphyton | | 4–12.5 | Havens et al. 1999 |
| Silver Springs, FL, epiphytes | 6.2 | | Odum (1957) |
| Michigan lake epiphytes | 2 | | Wetzel (1972) |
| Tundra pond epipelon | 0.02 | 0.01–0.03 | Stanley (1976) |

*Notes*: We refer to these rates as simply "primary production" because they were measured with both oxygen change and $^{14}C$ methods. Oxygen was converted to carbon for several studies assuming a photosynthesis coefficient of 1. These data are far from exhaustive, but they give the range of rates an investigator might encounter.

part in chapter 9. We concentrate on methods that are new or have been substantially improved since the 1980s (e.g., free-water methods, micro-oxygen methods). We do not consider production of macrophytes in wetlands or littoral sections of lakes, because methods for estimating that production should more closely follow those for grasslands (chap. 3) and the marine intertidal (chap. 7) in this volume.

## General Considerations and Concepts

A key difference between aquatic and terrestrial plant-based ecosystems is that it is operationally difficult to measure net primary production (NPP), because most of the autotrophic production in freshwater is by unicellular algae (or small colonies of cells) that have high biomass turnover due to grazing, release of extracellular dissolved organic matter (DOM), sinking in pelagic algae, and, in the case of benthic algae, sloughing. Production can be high while biomass remains low, and constant such that the change in biomass divided by change in time (the approach for terrestrial vascular plants) will appear to be zero. Therefore, aquatic ecologists typically measure photosynthetic rates without converting these values to NPP, because doing so requires estimates of autotrophic respiration. However, almost all respiration measurements include a substantial amount of heterotrophic activity, and this activity is difficult to separate from that of the autotrophs (Robinson 1983). An additional concern in defining NPP is that much fixed C may not be converted to plant biomass or respired, but rather excreted as DOM or exopolymers that may fuel the closely associated heterotrophic bacteria, which also contribute to biomass accrual (e.g., McFeters et al. 1978; Kühl et al. 1996; Glud et al. 1999, 2002). The tight coupling between photosynthetic and heterotrophic metabolism inherent in microbial communities makes NPP very difficult to conceptualize and measure (Flynn 1988; Jensen and Revsbech 1989). Hence, production measures usually refer to gross primary production (GPP), which is the total amount of C fixed by photosynthesis per unit time. The most common approach to measure metabolism is via the change in $O_2$ concentrations in water per unit time in light and dark situations. In the dark, one measures community respiration (CR), that is, respiration of autotrophs and heterotrophs. In the light, one measures net ecosystem production (NEP), which reflects the combined processes of GPP and CR. NEP nearly always includes a substantial heterotrophic component (del Giorgio et al. 1997). GPP is then calculated as the difference between NEP and CR, where GPP is considered to be a positive flux of organic C, CR is a negative flux, and NEP can be positive or negative, depending on the degree of autotrophy in the ecosystem.

A central assumption of this approach is that CR in light is the same as in the dark, which is likely not true; light respiration is probably higher than dark respiration. Light stimulation of coupled heterotrophic activity, increased cellular respiration during light, and photorespiration may all contribute to increased daytime CR. For example, periphytic respiration is especially light-driven: respiration by both algae and bacteria in the dark can be half the levels that occur in the light (e.g., Lindeboom et al. 1985), allowing fairly accurate assessments of NEP and dark respiration but an underestimate of GPP. Sunlight may increase DOM lability, which may also fuel higher heterotrophic respiration (Wetzel 2001). Currently, approaches are being developed to measure light CR using oxygen $O_2$ isotopes (Roberts 2004).

The biogeochemical transformations involved in the steps of algal photosynthesis allow rates to be assessed directly via four types of measurements which are incorporated in the methods presented in this chapter. These include, from the photosynthetic light reaction, (1) $O_2$ released during the oxidation of water ($H_2O$) at the PSII ($P_{680}$) reaction center in the algal chloroplast; (2) ATP formation via electron transport at the PSI ($P_{700}$) reaction center; (3) the transmission rate of electrons between PSII and PSI; and (4) via the dark reaction, the uptake of $CO_2$ and reduction to photosynthates (carbohydrate). We discuss application of existing methods to both planktonic and benthic algal communities in lakes, streams, and wetlands at a variety of scales, beginning at the largest (the whole ecosystem) and ending at the smallest (microscale $O_2$ distribution within periphyton communities), and provide precautions and suggest modifications for particular situations which we have encountered in our own work and in the literature.

## Recommended Methods for Freshwater Ecosystems

### Whole-Ecosystem Methods

#### General Approach

In some cases the ecosystem may be treated as a chamber where $O_2$ flux during the day (GPP and CR) and at night (CR only) is measured while correcting for the air-$H_2O$ flux of $O_2$ (the free-water method). It is a method originally developed by Odum (1956), but has been updated for rivers (Marzolf et al. 1994; Young and Huryn 1998). Its advantages are that it provides an ecosystem-scale estimate of metabolism that includes benthic and pelagic zones (lakes) or surface and hyporheic zones (streams; e.g., Fellows et al. 2001), and that it measures integrated daily production, so there is no need for modeling and scaling chamber-estimated production values measured at various times of the day. Disadvantages are that it has large error in oligotrophic ecosystems where both diel shifts in $O_2$ are small and concentration is close to saturation, and that it will not work in streams that have high rates of reaeration (McCutchan et al. 1998). In large rivers and lakes, air-$H_2O$ exchange of $O_2$ becomes difficult to measure empirically, and one must use published approximations of air-$H_2O$ exchange.

In rivers, oxygen budgets are created using the two-station free-water method (Odum 1956). Oxygen concentrations are measured at an upstream station and a downstream station at 5–20 min intervals throughout the day and night. These stations are typically placed such that the travel time of water between stations is about 20–30 min. Flux of $O_2$ across the surface of the water is estimated, and metabolism is calculated by difference. Net ecosystem production is measured during the day and CR during the night. Instantaneous metabolism is calculated as

$$M = \left( \frac{C_t - C_0}{\Delta t} - K_{o2}D \right) z , \tag{1}$$

where $M$ is metabolism (g $O_2$ m$^{-2}$ min$^{-1}$), $C$ is concentration of $O_2$ at time $t$ and time 0 (g $O_2$ m$^{-3}$); these correspond to the upstream and downstream stations with a travel time $\Delta t$ (min). $K_{O2}$ is the reaeration coefficient of $O_2$ (min$^{-1}$) that has been corrected to the ambient temperature (T), $D$ (g $O_2$ m$^{-3}$) is the saturation deficit of $O_2$—that is, ($C_{sat} - C_{ave}$), where $C_{sat}$ is the saturation concentration at that temperature and atmospheric pressure, and $C_{ave}$ is the average of the upstream and downstream $O_2$ concentrations, and $z$ is mean stream depth (m).

Instantaneous metabolism during the night is scaled to 24 hr to estimate CR. Gross primary productivity is calculated as the area under the curve of the diel $O_2$ increase (fig 10.1). Using this approach, community respiration is assumed to be constant and similar to nighttime values. Oxygen and temperature are measured with recording $O_2$ electrodes every 5–10 min, using, for example, multiprobes from YSI or

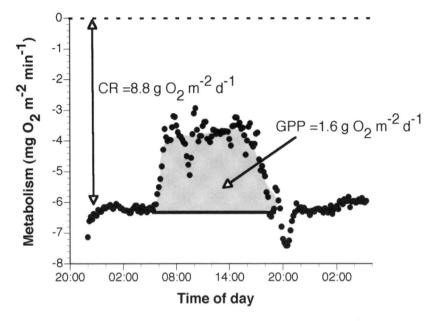

Figure 10.1. Example data for two-station, free-water method of measuring metabolism in a stream. This stream is highly heterotrophic with high rates of CR relative to GPP. Data are from Two Ocean Lake outlet, Grand Teton National Park (Hall and Tank 2003).

Hydrolab which have self-contained recorders, stirrers, and power. It is also possible to record $O_2$ concentrations manually, or, if higher accuracy and precision are required, by Winkler titration, but labor increases dramatically with this approach and is probably not worth the cost, given other sources of error with this method.

An assumption of the free-water methods is that groundwater inputs to the reach are minimal, since high groundwater fluxes with low $O_2$ can cause overestimation of CR (i.e., more negative) and underestimate GPP. If groundwater flux and $O_2$ concentration in groundwater are known, then it is possible to correct the metabolism estimate using the following equation (Hall and Tank 2005):

$$M = z\left(\frac{C_t - C_0}{\Delta t} - K_{O2}D\right) - (C_g - C_t)\frac{Q_g}{A}, \tag{2}$$

where $C_g$ is concentration of $O_2$ in groundwater (g m$^{-3}$), $Q_g$ is groundwater discharge into the study reach (m$^3$ min$^{-1}$), and $A$ (m$^2$) is the area of the study reach. Groundwater discharge is easy to estimate on the basis of the degree of dilution of the conservative tracer used for the reaeration estimation (see below). Oxygen concentration of groundwater can be measured on samples drawn from wells or, as a worst-case scenario, anoxic groundwater could be assumed. It is important that the water in fact be new to the stream and not shallow hyporheic water which may be rapidly exchanging with water in the channel.

In lakes, a single-station approach is used; this method is the same as the two-station approach above, except that the changes in $O_2$ concentrations are recorded from one station and $\Delta t$ is the time between sequential measurements at that station. Depth is calculated based on mixing; if the lake is stratified, then $z$ = epilimnion depth, and if the lake is mixed to the bottom, then $z$ = mean depth of the lake. The single-station method can also be used in rivers under the assumption that there is not much spatial variation upstream, with respect to both metabolism and reaeration.

Although both single-station and two-station methods have been used to measure metabolism in rivers, there has been no comparison of the methods. Young and Huryn (1999) compared the two methods, but they used a different method to estimate reaeration which may have driven variation between the two approaches. In theory, a two-station method will be better because it measures metabolism for a defined reach of stream and not an undefined reach upstream of the $O_2$ electrode. Chapra and DiToro (1991) suggested that the distance upstream measured by the single-station method is $<3V/K_{O2}$, where $V$ is stream velocity. This equation suggests that reaches with longer travel time and high reaeration will have similar results for the two-station and one-station approaches. A comparison of one- and two-station approaches was conducted using data from Hall and Tank (2003; unpubl. data) for 12 streams for which continuous upstream and downstream data were available. Overall, the results agreed closely; average CR of one-station was 91% of two-station, but there was high variability, with a range of 59–122%. Average GPP of one-station was 108% of two-station, with a range of 72–154%. What controls this variation? Multiplying $K_{O2}$ by travel time ($\Delta t$) yields a unitless coefficient that describes the amount of exchange due to reaeration along the reach. Low reaeration and low travel time mean little exchange with the atmosphere, so that the one-station method will effectively integrate a long distance upstream. High reaeration and high travel time lead to fast exchange with the atmosphere, and the station will be measuring a short distance above the reach. Thus, as $K_{O2}*\Delta t$ gets large, the one-station results should converge on the two-station, as noted for both GPP and CR (fig 10. 2).

## Measuring Reaeration

Two improvements in the free-water method have been the development of easy methods to measure river reaeration values (Wanninkhof et al. 1990; Marzolf et al. 1994) and better predictive equations for lakes (Cole and Caraco 1998). Metabolism estimates are greatly affected by reaeration (McCutchan et al. 1998), and for decades, published equations for estimating metabolism rates were used with no way of knowing if the estimated reaeration was accurate (Hall and Moll 1975). For small streams with high reaeration, published equations can deviate strongly from measured rates (Mulholland et al. 2001). Therefore, free-water metabolism estimates in rivers should be corrected empirically for reaeration. In some cases it is possible to measure reaeration a few times at different river stages, which can serve many metabolism estimates for a particular site (e.g., Uehlinger and Naegli 1998). Two commonly used tracer gases are propane and sulfur hexafluoride ($SF_6$). Either can produce good estimates of reaeration (Wanninkhof et al. 1990; Marzolf et al. 1994). Propane is easier to acquire (tanks from gas stations work fine), and it is easy to

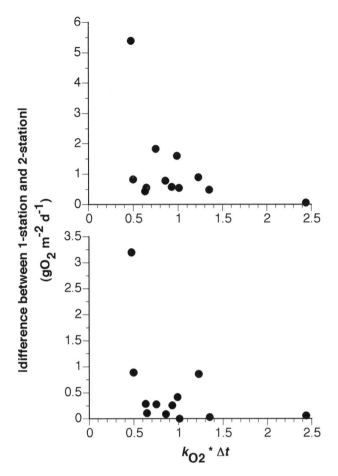

Figure 10.2. Relation between reaeration * reach travel time and absolute value of the difference between 1-station and 2-station metabolism. As $k_{O2}*\Delta t$ increases, the difference between conducting a 1-station and a 2-station metabolism estimate decreases (based on data from Hall and Tank 2003).

analyze (flame ionization detector on a gas chromatograph), but may not be effective in large rivers. $SF_6$ is more expensive initially, but one tank can last for many measurements. It also requires an electron capture detector to analyze on a gas chromatograph. It can be detected to picomolar quantities; thus, little gas needs to be added to a stream, and reaeration on large streams ($>> 1$ $m^3$ $s^{-1}$) can be estimated. Tracer gas is bubbled into a stream of known temperature along with a conservative tracer of water (e.g., NaCl or NaBr) that corrects for dilution of groundwater or variations due to mixing. Eight to 50 water samples are collected along the study reach from which gas is extracted into a headspace by shaking for 10 min, and the remaining water is analyzed for the conservative tracer. NaCl is easy to measure with a conductivity meter if added concentrations are $> 5$ mg $L^{-1}$ in a small stream. NaBr is better for larger rivers because it can be precisely measured in a range of

20–100 μg Br L$^{-1}$, using ion chromatography. Reaeration rate for the tracer gas ($K_{tracer}$) is estimated as the per-unit time decline of gas concentration relative to the salt tracer, using the following equation:

$$\ln C_t = \ln C_0 - K_{tracer}\,\Delta\,t, \tag{3}$$

where $C_t$ and $C_0$ are dilution-corrected tracer gas concentrations at times $t$ and 0. In practice, gas concentrations are measured at known distances downstream, but these can be converted to time-based average stream velocity measured using a salt tracer addition. Gas tracer reaeration rate ($K_{tracer}$) is converted to $O_2$ reaeration rate ($K_{O2}$) on the basis of the ratio of their Schmidt numbers, which is near $1.4 \times K_{tracer}$ for both $SF_6$ and propane (Wanninkhof et al. 1990). Oxygen reaeration rates vary as a function of temperature, so that $K_{O2}$ at temperature $T$ ($K_{O2\,T}$) is normalized to that at 20 °C, using the following equation:

$$K_{O220} = \frac{K_{O2T}}{1.024^{(T-20)}}. \tag{4}$$

Gas exchange rates are estimated similarly for lakes, though there are many fewer measurements than for streams. A tracer gas (e.g., $SF_6$) is added to the lake and its decline is measured through time (e.g., Cole and Caraco 1998). Gas exchange in lakes is largely a function of wind speed above a certain threshold (e.g., 3.7 m s$^{-1}$; Crusius and Wanninkhof 2003), and thus empirical measurements of reaeration often are not necessary if wind speed is known. Cole and Caraco (1998) synthesized gas exchange rates from several studies to show that those for lakes are a power function of wind speed:

$$k_{600} = 2.07 + 0.215U_{10}^{1.7}, \tag{5}$$

where $k_{600}$ is the piston velocity of gas exchange (cm h$^{-1}$) for a gas with a Schmidt number of 600, and $U_{10}$ is average wind speed (m s$^{-1}$) at 10 m above the lake surface. Data from subsequent studies (Frost and Upstill-Goddard 2002; Crusius and Wanninkhof 2003) fit this model reasonably well, suggesting that this power function is a good approximation of actual exchange rates. Given a Schmidt number for $O_2$ of 530 at 20°C, $k_{O2\,20}$ will be $1.06*k_{600}$. The piston velocity $k_{O2\,20}$ can be converted to a per unit time rate (i.e., $K_{O2\,20}$) by dividing it by the depth of the mixed layer of the lake.

## Error Estimation

There is high potential for error with this method because the final metabolism value depends on several independent measurements, including reaeration and $O_2$ concentrations. A thorough error analysis was carried out by McCutchan et al. (1998) for stream studies. They found that at $K_{O2}$ >100 day$^{-1}$, error for metabolism estimates is very high (fig. 10.3). Additionally, accurate calibration of recording electrodes is very important. McCutchan et al. (1998) assumed that electrodes could be calibrated within 0.02–0.03 mg $O_2$ L$^{-1}$, which is at the limits of tolerance for these

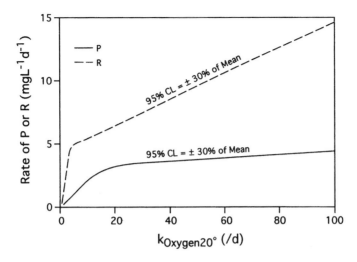

Figure 10.3. Relative error in respiration estimates increases as a function of the reaeration rate of $O_2$. Reproduced from McCutchan et al. (1998) with permission from the Journal of the North American Benthological Society.

instruments (R. Hall, personal observation). However, calibration errors can lead to large error in CR estimates, especially if CR is low and concentrations are near saturation. Because GPP is estimated as a change in $O_2$ concentration, it is less susceptible to calibration errors. Therefore the ideal conditions to minimize error with this method are low (but measurable) $K_{O2}$ and high metabolism. As reaeration rates increase and metabolism decreases, error will increase; thus, this method may not be useful for oligotrophic mountain streams with $K_{O2}$ greater than 100 day$^{-1}$.

### Chamber Methods

### $O_2$ Change with Pelagic Chambers

An alternative to the free-water approach that avoids the problem of estimating oxygen exchange across the air-$H_2O$ interface is to contain $H_2O$ in an enclosure and measure $O_2$ change (or $CO_2$ uptake) through time. The most common approach is to incubate $H_2O$ in light and dark biological $O_2$ demand (BOD) bottles, a technique first described by Gaarder and Grann (1927) for planktonic communities and used extensively since then (Wetzel and Likens 2000). Paired light and dark BOD bottles (generally 300 ml) are filled with $H_2O$ sampled in situ, and the initial dissolved $O_2$ concentration is measured. Optionally, the water is filtered through a mesh of determined pore size that excludes larger zooplankton, thus narrowing the lifeforms in the chamber to microphytes, bacteria, and viruses. The bottle is then capped and incubated for time $t$ (generally 2–4 hr), which is set to be (1) long enough that dissolved $O_2$ concentration changes measurably, which is linked to the accuracy of the $O_2$ measurement method; (2) short enough to avoid formation of $O_2$ bubbles on the BOD wall due to supersaturation; and (3) short enough to limit any excess of bacterial activity related to containment. Bottles are incubated at the depth at which they are collected. Calculations are the following:

$$NEP = \Delta O_2 \text{ concentration in light bottle}/t \qquad (7)$$

$$CR = \Delta O_2 \text{ concentration in dark bottle}/t \qquad (8)$$

$$GPP = NEP - CR \qquad (9)$$

Commonly, $O_2$ concentration is measured using Winkler titration with an optimal precision of $\pm 0.05$ mg $L^{-1}$ (Vollenweider 1969), though higher precision (up to 0.002 mg $L^{-1}$), often called for in oligotrophic environments, is possible using automatic titrators with a potentiometric end point (Carignan et al. 1998). Dissolved $O_2$ probes can also be used; they have a similar or slightly better precision than the standard Winkler titration.

## $O_2$ Change with Benthic Chambers

Benthic chamber approaches are old, but still very common, approaches to estimating primary production (Odum 1957; McIntire et al. 1964). Theoretically, using benthic chambers to estimate production of attached algae yields results that are similarly simple to pelagic estimates, but in reality it is more complicated because of necessary disruptions of attached communities from their substrate and complications involving the diffusive boundary layer (DBL) between the periphyton and the $H_2O$. In the crudest approach, periphyton is removed from the substrate and transferred into paired light and dark chambers filled with either overlying $H_2O$ (which may be filtered to exclude phytoplanktonic production; e.g., Brandini and da Silva 2001) or artificial $H_2O$ mimicking the natural chemical matrix of the real $H_2O$ (e.g. Thomas et al. 2006). Changes in $O_2$ due to activity in the incubation $H_2O$ are accounted for in an additional set of light/dark bottles without periphyton (= control).

Perhaps the major assumption in using chambers to measure periphyton metabolism is that the interstitial $H_2O$ of the periphyton and the overlying $H_2O$ where $O_2$ is being measured are at equilibrium (Carlton and Wetzel 1987). Periphyton communities can be thick and layered, and therefore metabolic processes contained within the periphyton matrix would not be measured in overlying $H_2O$. Additionally, biofilms can have a thick diffusive DBL, which lowers the diffusion of respiratory gas exchanges between the periphyton and the enclosed $H_2O$. Enclosing periphyton changes the hydrodynamics at the DBL. Stagnating conditions in chambers will increase DBL thickness such that $O_2$ measurements in overlying $H_2O$ will underestimate photosynthesis. Moreover, when working with productive mats under high light intensities, $O_2$ can be supersaturated in the chamber, often forming bubbles at the surface of the periphyton, again resulting in an underestimation of photosynthesis because $O_2$ is not fully dissolved in the $H_2O$. Additionally, the change in $O_2$:$CO_2$ ratio enhances photorespiration, and high $O_2$ concentrations can deactivate PSII (Long et al. 1994; Dodds et al. 1999) and reduce photosynthesis.

Problems with gas exchange between the periphyton and overlying $H_2O$ may be minimized by agitation, though this process may disrupt the periphyton matrix, which is often highly structured and, thus, functionally consequential. While stagnation may limit coupling between the photosynthetically active and anoxic layers, mixing in excess of natural turbulence may introduce an unnatural association between periphyton layers. To compromise, for incubations of periphyton from lentic environments, where the DBL is thick and subject to supersaturation, mixing can be delayed before measuring dissolved $O_2$ at the end of the incubation. Studies have shown that very little agitation is necessary for complete homogenization of overlying $H_2O$ (e.g., Bott et al. 1997). For assessments in lotic environments, agitation is mechanized to mimic the natural stream hydrodynamics by the use of stirrers or an $H_2O$ pump with a small dead volume (e.g., Dodds and Brock 1998). Velocity adjustments are often done empirically, although modeling can help (e.g., Broström and Nilsson 1999).

Finally, sampling and transferring the periphyton community into the incubation chamber alters the structure, thickness, and association with and orientation to the attachment substratum. The extent of the structural disruption is somewhat linked to the cohesiveness of the periphyton. A cohesive periphyton mat may retain its overall structure after sampling, while a loosely structured periphyton aggregation or biofilm may not. Loss of organization in the periphyton matrix, if any, potentially changes gradients ($O_2$, $CO_2$, nutrients) and exposes normally buried algae to direct light, potentially altering photosynthesis (e.g., Jonsson 1991). Sampling also may expose edges of thicker benthic mats, leading to a reduction in $O_2$ in initial measurements in the bottle (Bott et al. 1997).

Chamber design is an important consideration for measuring benthic productivity. Creating a unique enclosure for a given benthic community is encouraged if a chamber design is warranted, and there are many different chamber designs. Typically bell jars or semi-enclosed dome boxes are used that tightly isolate a periphyton community in situ (cf. Glud and Blackburn 2002; fig. 10.4 in this volume), thus minimally disturbing the mat. Alternatively, the biofilm and attachment substrate are removed and placed in sealed tubes or box-shaped chambers (e.g., Tank and Musson 1993; Dodds and Brock 1998). Other chambers have been designed to measure metabolism of subsurface sediment (Uzarski et al. 2001). Hydrodynamics can be mimicked in a homemade chamber, and enclosure size can be scaled to the productivity of the community to avoid "edge effects" and supersaturation. Another consideration when making incubation chambers is the effect of the enclosing material on incident irradiance (e.g., Hall and Moll 1975). Optical interference is minimized by quartz, followed by clear Plexiglas and acrylic. These chambers can greatly improve the standard BOD incubation method, but some enclosure artifacts will invariably remain.

## [14]C Methods

Isotopically labeled $CO_2$ can be used to measure inorganic C uptake by autotrophs in aquatic ecosystems. The method, introduced by Steemann Nielsen (1951), consists of the introduction of a known amount of labeled [14]C-$CO_2$ into light and dark bottles, filled with $H_2O$ containing phytoplankton, of known dissolved inorganic C (DIC) concentration. Linked to increasingly restrictive use of radioactive isotopes in natural environments, or to allowing a simultaneous uptake estimate of DIC and nitrogen (N), stable isotopes [13]C-$CO_2$ and [15]N, respectively, are also employed (Slawyk et al. 1977). The [14]C and [13]C methods agree (e.g., Sakamoto et al. 1984),

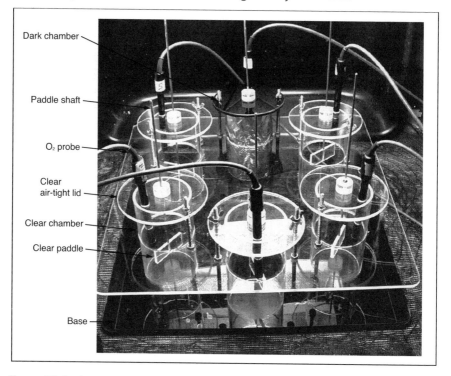

Figure 10.4. A prototype set of six incubation chambers designed for measuring metabolism of benthic periphyton mats. Each chamber is made of a clear Plexiglas tube (15.24 cm high by ø 8.255 cm; 815 cm³) and is covered with a clear, airtight lid. A chamber can be darkened with aluminum foil or black electrical tape, and capped with a dark lid. The chambers are equipped with stirring paddles, which can be rotated by hand at steady time intervals or connected to a submersible rotary motor (not shown in the picture). The bottom of the chamber can be sealed to the base containing the biofilm if incubations are performed in the laboratory or when the biofilm is too thin or on an unlevel/tough substratum that does not allow a watertight closure of the chamber. When the biofilm is thick enough, incubations are performed in situ: the base is removed and the chambers are pushed down to the sediment. In this design, the $O_2$ probes are connected to a continuously logging meter.

and the same formula for computing net transport rate of labeled C is employed for both (cf. Legendre and Gosselin 1996). The difference lies in the equipment used to measure radioactive vs. nonradioactive isotopes and also in the fact that, to our knowledge, no results involving $^{13}C$ incubations with benthic algal populations have been published. Thus, we address only the $^{14}C$ method below.

The labeled C removed by the phytoplankton is proportional, after correction of isotope uptake discrimination, to the photosynthetically incorporated DIC. Depending on the amount of light received during the incubation, which is a function of the duration of the incubation and the light intensity, the labeled C is either still within the phytoplankton biomass or is already partially excreted into the $H_2O$ column as dissolved organic C (DOC). The DIC uptake in the dark bottle (= dark uptake) can be subtracted from that in the light bottle (= light uptake) or omitted (dark uptake being generally very low). The specifics of this method are described in detail for lakes by Wetzel and Likens (2000) and for marine systems in chapter 9 of this volume.

The primary advantage of this method is that it accurately measures production, especially in unproductive systems where changes in $O_2$ concentration are too small to measure. Its main drawback is that it is difficult to determine the degree to which it measures GPP or NPP (Peterson 1980). The method optimally measures GPP, provided DOC has not already been excreted and the incubation time is very short (Dring and Jewson 1982). More likely, the productivity estimate lies somewhere between GPP and NPP (e.g., Vollenweider 1974; Peterson 1980), which complicates comparison among ecosystems because, given a constant incubation time, variation in algal turnover rates will affect the degree to which GPP or NPP is measured.

The [14]C uptake method has also been adapted for periphyton, where it has an advantage over the $O_2$ method by assessing both anoxygenic photosynthesis and chemolithotrophy (e.g., Epping and Jørgensen 1996). It is usually performed in the laboratory (e.g., Hill et al. 1995), in tight enclosures with no headspace, to avoid radioactive contamination, which causes the same enclosure problems described above for the $O_2$ exchange method.

One major problem of the [14]C method is that the labeled DIC needs to be rapidly and equally available to the whole photosynthetic periphytic layer (Jonsson 1991) before the light incubation is started. One solution is to mechanically suspend the photosynthetically active layer in the incubation $H_2O$ containing the labeled C (slurry technique; Grøntvoed 1960), but this method disrupts the natural organization of the periphyton, leading to high light exposure of algae normally found deep in the biofilm and interruption of natural fluxes occurring between the photosynthetically active and the heterotrophic layers (Revsbech et al. 1981). Alternatively, interstitial $H_2O$ can be replaced by overlying $H_2O$ containing the homogeneously distributed [14]C, using a percolation tube (the P-tube, Jonsson 1991). This method is, however, limited to epipelon. Other studies have effectively equilibrated the [12]C:[14]C ratio between the periphyton and overlying $H_2O$ by simply allowing the [14]C to diffuse into the photosynthetically active layer in the dark for a couple of hours (Revsbech et al. 1981). If equilibration is not complete, the [12]C:[14]C ratio may not vary much in the overlying $H_2O$, but interstitial $H_2O$ may be altered by labeled C fixation, production of unlabeled DIC, and pH alterations under illumination modifying the carbonate equilibrium (Revsbech et al. 1981). In these conditions, [14]C fixation must be measured by the [12]C:[14]C ratio change in the pore $H_2O$ rather than in the overlying water (Revsbech et al. 1981).

## Scaling Up from Chambers

Bottle methods give only instantaneous estimates of photosynthesis for a particular point in a lake. In order to estimate daily production for the entire pelagic zone, it is necessary to scale through space and time. For example, to convert volumetric production rates to an area-specific rate, production must be measured at various depths in the $H_2O$ column and integrated across these depths, given the area of the lake at a given depth. Because incubations are much shorter than one day (e.g.,

4 hr), it is necessary to scale to an entire day. The best approach is to measure production at 4-hr intervals throughout the day, which will account for varying light conditions. It is also possible to assume a linear relationship between productivity and available light, and to scale midday productivity to the entire day, though there is obvious potential for bias with this approach, since assuming a linear relationship may be incorrect (Bower et al. 1987). Another method is to measure production from each depth under controlled light environments and model daily production (Fee 1969; Lewis 1988). It is also possible to derive photosynthetic-irradiance curves (PI curves; Platt et al. 1980) from in situ incubations and to model daily production based on the varying irradiances at lake depths. Scaling up bottle production measurements to the entire lake can work; production estimated from a whole-lake radiocarbon addition agreed closely with modeled estimates based on bottle production measurements (Bower et al. 1987).

Scaling up from benthic chambers to an area of stream or marsh involves estimating the area of substrate in the chamber and converting to an area-specific rate. In simple systems, scaling the measurement can work. For example, primary production estimates scaled up from chambers were similar to estimates from entire laboratory streams (McIntire et al. 1964). However, because of spatial heterogeneity of substrate and the potential for subsurface metabolism, chamber estimates may be difficult to scale to whole-ecosystem rates; thus, free-water methods of measuring CR often yield much higher figures than chamber estimates (Bott et al. 1978). For example, in two New Mexico streams, chamber estimates of surface sediment respiration were much smaller than whole-stream estimates because the hyporheic zone contributed 40% to 93% of total ecosystem respiration (Fellows et al. 2001).

To estimate production on a seasonal or yearly scale, frequent measurements are necessary. Whole-system methods can be scaled by deploying recording probes on a daily or less frequent basis (e.g., Uehlinger and Naegli 1998; Cole et al. 2000), though it is important to periodically (every day to every week) recalibrate these instruments and make sure that the electrode is clean, since small errors in the calibration caused by electrode drift may cause large errors in the production estimate (McCutchan et al. 1998). Despite electrode maintenance, it is much less work to measure production frequently, using whole-system methods with recording sondes, rather than chamber methods.

### Microscale Methods

#### Overview

The main problem in adapting chamber methods to accommodate benthic periphyton is the heterogeneity inherent within benthic microbial mats, which cause $O_2$ gradients both within the biofilm or mat and above it in the overlying $H_2O$. Gas exchange in very thick biofilms may be limited by diffusion, and therefore the chamber methods described above may underestimate metabolism. However, the fact that gas exchange is diffusion-limited in these biofilms means that metabolism can be estimated by measuring the $O_2$ gradient at very small intervals through the biofilm. Techniques that measure the microscale distribution of $O_2$

within sediments have been modified for periphyton mats and biofilms in order to measure benthic community metabolism (Sørensen et al. 1979; Revsbech et al. 1981; Revsbech and Jørgensen 1983). The main advantage of the technique over the chamber method is that $O_2$ change is measured at the biofilm/DBL interface rather than in the overlying $H_2O$, where hydrodynamics are difficult to mimic. Moreover, because $O_2$ concentration is measured within the biofilm, vertical GPP integrated for the photosynthetically active layer can be assessed. Finally, because $O_2$ is measured rapidly in the light, GPP is addressed directly (i.e., not from the difference between the dark respiration and the NEP), and respiration in the light can thus be deduced (see below).

There are two complementary methods to measure metabolism of biofilms. The micro-$O_2$ exchange method is based on $O_2$ flux calculations at the DBL/biofilm interface and within the biofilm, using Fick's first law of diffusion, and after establishment of a stable $O_2$ microprofile under constant light or in the dark. The second, the light-dark shift method (Revsbech et al. 1981), is based on measurements of the rate of $O_2$ decrease, seconds after the light has been turned off, at each depth of a biofilm exhibiting a stable $O_2$ microprofile under constant light.

## Choice of Microelectrodes

Microscale investigations of biofilm photosynthesis rely on $O_2$ microelectrodes developed in the late 1960s (Whalen et al. 1967). Three main types of $O_2$ microelectrodes exist: the cathode-type and Clark-type microelectrodes, plus, more recently, the micro-optrode. Despite its ruggedness and potential, the actual design of the micro-optrode is marginally utilized and is limited to the assessment of NEP and dark respiration. The sensor does not measure $O_2$ concentration rapidly enough to directly estimate GPP (cf. Klimant et al. 1995 for more details). Thus it will not be described in the following.

The cathode- and the Clark-type microelectrodes function similarly. They are composed of a reference anode (Ag/AgCl) and a sensing cathode (Pt) powered with a low negative voltage that maintains a steady polarization in an electrolyte solution, providing the ionic conduction of electricity and $Cl^-$ ions at the anode (Ag + $Cl^- {\rightarrow} AgCl + e^-$) following $O_2$ reduction at the cathode ($O_2 + H_2O + 4e^- {\rightarrow} 4OH^-$). These microelectrodes are called amperometric because the $O_2$ concentration where they are plunged is proportional to the extent of $O_2$ reduction at the cathode. For technical reasons, despite the fact that the cathode type has a faster dissolved $O_2$ measurement (90% response time), the Clark-type microelectrode, with 1-2 sec 90% response time, is preferred, mainly because it does not need to be calibrated in the $H_2O$ where the $O_2$ measurement is performed, and thus can be used in different liquid media upon calibration (e.g., Carlton and Wetzel 1987). Moreover, since the improvement of the Clark-type microelectrode (Clark-type microelectrode with a rugged cathode; Revsbech 1989), this microelectrode has a more stable signal than the cathode type.

The Clark-type microelectrode often has a glass tip of 15–40 μm diameter, capped with a rubber membrane permeable to salts and dissolved $O_2$. The microelectrode can be enclosed in a syringe needle to improve the ruggedness of the overall microelectrode for tough biofilms. Because the microelectrode consumes $O_2$ in the measuring media, the rate of $O_2$ supply is different in stagnant ($O_2$ supply through diffusion) and stirring conditions ($O_2$ supply through advection). This could present a problem if measurements are performed under different hydrologic conditions, such as those encountered while moving the electrode from the overlying $H_2O$, through the DBL, and into the biofilm. However, with a low 1–2% stirring sensitivity, this problem is negligible. The spatial definition of the microelectrode technique is restricted by the response time of the microelectrode itself and the spatial diffusion of the $O_2$ (cf. Revsbech and Jørgensen 1983). Practically speaking, spatial definition of 0.1 mm can be reached with the Clark-type microelectrode (Revsbech and Jørgensen 1983), which is moved at an angle to avoid shading with a micromanipulator that can be computer controlled. The life duration of a microelectrode is limited mostly because it is quite fragile and also because the membrane loses its permeability through accumulation of salts. Aside from breakage, it lasts about 1–2 mos (e.g., Carlton and Wetzel 1987).

## Micro-$O_2$ Exchange Method

Metabolism measurement is based on the steepest $O_2$ fluxes at the DBL/biofilm interface after establishment of steady-state dissolved $O_2$ profiles in the light (NEP) and in the dark (dark respiration, $R_{dark}$). Flux of $O_2$ (J) is expressed following Fick's first law:

$$J = D \ dC(z)/dz \ (\text{mol cm}^{-2} \text{ s}^{-1}), \tag{11}$$

where D is the coefficient of diffusion ($\text{cm}^2 \text{ s}^{-1}$), $C(z)$ is the concentration of solute at depth $z$ (cm), and $dC(z)/dz$ is the gradient of concentration between the two depths ($\text{mol L}^{-1} \text{ cm}^{-1}$).

At steady state in the dark (fig. 10.5–A),

$$R_{dark} = J_{down} = - D_{DBL} dC(z)/dz, \tag{12}$$

where the dark respiration $R_{dark}$ equals $J_{down}$, the downward flux of $O_2$ at the interface; $D_{DBL}$ is the $O_2$ coefficient of diffusion in the DBL; and $dC(z)/dz$ is the $O_2$ gradient.

At steady state in the light (fig. 10.5–B),

$$NEP = J_{up} = - D_{DBL} dC(z)/dz, \tag{13}$$

where $J_{up}$ is the upward flux of $O_2$ that equals NEP. Additionally, the downward flux of $O_2$ is estimated as

$$J_{down} = - \Phi D_s dC(z)/dz, \tag{14}$$

where $D_s$ is the $O_2$ coefficient of diffusion in the pore $H_2O$ and $\Phi$ is the sediment porosity. Sediment porosity is generally determined by the dry:wet weight ratio (e.g., ~0.9 for a cyanobacterial mat; Epping and Jørgensen 1996; see more accurate method in Epping et al. 1999).

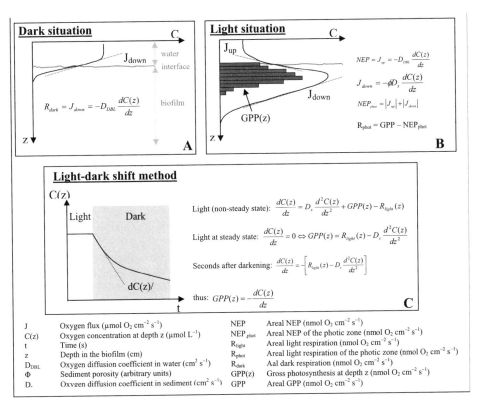

Figure 10.5. Assessment of the different metabolic rates of biofilms using dissolved $O_2$ microelectrodes. Curves refer to $O_2$ concentrations, and bars represent GPP at a given depth. A: Calculation of the dark areal respiration at equilibrium. B: Calculations of total areal NEP and of areal NEP of the photic zone at equilibrium. C: Theory behind the light–dark shift measurements to directly assess GPP at a given depth. Partly redrawn from Kühl et al. 1996.

It is then possible to calculate the areal NEP of the photic zone:

$$NEP_{phot} = |J_{up}| + |J_{down}|. \tag{15}$$

At this stage, GPP could be calculated as the sum of NEP and $R_{dark}$, but this would result in an underestimation because respiration in the light, $R_{light}$, which should be used to calculate GPP but cannot be assessed directly, is higher than $R_{dark}$ (cf. above, on the chambers).

## Light–Dark Shift Method

Due to the rapidity of $O_2$ measurements recorded with a Clark-type microelectrode, it is possible, paradoxically, to directly assess GPP in the dark within seconds after the light has been turned off (fig. 10.5-C). This method is based on the fact that a biofilm with a steady $O_2$ microprofile under constant light has a fixed rate of decline in $O_2$ for a considered depth in the biofilm. The fixed decrease in $O_2$ lasts at least 4-4.6 sec (Glud et al. 1992; Lorenzen et al. 1995), and during that time frame, $R_{light}$ remains fairly unchanged (e.g., Peterson 1983).

In the light, variation in dissolved $O_2$ concentration (C) for a given depth $z$ at time $t$ in the periphyton results from diffusion of $O_2$ within the periphyton matrix following Fick's second law of diffusion and the balance of the production-consumption of $O_2$ at this depth through GPP and $R_{light}$, respectively.

In the light and within the photic zone,

$$\frac{dC(z)}{dt} = D_s \frac{d^2C(z)}{dz^2} + GPP(z) - R_{light}(z). \tag{16}$$

After a period of illumination and at each depth, the $O_2$ concentration reaches a steady state and the $O_2$ production counterbalances (1) the biological and chemical $O_2$ consumption and (2) the diffusional import and export of $O_2$ from this given depth. Thus, from the equation above, we have at steady state in the light within the photic zone:

$$D_s \frac{d^2C(z)}{dz^2} + GPP(z) - R_{light}(z) = 0 \Leftrightarrow GPP(z) = R_{light}(z) - D_s \frac{d^2C(z)}{dz^2}. \tag{17}$$

When the light is switched off, $GPP(z)$ instantaneously stops, and both rates of $O_2$ consumption as $O_2$ diffusional transports are supposed to remain unchanged. Thus, after darkening,

$$\frac{dC(z)}{dt} = -\left[ R_{light}(z) - D_s \frac{d^2C(z)}{dz^2} \right] \tag{18}$$

and

$$GPP(z) = -\frac{dC(z)}{dt}. \tag{19}$$

Practically speaking, $O_2$ measurements are made 1 sec after switching the light off (Revsbech and Jørgensen 1983), but longer waits are sometimes required (4 sec), especially in the case of periphyton of low productivity (10 sec; Revsbech and Jørgensen 1983). One light-dark cycle lasting 10–60 sec, depending on productivity rates, is required to calculate depth-specific rates. Thus, for a 2-mm-thick biofilm, a GPP($z$) profile is established every 0.1 mm within 3.3 to 20 min. GPP($z$) is a pore $H_2O$ activity, and must be multiplied by the biofilm porosity, $\Phi$, to obtain the activity per volume of wet biofilm at a given depth $GPP_{vol}(z)$. Then, the areal GPP can be assessed by integrating all the $GPP_{vol}(z)$ over the photic zone ($Z_{phot}$):

$$GPP = \sum_{z=0}^{z_{phot}} GPP_{vol}(z).$$   (20)

## Limitations

The main limitation of the micromethods is that metabolic rates are calculated from single point measures through the periphyton (Glud et al. 1996, 2002). Thus, given the substantial spatial heterogeneity of most biofilms, many microprofiles are necessary to assess community metabolism. However, a very promising and complementary method involves the use of planar optrodes to assess the distribution of $O_2$ in a two-dimensional transverse section of periphyton (cf. Glud et al. 1996). The planar optrode can measure $O_2$ distribution along 25-mm transects, accounting for at least some of the spatial heterogeneity of the biofilm.

One implicit limitation of the microelectrode technique is the assumption of an $O_2$ diffusion gradient within the biofilm. Rapid advection of overlying $H_2O$, coupled with moderate porosity of the periphyton, can lead to a lack of an $O_2$ gradient, and thus a lack of diffusion, making the microelectrode techniques impracticable in situ (Carlton and Wetzel 1987; Kühl et al. 1996; Dodds et al. 1999). In this situation a chamber method should be used.

Moreover, changing or low light conditions in the field may not permit the required steady-state $O_2$ profile (Bott et al. 1997). Often, in these situations, a stable, high-intensity light source is supplied. However, excessive light intensities may elicit photoinhibition and algal migration, thus changing local photosynthetic rates (Kühl et al. 1996; Epping et al. 1999). High irradiance may also cause temperature at the surface of the biofilm to exceed that of the overlying water where the microelectrode is calibrated, leading to an overestimation of dissolved $O_2$. Practically speaking, however, this overestimation has been shown to be negligible (e.g., 2%; Revsbech et al. 1983). It is also supposed that the number of light-dark cycles performed to measure the photosynthesis rates within the biofilm (e.g., 20 for a 2-mm-thick mat) limits algal migration within the biofilm (Glud et al. 2002). Finally, light-enhanced $O_2$ bubble formation at the surface of or within the periphyton matrix lowers $O_2$ measurements and leads to localized high $O_2$ fluxes, thereby modifying local gradients (cf. Revsbech and Jørgensen 1983; Kühl et al. 1996). Thus, light intensity should be controlled at a moderate level relative to the range encountered in the natural environment.

The physical intrusion of the microelectrode from above compresses the DBL thickness, and thus increases $O_2$ exchange across the DBL. This perturbation can overestimate the $O_2$ flux across the DBL (Lorenzen et al. 1995). Moreover, decreasing thickness of the DBL can potentially change the $O_2:CO_2$ ratio, altering periphyton physiology (e.g., photorespiration intensity). This bias is important for smooth-surfaced periphytic biofilms (cf. Lorenzen et al. 1995).

### *Pulse Amplitude Modulation (PAM) Fluorometry*

This relatively new, noninvasive method calculates photosynthesis on a scale of relative units. Thus, it must be performed as a complement to the methods above if absolute numbers are necessary. However, it does not necessarily require calibration with other methods ($O_2$, $^{14}C$) if it is used as an assay of photosynthesis for comparative purposes (e.g., toxicology). The great advantage of the method is that photosynthesis and relative production can be assessed for specific algal groups (i.e., cyanophytes, green algae, and diatoms) in a matter of seconds. Coupled with an ease of use upon calibration, it is very often employed for repeated photosynthesis measures in studies of large spatiotemporal scales that require rapid assessment.

## The PAM Method

The method is based on the fact that not all the energy received at the antennae of the photosystems is actually transmitted to the reaction centers. Each quantum of light absorbed by a chlorophyll molecule in the antennae excites an electron from ground state to an excited state. The de-excitation of the chlorophyll molecule from this excited state to ground state can be achieved through three different pathways: (1) by excitation of the next neighboring molecule until the final energy discharge in the reaction center that fuels photochemical reactions, (2) by the energy dissipation through heat, and (3) by red (because some energy is lost in the process, and thus the wavelength is longer) fluorescence (generally 3–5% in vivo). Depending on the degree of light adaptation of the sample (i.e., biofilm or the pelagic water) and on the intensity of the light received at the moment, the total energy received will be distributed differently within the three pathways (cf. Krause and Weis 1991 for more details).

  The principle of the PAM method (Schreiber et al. 1986) is to measure the easily quantifiable PSII fluorescence yield of, ideally, a dark-adapted sample (all the reaction centers are opened) at two light intensities including, necessarily, one measurement at a sudden saturated light, which closes the reaction centers (= photosynthesis stopped). The fluorescence yield of PSI is distinctly lower than that of PSII and is negligible. The exposure to saturating light has to be done quickly to avoid electron accumulation at the acceptor side of the photosystem and to allow multiple PAM measurements on the same sample. This is performed using μsec pulses of light through light-emitting diodes (LED) and a 0.5 sec pulse of red 660 nm light for the saturating light pulse (SP) measurement. The fluorescence pulses following each excitation wavelength are detected by a photomultiplier with high red sensitivity (>710 nm).

  The photosynthetic capacity of the sample (the maximum energy conversion efficiency, $\Phi_p^0$) is assessed first. The fluorescence yield is measured by exciting chlorophyll of a dark-adapted sample with low intensity light ($<<1$ μmol m$^{-2}$ s$^{-1}$). In that case, the conduction of energy efficiency to the PSII reaction center or quantum yield is maximal because the fluorescence yield is minimal ($F_0$) and there is no heat dissipation. The SP that follows, closes all PSII reaction centers (= no photochemical reactions). In that case, heat dissipation increases and fluorescence yield is maximum ($F_m$). It is then possible to calculate $\Phi_p^0$:

$$\Phi_p^0 = (F_m - F_0)/F_m. \tag{21}$$

Thereafter, fluorescence under various actinic irradiances is measured and the effective energy conversion efficiency, $\Phi_p$, is calculated:

$$\Phi_p = (F_m'-F_s)/F_m', \tag{22}$$

where $F_m'$ is the maximal fluorescence in actinic light and $F_s$ is the steady-state fluorescence in actinic light. $\Phi_p$ is used to calculate the linear rate of photosynthetic electron transport (ETR) between the PSII and PSI (= photosynthesis at a certain actinic irradiance):

$$ETR = \Phi_p \times PPFD \times a_L \times r_{(PSII/PSI)}, \tag{23}$$

where PPFD is the photosynthetic photon flux density, $a_L$ is the fraction of PPFD absorbed (generally 0.84 is used by default, the remaining PPFD being reflected), and $r_{(PSII/PSI)}$ is the ratio between light absorptions of PSII and PSI (generally 0.5). ETR is thus expressed in $\mu$mol electrons $m^{-2}$ $s^{-1}$ and can be plotted against actinic irradiance. Contrary to classic photosynthesis vs. irradiance (PI) curves requiring long incubations, the light curve resulting from the PAM method is measured within 10–20 sec, and thus no photoacclimatation of the algal assemblage over time is possible. The PI curves are thus called rapid light curves (RLC), and give a snapshot of the photosynthetic capabilities of the sample at a given time (Glud et al. 2002). However, the ETR is dimensionless, and thus requires cross comparisons with other methods (see below).

## Apparatus

The PAM method is linked to the progress in light-emitting diode (LED) optics capable of delivering $\mu$sec pulses of actinic light, thus exciting the pigments against which pulse fluorescence is measured. Furthermore, because accessory pigments can characterize different algal groups, it is then possible to discriminate these spectral groups by exciting the pigment pool at different wavelengths within the visible spectra. These wavelengths are selected to enhance optimal differences in the excitation of the various antenna pigments in PSII. The contribution of each spectral group to the overall fluorescence is then determined through deconvolution, using the spectral signature or reference excitation spectra of each dominant alga representing each algal group present in the sample. The reference excitation spectra correspond to the yield exhibited at the four excitation wavelengths for a pure algal culture. A number of these multi-wavelength fluorometers are commercially available today, and they use trichromatic, quadrachromatic, and even pentochromatic excitations. The most popular are the Fluoroprobe bbe Moldaenke (Kiel, Germany; http://www.bbe-moldaenke.de) and PHYTO-PAM (Walz; Effeltrich, Germany, www.walz.com). In the following, the PHYTO-PAM, which offers a large range of apparatus, is described succinctly.

The PHYTO-PAM is a quadrachromatic fluorometer exciting the pigment pool at the PSII antenna at 645 nm for phycocyanine in cyanobacteria, at 470 nm and 665 nm for chlorophyll $b$ in green algae, and at 520 nm for chlorophyll $c$ and carotenoids in diatoms/dinoflagellates. The emitter detector unit (ED) of the PHYTO-PAM encapsulates several LEDs for the four excitation wavelengths (maximum

PPFD 600 μmol quanta $m^{-2}$ $s^{-1}$), as well as some for the actinic light (maximum PPFD 1800-2000 μmol quanta $m^{-2}$ $s^{-1}$, 3500-4000 μmol quanta $m^{-2}$ $s^{-1}$ for the saturating pulse) and a photomultiplier detector with high red sensitivity (>710 nm) for detection of the fluorescence signal.

The apparatus is operated with the Phytowin software, which not only controls the apparatus but also is used to save the bulk fluorescence data in a report. The report will be read by the software, and the stored fluorescence data will be deconvoluted, using the proper reference excitation spectra. The software also determines the RLC parameters such as Ik, alpha, and $ETR_{max}$, following a curve-fitting using the Eilers and Peeters model (1988). Finally, the deconvoluted data (i.e., the yield per group of algae and the yield at various light intensities) can be exported. The RLC parameters are, however, not saved with the current software version, and have to be recorded manually.

In order to deconvolute the fluorescence correctly, it is essential to use the "right" reference excitation spectra of the main alga representing each algal group present in the sample investigated. The reference spectra used for the deconvolution of the signal can be determined after the investigation of the sample with the PHYTO-PAM. The "bulk" fluorescence data generated for each sample can indeed be interpreted, and even reinterpreted later, once the taxonomic composition of the sample is known or when new information arises.

A wide range of apparatus specifics are available for different application types: the PHYTO-PAM series is dedicated to $H_2O$ samples but can be customized into a fiber-optic version for attached algae. A submersible fiber-optic unit is also available (DIVING-PAM). For microscopic resolutions, the MICROSCOPY-PAM (definition down to the cell and even to a single chloroplast) and the MICROFIBER-PAM (definition down to a single cell level in the cell layers) are offered by the company.

## Limitations

The main problem with the PAM method is that it requires calibration with other techniques. Very little is known about the relationships between the $O_2$ or DIC uptake and linear ETR, and more calibrations are necessary to validate the method (Glud et al. 2002).

However, the relationships between ETR and photosynthesis issuing from other methods performed with phytoplankton are mostly linear (Glud et al. 2002). These relationships are controversial with periphyton samples where nonlinear relationships are observed (Barranguet and Kromkamp 2000, linear; Glud et al. 2002, nonlinear). This is linked to the fact that all the phytoplankton in a cuvette is excited, and overall fluorescence is analyzed. However, in the case of a benthic or layered sample, if the working surface area excited is fairly well defined, it is not known how deeply the sample is actually excited and what part of the fluorescence yield is lost. Thus, the method with layered samples more likely provides an integrated ETR of poorly defined sample volume (Glud et al. 2002), and the calculated ETR reflects mostly photosynthesis of only the superficial upper layer.

## Summary and Recommended Methods

The recommended methods will depend upon the type of ecosystem (lotic vs. lentic) and the spatial scale of the research question. For example, grazing studies usually occur at a small scale (e.g., bottle or small enclosure); hence, a chamber production measurement will be best. Questions focusing on whole-system metabolism are best addressed by using whole-ecosystem diel $O_2$ change methods, if that approach is possible. Below we summarize conditions and ecosystems in which an investigator will choose a certain method.

### Lake or Large River, Pelagic Only

Bottle methods work well, either $^{14}C$ or $O_2$ change, if production is high enough (but see Carignan et al. 1998 for sensitive $O_2$ measurements). To obtain accurate daily estimates, water must be collected and incubated at different depths several times throughout the day. If diel measures are not possible, a photosynthesis vs. irradiance model (preferably derived empirically from the studied community) must be used to estimate production as light intensity varies throughout the day (Fee 1969). Epilimnetic production can also be estimated by free-water methods if littoral areas are small relative to pelagic areas, or if littoral areas are to be calculated as part of overall epilimnetic production.

### Lake or Large River, Benthic Only

Chamber methods and microscale methods are employed. These estimates must be scaled to estimate metabolism for the entire ecosystem, which means that they have to be replicated at varying depths and/or habitats (substrates), and weighted for the fraction of the benthos at various depths and/or habitats.

### Lake or Large River, Benthic and Pelagic

Free-water methods are the best here, because scaling both benthic and pelagic measurements is difficult. For lakes and rivers, it is possible to predict reaeration based on published equations, but it is possible (and better) to empirically measure reaeration if the ecosystems are not too large (though gas exchange has been measured in the ocean; e.g., Wanninkhof et al. 1990). If production is very low, then it may not be possible to estimate using whole-system approaches, and $^{14}C$ methods should be used instead (e.g., Goldman 1988).

### Streams and Small Rivers

Methods here will depend on spatial scale. If small-scale measurements are desired, then chambers work well. For example, if one is measuring the response of epilithic periphyton to grazing by invertebrates, then chamber methods will give an accurate assay of metabolic differences among grazer treatments. Like bottle methods, in order to scale for the whole day, either 24-hr chambers must be used (Bott et al. 1978) or production should be assessed throughout the day to account for varying light intensity.

On the other hand, if a researcher wants a reach-scale estimate of production, then the free-water technique is by far the best, provided it is employable at the site. Free-water methods integrate spatially over the substantial heterogeneity inherent in small streams and include hyporheic metabolism, which is less amenable to chamber measurements (Fellows et al. 2001). If reaeration is low, then 2-station estimates may give less error than 1-station estimates. Assuming minimal planktonic metabolism in streams and small rivers, this approach measures benthic metabolism.

Last, we echo the suggestion of Bott et al. (1997) that methods be chosen with knowledge of their advantages, disadvantages, and potential errors. Method choice minimizes both logistic constraints and measurement error and bias. We have attempted to inform investigators of some of the advantages and shortcomings of each method, but application to the investigators' questions and ecosystems should be made with care.

*Acknowledgments*   Thanks to Steve Thomas for reading and commenting on this chapter. We thank the National Science Foundation's Florida Coastal Everglades Long-Term Ecological Research program for providing support for establishing appropriate methods for benthic algal production assessments.

References

Bachmann, R. W., M. V. Hoyer, and D. E. Canfield, Jr. 2000. Internal heterotrophicity following the switch from macrophytes to algae in Lake Apopka, Florida. Hydrobiologia 418:217-227.

Barranguet, C., and J. Kromkamp. 2000. Estimating the primary production rates from photosynthetic electron transport in estuarine microphytobenthos. Marine Ecological Progress Series 204:39-52.

Bott, T. L., J. T. Brock, A. Baatrup-Pedersen, P. A. Chambers, W. K. Dodds, K. T. Himbeault, J. R. Lawrence, D. Planas, E. Snyder, and G. M. Wolfaardt. 1997. An evaluation of techniques for measuring periphyton metabolism in chambers. Canadian Journal of Fisheries and Aquatic Sciences 54:715-725.

Bott, T. L., J. T. Brock, C. E. Cushing, S. V. Gregory, D. King, and R. C. Petersen. 1978. A comparison of methods for measuring primary production and community respiration in streams. Hydrobiologia 60:2-12.

Bower, P. M., C. A. Kelly, E. J. Fee, J. A. Shearer, D. R. DeClerq, and D. W. Shindler. 1987. Simultaneous measurement of primary production by whole-lake and bottle radiocarbon additions. Limnology and Oceanography 32:299-312.

Brandini, F. P., and E. T. da Silva. 2001. Production and biomass accumulation of periphytic diatoms growing on glass slides during a 1-year cycle in a subtropical estuarine environment (Bay of Paranaguá, southern Brazil). Marine Biology 138: 163-171.

Broström, G., and J. Nilsson. 1999. A theoretical investigation of the diffusive boundary layer in benthic flux chamber experiments. Journal of Sea Research 42:179-189.

Carignan, R., A.-M. Balis, and C. Vis. 1998. Measurement of primary production and community respiration in oligotrophic lakes using the Winkler method. Canadian Journal of Fisheries and Aquatic Sciences 55:1078-1084.

Carlton, R. G., and R. G. Wetzel. 1987. Distribution and fates of oxygen in periphyton communities. Canadian Journal of Botany 65:1031-1037.

Chapra, S. C., and D. M. DiToro. 1991. Delta method for estimating primary production, respiration and reaeration in streams. Journal of Environmental Engineering 117:640–655.

Cole, J. J., and N. F. Caraco. 1998. Atmospheric exchange of carbon dioxide in a low-wind oligotrophic lake. Limnology and Oceanography 43:647-656.

Cole, J. J., S. R. Carpenter, J. F. Kitchell, and M. L. Pace. 2002. Pathways of organic carbon utilization in small lakes: Results from a whole-lake $^{13}C$ addition and coupled model. Limnology and Oceanography 47:1664-1675.

Cole, J. J., M. L. Pace, S. R. Carpenter, and J. F. Kitchell. 2000. Persistence of net heterotrophy in lakes during nutrient addition and food web manipulations. Limnology and Oceanography 45:1718-1730.

Crusius, J., and R. Wanninkhof. 2003. Gas transfer velocities measured at low wind speed over a lake. Limnology and Oceanography 48:1010-1017.

Del Giorgio, P. A., J. J. Cole, and A. Cimberlis. 1997. Respiration rates in bacteria exceed phytoplankton production in unproductive aquatic systems. Nature 358: 148-151.

Dodds, W. K., J. F. Biggs, and R. L. Lowe. 1999. Photosynthesis—irradiance patterns in benthic microalgae: Variation as a function of assemblage thickness and community structure. Journal of Phycology 35:42-53.

Dodds, W. K., and J. Brock. 1998. A portable chamber for in situ determination of benthic metabolism. Freshwater Biology 39:49-59.

Dring, M. J., and D. H. Jewson. 1982. What does $^{14}C$ uptake by phytoplankton really measure? A theoretical modelling approach. Proceedings of the Royal Society of London B214:351-368.

Eilers, P. H. C., and J. C. H. Peeters. 1988. A model for the relationship between light intensity and the rate of photosynthesis in phytoplankton. Ecological Modelling 42: 199-215.

Epping, E. H. G., and B. B. Jørgensen. 1996. Light-enhanced oxygen respiration in benthic phototrophic communities. Marine Ecology Progress Series 139:193-203.

Epping, E. H. G., A. Khalili, and R. Thar. 1999. Photosynthesis and dynamics of oxygen consumption in a microbial mat as calculated from transient oxygen microprofiles. Limnology and Oceanography 44:1936-1948.

Fee, E. J. 1969. A numerical model for the estimation of photosynthetic production, integrated over time and depth, in natural waters. Limnology and Oceanography 14: 906-911.

Fee, E. J. 1973a. Modelling primary production in water bodies: A numerical approach that allows vertical inhomogeneities. Journal of the Fisheries Research Board of Canada 10:1469-1473.

Fee, E. J. 1973b. A numerical model for determining integral primary production and its application to Lake Michigan. Journal of the Fisheries Research Board of Canada 30: 1447-1468.

Fellows, C. S., H. M. Valett, and C. N. Dahm. 2001. Whole-stream metabolism in two montane streams: Contributions of the hyporheic zone. Limnology and Oceanography 46:523-531.

Flynn, K. J. 1988. The concept of "primary production" in aquatic ecology. Limnology and Oceanography 33:1215-1216.

Frost, T., and R. C. Upstill-Goddard. 2002. Meteorological controls of gas exchange at a small English lake. Limnology and Oceanography 47:1165-1174.

Gaarder, T., and H. H. Grann. 1927. Investigations of the production of plankton in the Oslo

Fjord. Rapport et Procès-Verbaux des Réunions. Conseil Permanent International pour l'Exploration de la Mer 42:3-31.

Glud, R. N., and N. Blackburn. 2002. The effects of chamber size on benthic oxygen uptake measurements: A simulation study. Ophelia 56:23-31.

Glud, R. N., M. Kühl, O. Kohls, and N. B. Ramsing 1999. Heterogeneity of oxygen production and consumption in a photosynthetic microbial mat as studied by planar optodes. Journal of Phycology 35:270-279.

Glud, R. N., M. Kühl, F. Wenzhöfer, and S. Rysgaard. 2002. Benthic diatoms of high arctic fjord (Young Sound, NE Greenland): Importance for ecosystem primary production. Marine Ecology Progress Series 238:15-29.

Glud, R. N., N. B. Ramsing, J. K. Gundersen, and I. Klimant. 1996. Planar optrodes: A new tool for fine scale measurements of two dimensional $O_2$ distribution in benthic communities. Marine Ecology Progress Series 140:217-226.

Glud, R. N., N. B. Ramsing, and N. P. Revsbech. 1992. Photosynthesis and photosynthesis-coupled respiration in natural biofilms quantified with oxygen microsensors. Journal of Phycology 28:51-60.

Glud, R. N., S. Rysgaard, and M. Kühl. 2002. A laboratory study on $O_2$ dynamics and photosynthesis in ice algal communities: Quantification by microsensors, $O_2$ exchange rates, [14]C incubations and PAM fluorometer. Aquatic Microbial Ecology 27:301-311.

Goldman, C. R. 1988. Primary productivity, nutrients, and transparency during the early onset of eutrophication in ultra-oligotrophic Lake Tahoe, California-Nevada. Limnology and Oceanography 33:1321-1333.

Grønved, J. 1960. On the productivity of microphytobenthos and phytoplankton in some Danish fjords. Meddelelser fra Danmarks Fiskeri-Og Havundersogelser 3:1-17.

Hall, C. A. S., and R. Moll. 1975. Methods of assessing aquatic primary productivity. Pages 19-53 in H. Lieth and R. H. Whittaker (eds.), Primary Productivity of the Biosphere. Springer-Verlag, New York.

Hall, R. O., and J. L. Tank. 2003. Ecosystem metabolism controls nitrogen uptake in streams in Grand Teton National Park, Wyoming. Limnology and Oceanography 48: 1120-1128.

Hall, R. O., and J. L. Tank. 2005. Correcting whole-stream estimates of metabolism for groundwater input. Limnology and Oceanography: Methods 3:222–229.

Hanson, P. C., D. L. Bade, S. R. Carpenter, and T. K. Kratz. 2003. Lake metabolism: Relationships with dissolved organic carbon and photosynthesis. Limnology and Oceanography 48:1112-1119.

Havens, K. E., T. L. East, S. Hwang, A. J. Rodusky, B. Sharfstein, and A. D. Steinman. 1999. Algal responses to experimental nutrient addition in the littoral community of a subtropical lake. Freshwater Biology 42:329–344.

Hill, W. R., M. G. Ryon, and E. M. Schilling. 1995. Light limitation in a stream ecosystem: Responses by primary producers and consumers. Ecology 76:1297–1309.

Ibelings, B. W., and L. R. Mur. 1992. Microprofiles of photosynthesis and oxygen concentration in *Microcystis* sp. scums. FEMS Microbiology Ecology 86:195-203.

Jensen, J., and N. P. Revsbech. 1989. Photosynthesis and respiration of a diatom biofilm cultured in a new gradient growth chamber. FEMS Microbiology Ecology 62: 29-38.

Jonsson, B. 1991. A [14]C-incubation technique for measuring microphytobenthic primary productivity in intact sediment cores. Limnology and Oceanography 36:1485-1492.

Klimant, I., V. Meyer, and M. Kühl. 1995. Fiber-optic oxygen microsensors, a new tool in aquatic biology. Limnology and Oceanography 40:1159-1165.

Krause, G. H., and E. Weis. 1991. Chlorophyll fluorescence and photosynthesis: The basics. Annual Review of Plant Physiology and Plant Molecular Biology 42:313–349.

Kühl, M., R. N. Glud, H. Ploug, and N. B. Ramsing. 1996. Microenvironmental control of photosynthesis and photosynthesis—coupled respiration in an epilithic cyanobacterial biofilm. Journal of Phycology 32:799-812.

Legendre, L., and M. Gosselin. 1996. Estimation of N or C uptake rates by phytoplankton using $^{15}N$ or $^{13}C$: Revisiting the usual computation formula. Journal of Plankton Research 19:263-271.

Lewis, W. M., Jr. 1988. Primary production in the Orinoco River. Ecology 69:679–692.

Lindeboom, H. J., A. J. J. Sandee, V. D. DeClerk, and H. A. J. Driessche. 1985. A new bell jar and microelectrode method to measure changing oxygen fluxes in illuminated sediments with a microalgal cover. Limnology and Oceanography 30:693–698.

Long, S. P., S. Humphries, and P. G. Falkowski. 1994. Photoinhibition of photosynthesis in nature. Annual Review of Plant Physiology and Plant Molecular Biology 45:633-662.

Lorenzen, J., R. N. Glud, and N. P. Revsbech. 1995. Impact of microsensor-caused changes in diffusive boundary layer thickness on $O_2$ profiles and photosynthetic rates in benthic communities of microorganisms. Marine Ecology Progress Series 119: 237-241.

Marzolf, E. R., P. J. Mulholland, and A. D. Steinman. 1994. Improvements to the diurnal upstream-downstream dissolved $O_2$ change technique for determining whole-stream metabolism in small streams. Canadian Journal of Fisheries and Aquatic Sciences 51: 1591-1599.

McCormick, P. A., M. B. O'Dell, R. B. E. Shuford III, J. G. Backus, and W. C. Kennedy. 2001. Periphyton responses to experimental phosphorus enrichment in a subtropical wetland. Aquatic Botany 71:119-139.

McCutchan, J. H., W. M. Lewis, Jr., and J. F. Saunders. 1998. Uncertainty in the estimation of stream metabolism from open-channel oxygen concentrations. Journal of the North American Benthological Society 17:155-164.

McFeters, G. A., S. A. Stuart, and S. B. Olson. 1978. Growth of heterotrophic bacteria and algal extracellular products in oligotrophic waters. Applied Environmental Microbiology 35:383-391.

McIntire, C. D., R. L. Garrison, H. K. Phinney, and C. E. Warren. 1964. Primary production in laboratory streams. Limnology and Oceanography 9:92-102.

Mulholland, P. J, C. S. Fellows, J. L. Tank, N. B. Grimm, J. R. Webster, S. K. Hamilton, E. Martí, L. Ashkenas, W. B. Bowden, W. K. Dodds, W. H. McDowell, M. J. Paul, and B. J. Peterson. 2001. Inter-biome comparison of factors controlling stream metabolism. Freshwater Biology 14:1503-1517.

Odum, H. T. 1956. Primary production in flowing waters. Limnology and Oceanography 1:103-117.

Odum, H.T. 1957. Trophic structure and productivity of Silver Springs, Florida. Ecological Monographs 27:55-112.

Peterson, B. J. 1980. Aquatic primary productivity and the $^{14}C$-$CO_2$ method: A history of the productivity problem. Annual Review of Ecology and Systematics 11:359-385.

Peterson, R. B. 1983. Estimation of photorespiration based on the initial rate of postillumination $CO_2$ release II. Plant Physiology 73:983-988.

Platt, T., C. L. Gallegos, and W. G. Harrison. 1980. Photoinhibition of photosynthesis in natural assemblages of marine phytoplankton. Journal of Marine Research 38:687–701.

Rejmankova, E., and J. Komarkova. 2000. A function of cyanobacterial mats in phosphorus-limited tropical wetlands. Hydrobiologia 431:135–153.

Revsbech, N. P. 1989. An oxygen sensor with a guard cathode. Limnology and Oceanography 34:474–478.

Revsbech N. P., and B. B. Jørgensen. 1983. Photosynthesis of benthic microflora measures with high spatial resolution by the oxygen microprofile method: Capabilities and limitations of the method. Limnology and Oceanography 28:749–756.

Revsbech, N. P., B. B. Jørgensen, and O. Brix. 1981. Primary production of microalgae in sediments measured by oxygen microprofile, $H^{14}CO_3^-$ fixation, and oxygen exchange methods. Limnology and Oceanography 26:717–730.

Roberts, B. J. 2004. Assessing diel respiration in pelagic ecosystems using oxygen stable isotopes: When do the highest rates occur and who is respiring under different light and nutrient regimes? Ph.D. dissertation, Cornell University.

Robinson, G. G. C. 1983. Methodology: The key to understanding periphyton. Pages 245–251 in R. G. Wetzel (ed.), Periphyton of Freshwater Ecosystems. Dr. W. Junk Publishers, The Hague.

Sakamoto, M., M. M. Tilzer, R. Gächter, H. Rai, Y. Collos, P. Tschumi, P. Berner, D. Zbaren, M. Dokulil, P. Bossard, U. Uehlinger, and E. A. Nusch. 1984. Joint field experiments for comparisons of measuring methods of photosynthetic production. Journal of Plankton Research 6:365-383.

Schreiber, U., U. Schliwa, and W. Bilger. 1986. Continuous recording of photochemical and non-photochemical chlorophyll fluorescence quenching with a new type of modulation fluorometer. Photosynthesis Research 10:51-62.

Slawyk, G., Y. Collos, and J. C. Auclair. 1977. The use of $^{13}C$ and $^{15}N$ isotopes for the simultaneous measurements of carbon and nitrogen turnover rates in marine phytoplankton. Limnology and Oceanography 22:925–932.

Sørensen, J., B. B. Jørensen, and N. P. Revsbech. 1979. A comparison of oxygen, nitrate and sulfate respiration in coastal marine sediments. Microbial Ecology 5: 105–115.

Stanley, D. W. 1976. Productivity of epipelic algae in tundra ponds and a lake near Barrow, Alaska. Ecology 57:1015–1024.

Steemann Nielsen, E. 1951. Measurement of production of organic matter in the sea by means of carbon-14. Nature 167:684-685.

Tank, J. L., and J. C. Musson. 1993. An inexpensive chamber apparatus for multiple measurements of dissolved oxygen uptake or release. Journal of the North American Benthological Society 12:406-409.

Thomas, S., E. E. Gaiser, M. Gantar and L. J. Scinto. 2006. Quantifying the responses of calcareous periphyton crusts to rehydration: a microcosm study (Florida Everglades). Aquatic Botany 84:317-323.

Uehlinger, U. 2000. Resistance and resilience of ecosystem metabolism in a flood-prone river. Freshwater Biology 45:319-332.

Uehlinger, U., and M. W. Nageli. 1998. Ecosystem metabolism, disturbance, and stability in a prealpine gravel bed river. Journal of the North American Benthological Society 17:165-178.

Uzarski, D. G., T. M. Burton, and C. A. Stricker. 2001. A new chamber design for measuring community metabolism in a Michigan stream. Hydrobiologia 455:137-155.

Vollenweider, R. A. (ed.). 1969. A Manual on Methods for Measuring Primary Production in Aquatic Environments. Blackwell Scientific, Oxford.

Vollenweider, R. A. (ed.). 1974. A Manual on Methods for Measuring Primary Production in Aquatic Environments, 2nd ed. Blackwell Scientific, Oxford.

Wanninkhof, R., P. J. Mulholland, and J. W. Elwood. 1990. Gas exchange rates for a first-order stream determined with deliberate and natural tracers. Water Resources Research 26:1621-1630.

Webster, J. R., and J. L. Meyer (eds.). 1997. Stream organic matter budgets. Journal of the North American Benthological Society 16:3–161.

Wetzel, R. G. 1972. The role of carbon in hard-water marl lakes. Pages 84–96 in G. E. Likens (ed.), Nutrients and Eutrophication: The Limiting Nutrient Controversy. American Society of Limnology and Oceanography Special Symposium 1. Allen Press, Lawrence, KS.

Wetzel, R. G. 2001. Limnology. Academic Press, San Diego.

Wetzel, R. G., and G. E. Likens. 2000. Limnological Analyses, 3rd ed. Springer-Verlag, New York.

Whalen, W. J., J. Rilay, and P. Nair. 1967. A microelectrode for measuring intracellular $pO_2$. Journal of Applied Physiology 23:798-801.

Young, R. G., and A. D. Huryn. 1998. Comment: Improvements to the diurnal upstream-downstream dissolved $O_2$ change technique for determining whole-stream metabolism in small streams. Canadian Journal of Fisheries and Aquatic Sciences 55:1784-1785.

Young, R. G., and A. D. Huryn. 1999. Effects of land use on stream metabolism and organic matter turnover. Ecological Applications 9:1359-1376.

# 11

# The Role of Remote Sensing in the Study of Terrestrial Net Primary Production

Scott V. Ollinger

Robert N. Treuhaft

Bobby H. Braswell

Jeanne E. Anderson

Mary E. Martin

Marie-Louise Smith

Even the most ambitious of field campaigns cover extremely small fractions of the earth's land area and capture limited samples of its ecological complexity. As a result, addressing regional-to-global environmental issues can be difficult or impossible without some means of extending field measurements to the appropriate spatial domain. Measurements that are stratified over a large number of ecological units are vital (chap. 1, this volume), but are still incomplete without knowledge pertaining to the distribution, variability, and spatial extent of each. Consequently, ecologists have become heavily invested in methods for relating observations of individual organisms and field plots to the broader regions in which they exist (e.g., Ehleringer and Field 1993; Cohen and Justice 1999).

Although a variety of scaling approaches have been investigated, there is widespread agreement that remote sensing holds a central and irreplaceable role. Remote platforms are the only means by which large and contiguous portions of the Earth's surface can be sampled, and the selective absorption and reflection of radiation by different plant tissues provide a unique basis for obtaining ecologically relevant information. However, remote observations also pose enormous methodological challenges, and to date, there is no single remote sensing method that offers an optimal approach to NPP measurement across all scales and for all research objectives. In this chapter, we review a number of approaches through which remote sensing data can be applied to terrestrial NPP, or some of its important constituents, and discuss the trade-offs of various methods. Analogous approaches for marine ecosystems are described in chapter 9 of this volume. By nature, the remote sensing determination of NPP is more indirect than other terrestrial methods described in this volume, because ecosystem properties are only inferred through their interactions with electromagnetic radiation.

Given the breadth of the topic, our goal in preparing this chapter was not to provide a working manual of all NPP remote sensing methods available. Instead, we sought to summarize important overall strategies for NPP detection into a framework that involves the types of instruments used, the ecological properties they can be designed to detect, and the manner in which those properties can be translated into estimates of NPP. For example, instrument types can be broadly categorized into (1) passive sensors, which record reflected radiation that originates from a natural source (most often the sun), and (2) active sensors, which emit a known form of radiation from their own source (e.g., laser or radar) and record what is returned from the target surface. These categories can be further subdivided by the nature of the radiation they detect, the spatial resolution of the recorded measurements, the temporal frequency of sampling, the number of spectral channels detected, and the degree of spectral aggregation within each channel.

Not surprisingly, the properties of different remote sensors have direct bearing on the types of ecological variables that can be estimated and the methods by which they can be used to estimate NPP. For example, data from passive optical sensors typically indicate the degree to which solar radiation is absorbed or reflected by the Earth's surface at different wavelengths or view angles. Because plant pigments associated with photosynthesis have unique absorption properties, this type of data can be related to a range of variables associated with the greenness, or physiological capacity, of vegetation. Such variables, however, are not commensurate with NPP, and can be used to derive NPP estimates only when combined with process models or empirically derived algorithms.

Finally, although the methods we discuss are relevant to NPP studies in terrestrial ecosystems generally, our focus will be more heavily directed toward forests. Because forests represent the largest fraction of terrestrial carbon (C) storage and because they confront us with some of the most difficult technical challenges concerning detection of biomass and productivity, they have received a disproportionate amount of attention in the ecological remote sensing literature.

## Vegetation Properties Using Passive Sensors

This section discusses methods in which vegetation properties that are observable using passive sensors can be combined with various models and empirical equations to estimate NPP. The approaches vary in terms of the complexity of the models used, their reliance on local field measurements, and the degree to which they can be applied to longer-term mechanisms of environmental and ecological change. Specifications of several passive sensors that are relevant to NPP are shown in table 11.1.

The simplest models used to derive NPP from remote sensing data are empirically derived productivity algorithms, which combine field-based relationships with

Table 11.1. Instrument characteristics for several passive sensors relevant to NPP

| Instrument | Spectral Channels | Swath (km) | Spatial Resolution (m) | Frequency of Coverage | Notes |
|---|---|---|---|---|---|
| AVHRR | 5 | 2400 | 1100 | Daily global coverage | Continuous data since 1981 |
| MODIS | 36 | 2330 | 250–1000 | Global coverage every 1–2 days | Algorithms developed for NPP prediction |
| Landsat 7 | 7 | 185 | 15–30 | 16 days | Landsat 1 launched in 1972 |
| SPOT | 4 | 60 | 10–20 | 25 days | Commercial sensor |
| Ikonos | 3 | 12 | 1–4 | 3 days off nadir (144 days at nadir) | High-resolution commercial sensor |
| Hyperion | 220 | 7.5 | 30 | 16 days | High spectral-resolution imaging spectrometer |
| AVIRIS | 224 | 2–11 | 4–18 | Irregular | High spectral-resolution aircraft instrument |

remotely sensed canopy properties that correlate strongly with rates of production. This method offers the benefit of generating productivity estimates that are constrained to known local patterns of growth, but does not consider ecological mechanisms by which estimates can be extended to broader spatial or temporal scales. Of intermediate complexity are the light-use efficiency (LUE) models, which use the remotely sensed fraction of absorbed photosynthetic radiation to estimate maximum C assimilation rates and then adjust for suboptimal climate conditions, using a series of simple climate response algorithms. These are more dynamic in that they can account for temporal variation in climate, but they focus only on current vegetation conditions and generally do not include the ecological or biogeochemical processes needed to simulate change over the course of vegetation development or longer-term environmental change. Of greatest complexity are ecosystem process models, which simulate a wider suite of ecological mechanisms, such as photosynthesis, respiration, litterfall, decomposition, and soil nutrient turnover. Their added complexity allows simulation of a greater range of environmental factors (e.g., pollution deposition, physical disturbance, climate change), often over long time scales, but this capacity comes with added demands in terms of vegetation parameters and environmental data inputs.

In the interest of simplicity, we discuss these 3 categories of remote sensing model linkages as being distinct from one another, but readers should be aware that the boundaries between them are often blurry and hybrid approaches are also available (e.g., LUE models that include mechanisms affecting long-term biogeochemical processes).

## Empirically Derived Production Algorithms

The simplest NPP models are those that consist of field-based empirical relationships between NPP and canopy properties that can be estimated using remote sensing. When such relationships are available, this approach offers a straightforward means of producing estimates that are constrained to known patterns of productivity. The resulting accuracy is dependent only on the strength of the observed trends and on the accuracy of the vegetation property estimates. The principal disadvantage is that these approaches include no mechanisms that would allow extrapolation in time or under varying environmental conditions. Most of the canopy properties that have been examined as potential scalars between plot-based measures of NPP and remote sensing based metrics can be generalized into two groups: (1) biophysical properties such as canopy biomass or leaf area index (LAI), which represent the vegetation surface area available for light absorption, and (2) biochemical variables such as chlorophyll or nitrogen (N) concentrations, which regulate the efficiency with which harvested light can be utilized for C assimilation.

## Detection of Biophysical Properties Using Broadband Sensors and Vegetation Indices

Methods for detecting biophysical properties have a longer history of development and are based on the distinct optical properties of live vegetation in the visible and near-infrared (NIR) regions of the solar spectrum. Whereas leaf reflectance in the visible region is typically low, due to the radiation absorption properties of leaf pigments (cholorphyll and caratenoids), reflectance in the NIR is high because plant cell walls strongly scatter NIR energy. Early research demonstrated that this difference in visible versus NIR reflectance could be significantly related to various properties of canopy "greenness" (e.g., Tucker 1979).

The advent of broadband Earth-observing satellites, such as the Landsat thematic mapper and the advanced very high resolution spectroradiometer (AVHRR) in the 1970s and 1980s resulted in efforts to produce simple metrics that captured variation in vegetation properties across broad spatial scales. The resulting vegetation indices (VIs), based on differing canopy reflectance at various visible and NIR wavelengths, are a composite property representing canopy cover, leaf area, and canopy architecture. Except in optically dense vegetation canopies (i.e., those with high leaf biomass and LAI), VIs tend to increase in a linear manner with increasing leaf area. The use of VIs, particularly NDVI (the normalized difference vegetation index; eq. [11.1] and SR (the simple ratio; eq. [11.2]), to predict LAI, and the application of both as estimators of productivity, has been most effectively demonstrated in forest monocultures and across large moisture gradients where substantial variation in LAI or canopy cover fraction has been correlated with field productivity measurements (e.g., Vose and Allen 1988; Gower et al. 1992; Matson et al. 1994; Fassnacht and Gower 1997; Luo et al. 2004).

$$NDVI = (NIR - red)/(NIR + red). \qquad (11.1)$$

$$SR = red/NIR \qquad (11.2)$$

Despite their advantages and widespread use, index-based methods are often challenged by factors that cause both vegetation indices and LAI to exhibit asymptotic relationships with canopy C assimilation. At high LAI, a decrease occurs in the incremental change in both VIs and C fixation capacity associated with a rise in LAI. This pattern has been well documented and stems from the fact that, as LAI increases, the amount of radiation intercepted by additional leaf layers declines exponentially due to increased self-shading (Gower et al. 1993; Reich et al. 1999a). The result is that relationships between NDVI and LAI eventually become saturated and, at LAI values above 3 or 4, the two variables become increasingly decoupled (Turner et al. 1999). In some systems this is a minor issue, but in closed-canopy forests, which often have mean LAI values of 4 or greater, this can be a substantial limitation. In such instances, wide variation in growth can still occur, but is driven instead by variation in leaf-level physiological capacity and the efficiency with which absorbed radiation is converted into $CO_2$ fixation (Reich et al. 1999b; Smith et al. 2002). The saturation in the NDVI application to LAI is exactly analogous to the saturation observed in the synthetic aperture radar application to biomass, discussed in a later section.

## Detection of Biochemical Properties Using High Spectral Resolution Sensors

More recently, development of methods that allow remote sensing of leaf biochemical properties offer additional means of characterizing spatial patterns in productivity and may provide a solution for areas where LAI-based approaches are most challenged. These methods make use of the more detailed spectral information provided by high-spectral-resolution sensors or imaging spectrometers (e.g., Ustin et al. 2004). An advantage of these instruments over more conventional sensors is that instead of measuring reflected radiation in a small number (typically from 1 to 6) of broad spectral bands, they record reflected radiation over hundreds of narrow and contiguous bands, often covering a spectral range from 400 to 2500 nm. Because of their enhanced spectral coverage, the data they record have been used to detect more subtle forms of ecological variation, including leaf pigments (Fuentes et al. 2001), species composition (Martin et al. 1998; Roberts et al. 1998), the fraction of photosynthetic versus nonphotosynthetic vegetation (Asner et al. 2003), and chemical constituents such as lignin and N concentrations (e.g., Martin and Aber 1997; Smith et al. 2002). At the time of this writing, there are at least a dozen aircraft-based imaging spectrometers in operation, including NASA's airborne visible and infrared imaging spectrometer (AVIRIS) and the commercial HyMap sensor, as well as one orbital sensor—NASA's Hyperion instrument, which is part of the EO-1 satellite. A principal limitation of all existing imaging spectrometers is their small spatial coverage (swath widths of approximately 2 to 10 km), a problem that may be overcome by future sensors.

The benefit of leaf biochemistry detection in studies of terrestrial productivity stems from the well-known relationship between leaf N and photosynthetic capacity in terrestrial plants (Field and Mooney 1986; Reich et al. 1999b). The relationship has its basis in the fact that foliar N is found primarily in cellular proteins and that the principal carboxylating enzyme, Rubisco, makes up a majority of total leaf protein. Evidence supporting the link between canopy N and ecosystem productivity comes from both theoretical and empirical studies. Because N is often the nutrient most limiting to plant growth, it has been argued that natural selection should favor individuals that allocate N in an efficient manner (Hirose and Werger 1987; Hollinger 1989; Field 1991). Sellers et al. (1992) extended this argument to show that C uptake is maximized when N is allocated optimally with respect to available solar radiation. A consequence of this relationship is that it should be possible to determine whole-canopy photosynthesis by knowing the leaf N concentration at the top of the canopy (Sellers et al. 1992). Data from stand-level studies in temperate forests support this notion, having demonstrated significant relationships among NPP, canopy-level N concentrations, and rates of N mineralization in soils (Reich et al. 1997; Smith et al. 2002; Ollinger et al. 2002).

Methods for estimating canopy N using high-spectral-resolution remote sensing have been tested by a number of investigators (Zagloski et al. 1996; Martin and Aber 1997; Townsend et al. 2003), and the usefulness of an N-based approach to estimating biomass production and soil N status has been demonstrated in studies of northern temperate forests (Smith et al. 2002; Ollinger et al. 2002). These analyses combined image data from NASA's AVIRIS instrument with extensive field measurements of canopy chemistry and related ecosystem properties. Field measurements demonstrated that patterns of aboveground NPP were more closely tied to canopy N than to several other commonly measured properties, including LAI and foliar biomass.

## Structural versus Biochemical Vegetation Properties and NPP

Despite the growing number of studies that point to either biochemical or biophysical vegetation properties as useful scalars for biomass production, the relative importance of structural versus biochemical sources of variability over broad spatial scales is still largely unresolved. This is partly due to the fact that individual studies tend to focus on one variable or the other, and rarely measure both over the same range of conditions. Another source of disparity lies in the different spatial scales to which various methods have historically been applied. Given the spatial limitations of imaging spectrometers, most studies of canopy biochemical properties have focused on relatively small landscape units, whereas studies carried out over broader scales necessarily look toward canopy properties predicted by broadband VIs. Future efforts to reconcile these relationships over a range of scales are greatly needed.

In absence of the field data needed to address this question directly, an interesting alternative method involves the combined application of models designed to simulate the behavior of light in forest canopies (radiative transfer models) and those designed to simulate canopy C assimilation (ecosystem process models). To investigate the potential of this approach, a pilot study was conducted using the PnET canopy photosynthesis model (Aber et al. 1995), coupled with the SAIL-PROSPECT model of canopy reflectance and leaf optics (Verhoef 1984; Jacquemoud and Baret 1990). The aim was to test the degree to which LAI versus leaf-level chemistry affects both C uptake and light reflectance over the range of climate- and forest- type conditions in the northeastern United States (using climate and vegetation data from Ollinger et al. 1998). Vegetation parameters that are central to both models include LAI and foliar chemistry, although SAIL-PROSPECT uses area-based chlorophyll concentrations and PnET uses mass-based foliar N concentrations. For this analysis, a linear N-chlorophyll relationship was used, based on data for deciduous tree species from Yoder and Pettigrew-Crosby (1995).

The results from this exercise indicate the combined effects of foliar chlorophyll concentrations and LAI on canopy net photosynthesis and NDVI (fig. 11.1). These results are consistent with earlier studies showing that variation related to LAI tends to saturate above LAI values of 3 or 4, whereas the effect of chlorophyll does not. Nevertheless, the responsiveness to LAI at lower LAI values is apparent, and suggests that methods for simultaneous detection of both variables would be very beneficial.

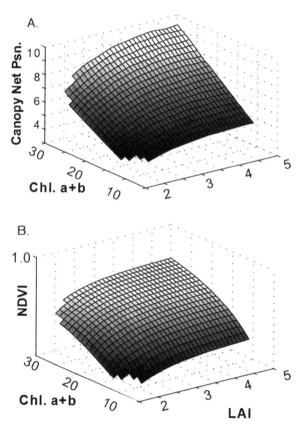

Figure 11.1.  Results of simulations using the PnET and SAIL/PROSPECT models, indicating predicted responses of (A) canopy photosynthesis and (B) NDVI across the northeastern United States to variation in LAI and chlorophyll $a+b$ concentrations ($\mu g/cm^2$). Climate inputs were sampled from a GIS-based climate model over the geographic range covered by northeastern U.S. deciduous forests.

## *Light-Use Efficiency Models*

The existence of time series of moderate resolution, multispectral reflectance data at the global scale, has enabled development of a family of techniques for estimating terrestrial productivity based on the concept of vegetation LUE. These models have evolved from the original arguments of Monteith (1972) that quantum yield of photosynthesis, the amount of C fixed per unit of incident radiation, can be used as an organizing principle for estimating overall canopy productivity. There are now a large number of efficiency models which differ in their details and complexity, but all are based on the idea that knowledge of incident radiation and the light-absorbing properties of the plant canopy can determine the maximum potential photosynthesis for that canopy. That these 2 quantities can be derived from satellite data has caused an increase in the application of LUE models, also called production efficiency models (PEM), as more remote sensing data have become available.

The first remote sensor to produce suitable large-scale data for LUE-based productivity modeling was AVHRR, which has been flown on board a series of National Oceanographic and Atmospheric Administration (NOAA) satellites from 1981 to the present, providing data at a relatively coarse spatial resolution (1 km). Since 1996, data have also been available through a commercial satellite called Systeme pour l'Observation de la Terre (SPOT), which has a radiometer (VGT) with similar properties as AVHRR but with more spectral bands, improved radiometric and geometric characteristics, and improved capacity for atmospheric correction. A recent mission led by NASA resulted in two additional satellites called Terra and Aqua, one with a morning overpass and one with an afternoon overpass. Terra and Aqua have been in orbit since 1999 and 2002, respectively, and both have an instrument called the moderate resolution imaging spectroradiometer (MODIS) that embodies similar improvements relative to AVHRR and VGT, but data are available at essentially no cost to researchers.

Monteith (1972) posited that terrestrial ecosystems are living machines whose metabolism and growth are driven primarily by the thermodynamic force of incoming solar radiation. In this framework, one needs only to consider the breakdown of efficiencies of use of this energy in order to model the individual or aggregate behavior of vegetated systems. He considered that there were seven factors, each with its associated efficiency ($\varepsilon$), and that the factors control all aspects governing the ratio between the amount of light incident at the top of the atmosphere and the eventual amount of fixed C. Thus, three of the factors did not relate to vegetation characteristics and are currently considered exogenous, for example, the amount of light transmitted by the atmosphere to the top of the canopy. The remaining four factors quantified the biochemical conversion and the effects of canopy structure. Most of the subsequent and current PEM approaches have been applied using coarse resolution data (1 km$^2$ or larger) across regional-to-global scales. These typically do not consider efficiency at the same level of detail as in Monteith's paper, primarily because of the lack of data at these scales to support the disaggregation of the canopy biochemical efficiency terms, although some recent research suggests the possibility of a unification of LUE and other approaches based on radiative transfer modeling.

Nearly all applications of PEM-type methods using satellite remote sensing data are based on the idea that the rate of C accumulation by plants ($P$) depends on environmental and biochemical factors in the following way:

$$P = \varepsilon * A * fAPAR * PAR,\tag{11.3}$$

where $\varepsilon$ represents the photochemical conversion efficiency of leaves under optimal conditions (g MJ$^{-1}$), and $A$ (dimensionless) represents the degree to which actual conditions are less than optimal. For example, the effects of moisture, temperature, or humidity could be represented by decomposing $A$ into a set of scalars that decrement photosynthesis by the appropriate amount. The quantity PAR (MJ) is the amount of photosynthetically active radiation incident on the canopy. In principle, all the quantities in equation (3) can be time-varying, but the parameter that is considered to be the most central for productivity modeling is fAPAR (dimensionless), the fraction of incoming PAR absorbed by the canopy. This variable reflects the changing capacity of the canopy to harvest available light, and its potential for estimation using multispectral satellite data has driven many applications of PEMs in large-scale productivity analyses.

The principal variations and uncertainties of LUE models can be explored by discussing the four terms of equation (11.3). A given LUE-based productivity model, in a given spatial or temporal context, must assign values for conversion efficiency, environmental factors, the amount of incident PAR, and the fraction of absorbed PAR. It is conventional to think of a PEM as being driven by satellite-derived estimates of fAPAR, with the remaining terms assigned on the basis of field studies, other models, or other satellite data. Nevertheless, each of the terms carries its own assumptions, data requirements, and set of possible approaches.

The biochemical efficiency parameter $\varepsilon$ represents the key link to plant physiology, but for the typically large spatial scale applications of PEMs (where variation in $\varepsilon$ is poorly known), this parameter is meant to generalize extremely broadly about the behavior of leaves and canopies. In equation (11.3), $P$ usually signifies NPP because $\varepsilon$ is conventionally determined by observing the amount of plant dry matter accumulated over time, relative to the total intercepted light. Some recent studies use a separate model for autotrophic respiration. The meaning of equation (11.3) remains the same, except that the conversion efficiency refers to photosynthesis only, or *gross* primary productivity. Xiao et al. (2004) use the subscripted forms $\varepsilon_n$ and $\varepsilon_g$ to indicate this important distinction between net and gross conversion.

A persistent challenge for efficiency models in general has been the lack of understanding of factors controlling variation in $\varepsilon$ both within and among vegetation types. Individual studies have suggested factors such as stand age, species composition, soil fertility, and foliar nutrients (Gower et al. 1999), but in the absence of a firm predictive understanding, most PEMs use either a global mean $\varepsilon$ value for all vegetation types (Potter et al. 1993) or rely on a lookup table that assigns single values for individual biomes (Running et al. 2000). Although these approaches are satisfactory in many circumstances, measured values of $\varepsilon$ vary considerably and, without some means of describing this variability, an important driver of C assimilation in real ecosystems remains undetected. A meta-analysis by Green et al. (2003) offers some promise for how this challenge might be overcome. The authors compiled published values of ε and a variety of leaf and canopy-level traits from a wide array of C3 plant communities, including deciduous and evergreen tree species, and herbaceous species consisting of grasses, forbs, and legumes. Their results showed that of all factors considered, the single variable that explained the majority of the observed variation in ε was the mass-based leaf N concentration. This result suggests a potentially promising synergy between PEMs and future high spectral resolution instruments.

The estimation of fAPAR from satellite data is the central remote sensing research question associated with the development of PEMs. Shortly after the launch of AVHRR, it was shown that the pattern of differential reflectance of NIR reflectance relative to red reflectance was broadly related to field- based data on foliar biomass, LAI, NPP, and the radiometric quantity fAPAR. These relationships all follow from the fact that photosynthesis uses energy in the PAR portion of the spectrum without affecting NIR reflectance. Thus, dense plant canopies appear brighter in the NIR and darker in the visible regions than sparse canopies (fig. 11.2). This differential reflectance can be summarized with one of several indices, but the most widely used are NDVI and the SR index (equations [11.1] and [11.2]). Data from sensors that have a blue band can also be used to calculate an "atmospherically resistant" index called the enhanced vegetation index:

$$EVI = G*(NIR - Red)/(NIR + (C1 * Red - C2*Blue) + L), \qquad (11.4)$$

where the constants G, C1, C2, and L are chosen to minimize the contaminating effects of soil and atmosphere variations.

Focusing on fAPAR as the canopy biophysical parameter of interest for PEMs, there is a theoretical basis for the correlation between these indices and the amount of PAR absorbed by vegetation (Sellers 1987; Myneni and Williams 1994), but because of the complexity of the radiation regime of canopies, there is no simple analytical formula. Sellers (1987) showed that under certain assumptions that probably generally hold true (e.g., that canopies are approximately twice as optically thick in the NIR as in the visible), SR is indeed proportional to fAPAR, given a simplified model of radiation transfer. The SR index is the preferred index for some PEM models (e.g., Potter et al. 1993), but algebraically it can be shown that SR and NDVI are approximately collinear except for extremely large values of NDVI. With more complicated models of radiative transfer (e.g., Myneni and Williams 1994), a conclusion similar to those based on field studies can be reached, which is that vegetation indices and fAPAR are strongly correlated but also strongly contingent upon several other local conditions. Those conditions include the state of the atmosphere, the type and color of soil or litter, the fractional ground cover of the canopy, and the geometry of the observation. Some LUE models (e.g., Running et al. 2000) therefore use fAPAR generated by mathematical inversion of the radiative transfer models, rather than by linear transformation of SR or NDVI.

Though PEM models are generally thought of as being at least potentially global in scope, most applications have been regional. As can be seen in a review of the literature, most PEM studies have been performed on crops, rangelands, and boreal forests, a focus that may reflect the economic value of these ecosystems.

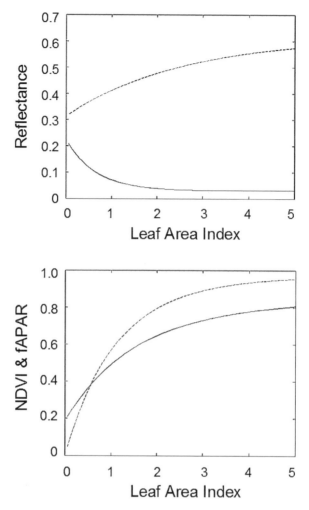

Figure 11.2. Results of a simple radiative transfer model that illustrate the underlying physical relationships that form the basis of LUE models. (Top): The differential effect of increasing leaf area index on red (solid line) and NIR (dashed line) reflectance. (Bottom): The resulting saturation curves for NDVI (solid line) and fAPAR (dashed line). It follows that NDVI and fAPAR will be approximately linearly correlated.

More recently, efforts to derive continuous NPP estimates at the global scale by combining fAPAR-based efficiency algorithms with data from the MODIS satellite have begun to take shape (Running et al. 2004; fig. 11.3). Major limitations of this approach are the relatively coarse spatial resolution (1 km), the fact that photosynthetic efficiency and foliar nutrient concentrations must be held constant within very general biome categories, and the challenge of validating predictions over such large spatial scales (e.g., Turner et al. 2003a). Nevertheless, the availability of continuous global data that can allow continuous monitoring of the NPP response to

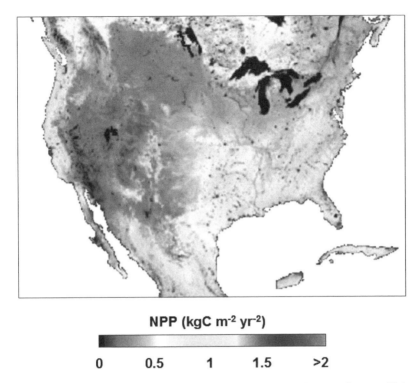

**NPP (kgC m$^{-2}$ yr$^{-2}$)**

0        0.5        1        1.5        >2

Figure 11.3. Predicted 2003 NPP over the continental United States, using an efficiency model designed to allow global monitoring of NPP with data from the MODIS satellite (Running et al. 2004).

factors that are accounted for (climate, LAI) is of obvious value for addressing a range of global-scale issues.

## Ecosystem Process Models

Ecosystem process models use remote sensing primarily to initialize important vegetation input variables and then simulate ecological processes—such as photosynthesis, C allocation, respiration, litterfall, decomposition, and water balances—that affect the NPP of an ecosystem. The added complexity in these models allows them to predict a range of additional variables and to examine responses to environmental factors such as rising $CO_2$, atmospheric pollution, and physical disturbance. Because they are often designed to be run over longer time scales, they are more suitable for considering changes in ecosystem components such as soil C and nutrient pools that have very long turnover times.

Rather than reviewing the structure and characteristics of individual models, the focus here is on the types of input variables on which these models rely that can be obtained from remote sensing. Methods and sensors available for deriving variables mentioned in the two previous subsections will not be repeated here (e.g., use of

VIs for estimating LAI). For a more thorough review of the linkages between ecosystem process models and remote sensing, we direct readers to the review provided by Turner et al. (2004).

## Vegetation Cover Type

As an initial step in parameterization, nearly all models that predict spatially distributed primary productivity in terrestrial ecosystems require some information about the vegetation or land cover type being simulated (e.g., Kimball et al. 2000; Turner et al. 2003b). Because vegetation functional classes are often adapted to specific sets of environmental conditions, they can differ greatly in terms of basic properties such as morphology, leaf life span, and C allocation patterns. Land cover classification maps provide at least a first step toward assigning appropriate values for the required parameters. Often, the challenge comes with determining how finely divided vegetation classes should be and how well remote sensing data sources can detect them.

The requirements of models in terms of specificity of vegetation classifications depends on the degree to which important parameters vary among different species and functional groups. Frequently, modelers face a trade-off between the spatial extent of a particular model application and the degree of parameter specificity that can be achieved. For instance, development of a broadly aggregated land cover map of the northeastern United States has allowed regional-scale NPP simulations to include differences in parameters such as foliar N, specific leaf weight, and leaf longevity in deciduous, evergreen, and mixed forest types (Ollinger et al. 1998). However, variation in these parameters within such broad classes can be quite large, and cannot be captured without more refined vegetation maps or independent methods of deriving these parameters directly. Additional difficulties are encountered when bringing together analyses from different regions that use different land cover classification schemes. At the global scale, efforts have been under way to reach a standardized classification for major functional types (Friedl et al. 2002), but landscape-to-regional efforts that require higher spatial resolution or greater vegetation specificity often need to derive land cover maps independently (Turner et al. 2003b).

## Leaf Area Index and Canopy Height

Because LAI provides a measure of the foliar surface area available for capturing solar radiation, models that base productivity on photosynthetic rates calculated over multilayered plant canopies often require an explicit LAI input (e.g., Running and Gower 1991; Liu et al. 1999). As described earlier, LAI estimates are often derived using simple VIs, but here, too, the problem of saturation in high LAI systems is an important challenge (Turner et al. 1999). Methods that involve use of multiple VIs and multiple image collection dates offer some improvement over single VI methods (Cohen et al. 2003), and new methods involving active sensors may offer further improvement still. For example, InSAR and Lidar instruments (described in later sections) offer the benefit of sampling the vertical foliage distribution, giving them the capacity to provide detailed information on LAI as well as canopy height

(Lefsky et al. 2002a; Treuhaft et al. 2004). Hurtt et al. (2004) demonstrated the utility of this type of data by combining information on vertical structure in a Costa Rican rain forest with a height-structured vegetation model (Moorcroft et al. 2001).

## Foliar N

The positive relationship between foliar N concentrations and maximum rates of net photosynthesis has been demonstrated for a large number of biomes and plant functional types, and has become a core component of a number of process-based ecosystem models (Running and Gower 1991; Comins and McMurtrie 1993; Aber et al. 1995). Spatial variation in foliar N is driven by a variety of interrelated factors, including species composition, site disturbance history, soil N, water availability, and climate (Yin 1993; Ollinger et al. 2002). Although foliar N variation is often assumed to be primarily of local importance, variation over broad spatial scales can be substantial, and is believed to be driven by the effects of climate and radiation on patterns of optimal N allocation in plants (Yin 1993; Haxeltine and Prentice 1996).

To date, only a small number of studies have made use of remotely sensed foliar N data as input for spatially distributed ecosystem models, but substantial improvements in prediction accuracy have been obtained over similar efforts where foliar N data were lacking. In one such example, Ollinger and Smith (2005) used the PnET ecosystem model to predict NPP at 18 m spatial resolution for the Bartlett Experimental Forest in north-central New Hampshire, and evaluated predictions using field measurements from a network of 39 inventory plots. When the model was run using mean foliar N values for deciduous, evergreen, and mixed forest types, agreement was reasonable in terms of the overall mean for the entire study area, but was poor on a plot-by-plot basis. When the model was run with foliar N inputs derived using the AVIRIS instrument, predictions showed a much higher degree of plot-level agreement and revealed landscape-scale spatial patterns associated with topography and forest management history (fig. 11.4).

## Vegetation Properties Using Active Sensors

An active sensor transmits electromagnetic radiation, which is reflected from elements of the Earth's surface. These reflected signals travel back to the sensor, where their detection enables estimation of properties of the vegetation or ground surface. The two types of active sensors discussed in this section are radar (microwave) and lidar (optical). Although passive microwave sensors have useful sensitivities to the quantities of interest in the NPP measurement (Pampaloni 2004) with 5- to 10-km resolution, active microwave sensors provide 5- to 100-m resolution. Another advantage of active over passive sensors is that both microwave and optical active sensors enable vegetation structure measurements via received phases or time delays, as will be described below. Active measurements, however, are generally more expensive and complex, because they require transmitting as well as receiving hardware on spacecraft or aircraft.

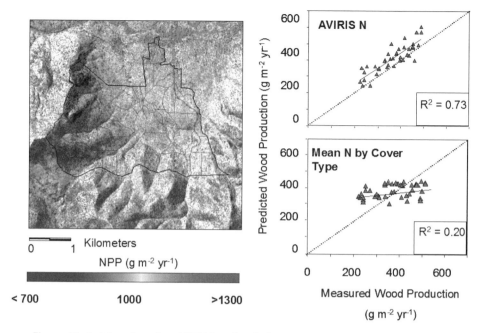

Figure 11.4. Map of predicted NPP (g m$^{-2}$ yr$^{-1}$) for the Bartlett Experimental Forest, NH, generated using AVIRIS-derived foliar N as input to the PnET ecosystem model. Shadowing indicates local topography. At right are comparisons of predicted and observed wood growth, first using the AVIRIS foliar N inputs (top) and then using mean foliar N values for deciduous, evergreen, and mixed cover types (bottom). Redrawn from Ollinger and Smith (2005). (See the cover for a color version.)

Whether microwave or optical, active remote sensing detects reflections of a transmitted beam from components of a vegetated land surface. Electromagnetic reflections arise whenever there is a discontinuity in a key electromagnetic property called the "dielectric constant." The dielectric constant of an object—a leaf, branch, trunk, or the ground—depends on its chemical composition. Its square root is inversely proportional to the speed with which an electromagnetic wave propagates in the medium, and directly proportional to how much of it is absorbed. Vegetation and ground surfaces reflect because their dielectric constants are higher than that of air, creating dielectric discontinuities at their surfaces. Figure 11.5 shows schematically that the signal reflected back to an active sensor depends on both the strength and the number of the discontinuities as well as on the organization of the reflecting objects in the scene. That is, it depends on what the objects are, how many there are, and where they are. The following subsections describe two conceptual ways in which radar and lidar sensors can contribute to estimating NPP in ecosystems (1) by determining biomass at multiple time periods and thereby estimating a biomass change rate (multiple-epoch biomass method), and (2) by determining other properties of the vegetation which, when combined with empirical production algorithms or process models, such as those described earlier, directly correlate with NPP (direct NPP correlate method).

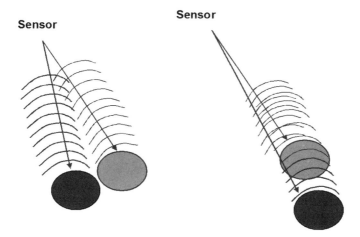

Figure 11.5. A schematic demonstration indicating that (Left) a strong reflection from the dark object, indicated by dark black wavefront lines, and a weaker one from the light object have an additive effect on the signal detected at the sensor. (Right) With the same 2 reflecting objects, the signal from the dark object is attenuated if it is vertically behind the light object. The total received signal will be weaker than that of the left portion of the figure, though the reflecting objects are the same; only their spatial locations have changed.

### Radar Measurements: Biomass and Its Accumulation over Time

The application of radar to estimates of NPP is in its infancy. Nevertheless, given the direct relevance of radar sensors to measurements of vegetation biomass, a review of the prospects for estimating NPP with products currently derived from radar is warranted. There are two types of radar that potentially can be applied to NPP measurements (fig. 11.6): (1) synthetic aperture radar (SAR) and (2) interferometric synthetic aperture radar (InSAR). SAR is primarily sensitive to the amount of vegetation in a scene, and InSAR is primarily sensitive to the vertical distribution of the vegetation.

Before proceeding, it should be stressed that vegetation biomass, biomass accumulation, and NPP are three distinct properties, and relationships among them vary over time and between ecosystems. Hence, the ability to detect biomass or its accumulation over time does not translate directly to an ability to estimate NPP. The relationship between NPP and biomass accumulation on a vegetated land surface can be generalized as follows:

$$NPP = \text{Increase in standing biomass}$$
$$+ \text{Biomass lost to mortality and litter production,} \qquad (11.5)$$

where the contributions of herbivory and C exudates are ignored. In young systems dominated by perennial vegetation, biomass accumulation can represent a large fraction of NPP because growth rates tend to be considerably higher than death rates. In maturing systems, however, biomass accumulation declines and typically

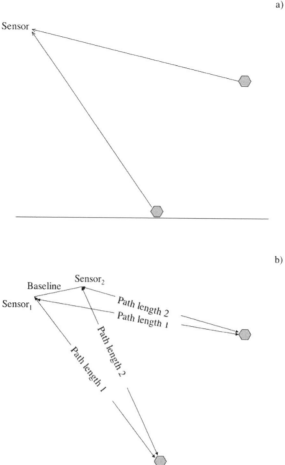

Figure 11.6. (Top): SAR uses 1 sensor to receive power reflected back to it by 2 green schematic vegetation components. SAR is sensitive to the powers induced by the components, but not to their location. (Bottom): InSAR uses 2 sensors and is sensitive to the powers as well as the vertical location of the scatterers via the differences in path length of each component to the 2 sensors.

becomes small with respect to NPP, due to increasing rates of both mortality and litterfall. In systems dominated by annual grasses and herbs, changes in biomass can be equal to NPP over seasonal time scales, but independent of NPP when viewed over multiple years. Despite the variability in NPP-biomass relationships, most NPP estimation methods require knowledge of biomass at some stage of their application. Hence, methods of direct biomass detection are often beneficial, particularly where estimates of other important properties are available. This section discusses how biomass measurements are made with radar and the prospects of using multiple-epoch measurements to estimate a rate useful to NPP determination.

## Biomass from SAR

Qualitatively, SAR measures the power of the returning signal due to vegetation using a single sensor (fig. 11.6a). SAR is directly sensitive only to the strength and number of the dielectric discontinuities, as suggested by figures 11.5 and 11.6. While SAR is affected by the location of components of reflecting surfaces—the signal at the sensor in figure 11.5 (right) is weaker than that of 11.5 (left)—SAR cannot tell the difference between changes in location and changes in component strength and number. In figure 11.6a, the received signal is independent of the vertical position of the 2 schematic vegetation elements shown.

For microwave radiation, the dielectric discontinuities depend principally on the water content of the vegetation or the ground surface. More discontinuities mean, for example, more foliage, which in turn can mean more biomass. SAR estimations of biomass (e.g., Dobson et al. 1995; Paloscia et al. 1999; Santos et al. 2003) assume that more power means more biomass. The biomass estimation procedure starts with the following general, often empirical, relationship between biomass and SAR powers:

$$f(biomass) = g(SAR\ power_1, SAR\ power_2 \ldots SAR\ power_{N;} \\ parameters_{1 \ldots M}) \tag{11.6}$$

where $f$ is some function of the biomass and $g$ is another function of SAR powers 1 through N and parameters 1 through M. The parameters must be determined by test plots, and then applied to other plots to estimate biomass, using equation (11.6). The internal consistency of equation (11.6) for a single set of plots is also used to demonstrate biomass estimation and gauge its error.

The different SAR powers in equation (6) arise from differences in polarization (Marion 1965) and frequency. Using SAR of different polarizations improves vegetation and surface characterization if there are "oriented" objects, such as the ground (Papathanassiou and Cloude 2001). Generally, lower frequencies (e.g., P-band at 80 cm wavelength) penetrate further into vegetation and scatter from larger objects than higher frequencies (C-band at 6 cm wavelength and X-band at 3-cm wavelength). Spaceborne demonstrations of biomass estimation have used JERS (Japanese Earth Resources Satellite, L-band, 25 cm), ERS (European Remote Sensing, C-band), RadarSAT (C-band), and SIR-C (Shuttle Imaging Radar, L-, C-, X-band).

Assigning a single accuracy to biomass estimation with SAR is difficult because virtually every study uses different forms of $f$ and $g$ in equation (6), and different numbers of parameters and input SAR powers. Forest biomass typically falls within the range of 0–700 Mg/ha, and is often less than 300 Mg/ha, except for tropical and old-growth forests, which populate the high end of the range. Average errors with 1–3 parameters in equation (6) have been reported in the 30% range with JERS-1 (Luckman et al. 1998). Some experiments which sort results by species, structure, or other metrics using more ground data show errors of less than 10% with SIR-C (Dobson et al. 1995). Many reports say that the estimation error may be due in large part to field measurement error, in which case the intrinsic radar error might be much smaller than the above. This hypothesis has yet to be carefully tested for SAR experiments.

Virtually all experiments agree with the implications of figure 11.5 regarding the ambiguity introduced by vegetation structure (e.g., Imhoff 1995). Although SAR powers respond to vegetation vertical structure, they cannot be used to uniquely specify the structure. A lack of knowledge regarding structure induces errors in the conversion of radar observations to biomass, as in equation (6). Figure 11.5, for example, would be interpreted as two different biomasses, though only the vertical organization of the vegetation differs between the figures.

Two salient features are used to describe biomass estimation accuracy, the location of the "saturation point," where the curve flattens out, and the scatter about a smooth trend (fig. 11.7, top). The saturation point generally occurs at lower biomasses for higher frequencies. The scatter indicates the biomass error. In the figure, below 100 Mg/ha, for example, scatters indicate that an approximate range of 10–20 Mg/ha of biomass, 20–40% of the biomass, could correspond to a single radar power.

## Biomass from InSAR

InSAR utilizes two receivers to view the vegetation from two different perspectives (fig. 11.6b). In addition to being sensitive to the reflecting strength and the number of vegetation components, InSAR is directly sensitive to the altitudes of the components of the vegetation above the ground surface via the path length difference from each component to the two sensors (Treuhaft et al. 1996). A single InSAR observation yields an InSAR phase, proportional to the average height of the vegetated surface, and an InSAR coherence, which decreases as the vegetation becomes more vertically distributed (Treuhaft et al. 2004). If one receiver is flown by a site at two different positions, a repeat-track baseline is formed. If there are two real receivers, an instantaneous baseline is formed. Biomass estimation has been reported a few times using repeat-track, spaceborne baselines of ERS and JERS-1 (Luckman et al. 2000; Santoro et al. 2002; Askne et al. 2003; Pulliainen et al. 2003; Wagner et al. 2003). The repeat-track studies, which derive their sensitivity to vegetation volumes from the changes in the scene during the observation epochs, generally show that C-band InSAR is much more useful for biomass estimation than C-band SAR. This conclusion is further supported for fixed-baseline InSAR by a simple model calculation (Treuhaft and Siqueira 2004). ERS repeat-track InSAR passes generally outperform JERS-1 because 1-day repeats could be obtained with the ERS-1 and ERS-2 satellites, whereas JERS repeated in 45 days. With longer repeat times, the degree of decrease in coherence no longer discriminates between different volumes of the forest, as it does for 1-day repeats (Askne et al. 2003), owing to excessive coherence loss induced by scene changes between repeat tracks. For frozen boreal forests, Santoro et al. (2002) and Askne et al. (2003) report the best biomass estimation precision using repeat-track InSAR coherence from ERS-1 and ERS-2 of about 25% on 2–20-ha stands of up to about 200 Mg/ha.

Biomass estimation has been reported only once, using instantaneous baselines from the airborne AirSAR (Treuhaft et al. 2003). In the instantaneous baseline experiment, the phases and coherences from six C-band baselines were used to estimate the vertical profile of leaf area density (LAD) for stands predominantly of ponderosa pine, grand fir, and larch in the Metolius River basin in central Oregon. Multiple baselines are required to estimate vertical profiles (Treuhaft et al. 2002) and, generally, multiple baseline, multiple polarization (Papathanassiou and Cloude 2001), multiple frequency, or ancillary information (Treuhaft et al. 1996; Kellndorfer et al. 2004) is required to quantitatively estimate vertical characteristics of forests. The profiles estimated in the multiple-baseline biomass experiment were normalized by LAIs estimated from airborne hyperspectral data from AVIRIS. Unlike equation (11.6) or the repeat-track InSAR experiments, biomass was correlated with structural variables estimated from the remote sensing rather than raw observations. Specifically, figure 11.7 (bottom) shows field biomass versus remote sensing biomass, calculated as

$$biomass = a + b \, (LAI^*\sigma_{LAD} + \langle z \rangle_{LAD}), \qquad (11.7)$$

where LAI is the leaf area index from hyperspectral data, $\sigma_{LAD}$ is the standard deviation of the LAD distribution from InSAR, $\langle z \rangle_{LAD}$ is the LAD-averaged vegetation height from InSAR, and $a$ and $b$ are best-fit parameters. The scatter of the remote sensing about the field measurements was 25 Mg/ha, or about 16% of the average biomass of the 1-ha stands. Note that biomass reported from C-band SAR power saturates at about 50 Mg/ha (Imhoff 1995), but there is little evidence of saturation based on C-band InSAR (fig. 11.7, bottom). This further suggests that InSAR saturation characteristics are more favorable than those of SAR for biomass estimation. Furthermore, the accuracy indicated by the scatter of figure 11.7 (bottom) is better than most two-parameter determinations from radar power, and is about the same as the best lidar determinations (e.g., Drake et al. 2002). Additionally, Treuhaft et al. (2003) estimate about a 15% error in the field biomass measurements, possibly implying very high performance for the remote sensing technique. As noted above, an experiment in which the field errors were reduced substantially would have to be conducted to claim much higher accuracy for InSAR-based remote sensing.

Though the results of figure 11.7 (bottom) are significant with 99.5% confidence, they are the first of their kind, sample only one conifer forest, are based on only 11 stands, and must be regarded as a suggestion of the potential of InSAR-determined, structure-based biomass, rather than a demonstration of reliable performance. Extensive tests of the InSAR-profiling approach in other vegetation types are needed. Although virtually every study using radar power to estimate biomass cites unknown structure ambiguities (e.g., fig. 11.5), studies in which estimated InSAR structure is used to interpret and correct radar power estimation have yet to be done.

## Multiple-Epoch Biomass Estimates

In young, rapidly aggrading ecosystems, where the accumulation of biomass over time represents an important fraction of NPP, radar-based biomass estimates have a clear application when acquired over two or more time periods. In more mature

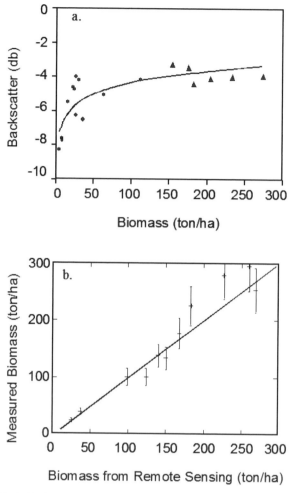

Figure 11.7. (Top): Radar P-band backscatter power versus field biomass in tropical forest stands of 0.1–0.25 ha, showing typical increases of power with biomass and a flattening out or saturation of power versus biomass at higher biomasses. Triangles are primary forest, and circles are secondary succession. The line is the function indicated. Redrawn from Santos et al. (2003). (Bottom): Biomass from C-band InSAR and hyperspectral remote sensing versus field biomass for 11 stands of ponderosa pine, grand fir, and larch in central Oregon. Relative vertical leaf area density profiles were estimated from multiple-baseline InSAR using AirSAR and normalized by a leaf area index from AVIRIS hyperspectral data. The line is y = x. Redrawn from Treuhaft et al. 2003.

systems, as standing biomass reaches its maximum, biomass accumulation declines toward zero, and the importance of combining biomass estimates with measurements of ingrowth, mortality, and litter production becomes paramount. In the maturing forests of eastern North America, for example, the incremental change in biomass is often less than 10% of NPP on an annual basis, and is an even smaller fraction of the standing biomass observed at any one time. In such cases, biomass must be remotely sensed with a few percent accuracy in any one year to contribute meaningfully to NPP determination using multiyear estimates. This can plausibly be accomplished with repeated measurements using InSAR and to a lesser degree with SAR. Multiple InSAR coherence-based experiments have yielded approximately 13% accuracy in biomass retrieval over boreal forests (Santoro et al. 2002), whereas Salas et al. (2002) suggested that repeated SAR measurements yielded accuracies at the 30% level. Higher accuracies could also be accomplished with fewer measurements if the fundamental performances of the SAR and InSAR approaches were improved. Quantifying the accuracy of the field measurements themselves will also be important to further the assessment of radar-based methods.

Virtually all radar biomass demonstrations that could be used in the multiple-epoch method measure aboveground biomass. Belowground NPP can be equal to 20%–50% of the aboveground NPP (Gower et al. 1992; Fahey et al. 2005). Therefore, to date, radar biomass measurements can be used only for the aboveground component of NPP, rather than the total NPP. The degree to which aboveground biomass rate can be reliably correlated with belowground biomass must be investigated, perhaps along with structural correlates from InSAR and lidar.

Biomass accumulation estimates using radar-based approaches are also possible in grasslands, crops, and herbaceous communities, although measurements are required over seasonal, rather than annual, time scales. In one example, Moreau and Le Toan (2003) show approximately 30%–50% biomass estimation accuracy for water-saturated grasslands observed at C-band with ERS. Ferrazzoli et al. (1997) obtained an accuracy of about 50% in the estimation of aboveground crop biomass using P-, L-, and C-band AirSAR over plots containing corn, sunflowers, sorghum, and wheat. These accuracies should translate to NPP accuracies of about the same order, but a correlative analysis of aboveground and belowground NPP would indicate the performance for the larger belowground component. Furthermore, observed accuracies depend somewhat on the availability of a so-called ground bounce due to the water in the ground. The task of correcting for standing dead tissues represents an additional challenge, though one which may be minimized in systems that are annually burned or harvested.

### Radar Measurements: Direct NPP Correlates

This section outlines the potential for estimating biophysical quantities from radar observations which correlate directly with rates of NPP. This link could be achieved through use of empirical algorithms or via process models. Because there are few examples of this approach in the literature, the discussion will be brief. As mentioned above, SAR correlates primarily with the amount of material, or the aboveground biomass, in a vegetated land surface. Under some circumstances, the biomass itself, rather than the biomass change between two or more time periods, could be considered as an NPP correlate. Across biomes, there is a coarse relationship between biomass and aboveground NPP (Webb et al. 1983). Within biomes, however, wide variation around this trend can be expected from factors such as stand age, species composition, disturbance history, and edaphic properties. Biomass from

SIR-C/X-SAR has been used with an allometric model of aboveground NPP (Bergen and Dobson 1999), but many additional field constraints were required. It is possible that with adequate ancillary information and sufficient diversity in frequency and polarization, this approach could represent a useful complement to other NPP detection methods.

Structural properties retrieved from InSAR are affected by factors such as species composition, stem density, stand age, and vegetation height, all of which relate to NPP in some way or another (Waring and Running 1998; Mencuccini and Grace 1996; Kicklighter et al. 1999; Smith et al. 1999). As an example, secondary forests that are in an early state of regrowth following harvesting contain less biomass, but are often more productive than more mature forests (Pregitzer and Euskirchen 2004). Hence, their vertical structure as measured by InSAR may well be useful as a correlate with NPP, although here, too, information regarding other properties of the sites and vegetation conditions would likely be required. The optimal means of direct-correlate estimation from InSAR and other techniques, perhaps combined with climate inputs that can also be derived from remote sensing (e.g., Running et al. 2004), are subjects for future research.

## Lidar Measurements

Laser altimetry, (or lidar, light detection and ranging) is an emerging remote sensing technology with a variety of applications of interest to terrestrial ecologists (Lim et al. 2003; Dubayah and Drake 2000; Wehr and Lohr 1999; Lefsky et al. 2001, 2002a). Lidar-derived metrics have proven effective for predicting ecological variables such as canopy height and structure, the density of forest cover, biomass, and light transmittance (Drake et al. 2002; Lefsky et al. 2002a; Means et al. 1999; Harding et al. 2001; Parker et al. 2001). In particular, the demonstrated capability of lidar to characterize the amount of standing biomass in an ecosystem provides a strong foundation for determining rates of primary productivity (but note the previous discussion on distinctions and relationships between biomass and productivity). After a description of the lidar measurement, the multiple-epoch biomass method from lidar data will be followed by the direct NPP correlate method, just as in the radar section.

## The Lidar Measurement

The basic measurement made by a lidar device is the distance between the sensor and the target, obtained by an accurate measurement of the time elapsed between a pulsed signal emission and the return signal. Lidar instruments currently in use can be described as either *discrete return* or *full waveform* lidar (Lim et al. 2003). Discrete return lidar instruments measure a single (or a few) vertical distance(s) within the lidar footprint, often the first and last signal returns. Discrete return lidar instruments typically operate at a high spatial frequency, with a small footprint, and are optimized to provide detailed information on ground elevation and canopy surface. In contrast, waveform recording lidar instruments record the time-varying intensity of the return signal, and thereby yield an increased amount of information on the vertical distribution of material within the footprint. Lidar sensors are used in combination with other instruments, such as global positioning system (GPS) receivers, to obtain the sensor position, and inertial navigation systems (INS), to measure the attitude (roll, pitch, yaw) of the lidar platform (Blair et al. 1999). This combination of instrumentation provides information on the vertical and horizontal location of the lidar data to a high level of accuracy. Lasers for terrestrial applications generally operate at wavelengths in the infrared range of 900–1064 nm, where vegetation reflectance is at a maximum. As with other optical remote sensing techniques, lidar is limited to operation during cloud-free conditions (Lefsky et al. 2002a).

Laser altimetry has been successfully deployed from a variety of airborne platforms, from near Earth orbit during the shuttle laser altimeter (SLA) missions, and in deep space via the Mars observer laser altimeter (MOLA; Flood 2001). Newly developed research instruments, including the Laser Vegetation Imaging Sensor (LVIS), developed at NASA's Goddard Space Flight Center in the 1990s (Blair et al. 1999), and its predecessor, the scanning lidar imager of canopies by echo recovery (SLICER; Harding et al. 2000) have expanded the capability of traditional laser altimeters to sample vegetation by recording a full waveform return signal. The spaceborne geoscience laser altimeter system (GLAS), launched in January 2003 onboard the ICESat satellite, also combines the function of accurately mapping land and ice surface elevations with the measurement of vegetation structure (Zwally et al. 2002). There are currently fewer than 200 lidar sensors operational for commercial applications worldwide (with the majority based on discrete-return systems), but growth in this sector remains rapid (Flood 2001).

Digitization of the lidar return signal with continuous, very high temporal resolution provides the capability of generating a number of metrics useful in determining vegetation structural characteristics. These include the mean elevation of the lowest detected mode (ground elevation), highest detected return (canopy height), heights (relative to ground elevation) at which the cumulative return energy equals a specified percentage of the total return—in particular, the height at which 50% of the cumulative return energy occurs, the "height of median energy" (Blair et al. 1999; Hofton et al. 2000; Drake et al. 2002; fig. 11.8).

## The Multiple-Epoch Biomass Accumulation from Lidar Observations

Relationships between lidar metrics and allometric estimates of aboveground biomass can be used to derive estimations of aboveground biomass which, if applied over multiple time periods and in conjunction with data for other important variables, can be used to estimate aboveground NPP. Canopy height distributions, derived from small-footprint, discrete return lidar, have been related to basal area, volume, and aboveground biomass in a number of studies (Patenaude et al. 2004; Naesset 2004, 2002; Nelson et al. 2003; Magnussen and Boudewyn 1998). Studies using full-waveform return lidar have estimated these same structural parameters in forested ecosystems, using a combination of metrics derived from waveform data. Several published waveform studies report that more than 78% of the variance in

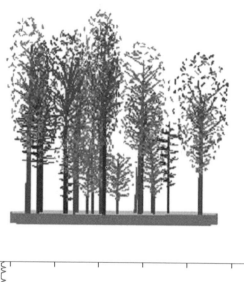

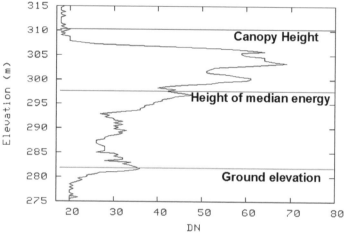

Figure 11.8.  Full waveform lidar example. (Top): Canopy simulation derived from field measurements. (Bottom): Waveform data corresponding to the canopy depicted at top, indicating the fully digitized waveform data and ground elevation, top of canopy, and height of median energy in the return signal. Data from a 2003 LVIS flight over the Bartlett Experimental Forest, Bartlett, NH.

aboveground biomass can be explained by lidar metrics (Lefsky et al. 1999a, 1999b, 2005a; Means et al. 1999; Nilsson 1996; Drake et al. 2002). These studies have been conducted in a number of different biomes, including temperate deciduous, temperate coniferous, tropical wet forest, and boreal coniferous biomes. In addition, Lefsky et al. (2002b, 2005b) and Drake et al. (2003) have begun exploring the generality of relationships between lidar metrics and allometric estimates of above-ground biomass across contrasting biomes and among sites of different productivities within a biome. Further investigations of these cross- and within-biome relationships will play a significant role in scaling from small-scale airborne lidar estimates of biomass to large-scale estimates from future spaceborne lidar instruments.

To date, few spatially coincident lidar data sets have been collected over multiple time periods. In a space-for-time substitution, Lefsky et al. (2005a) described a spatially extensive approach to estimating the woody component of aboveground NPP (NPP$_w$) based on deriving stand age and biomass in landscapes subject to stand-replacing disturbance regimes from Landsat and lidar data products. Stand age is derived by iterative unsupervised classification of a multitemporal sequence of images from a passive optical sensor (e.g., Landsat TM). Stand age is then cross-tabulated with estimates of stand height and aboveground biomass from lidar remote sensing. NPP$_w$ is calculated as the average increment in lidar-estimated biomass over the time period determined, using change detection. This approach compared well with forest inventory estimates, but contrasted significantly with estimates derived from a spatially distributed biogeochemistry model (Lefsky et al. 2005a).

## Model Parameters and NPP Correlates from Lidar Observations

A second approach uses lidar metrics as input data to ecosystem models that estimate NPP. Modeling studies have demonstrated the utility of lidar data in the production of input parameters such as tree height and vertical foliage distribution, which more accurately define current canopy conditions and relate directly to rates of productivity. Kotchenova et al. (2004) demonstrated the use of waveform lidar in modeling gross primary production (GPP) of deciduous forests. They parameterized a photosynthesis model using standard sunlit/shaded leaf separation (two-leaf) and multiple layer approaches. Model simulations using a uniform leaf distribution versus a vertical leaf distribution derived from waveform lidar resulted in large differences in the calculated GPP values, and demonstrated the importance of vertical canopy structure in determining both the distribution of direct and diffuse light within the canopy and the distribution of sunlit and shaded leaves. Hurtt et al. (2004) combined lidar-derived canopy height with a height-structured terrestrial ecosystem model called the ecosystem demography model (ED; Moorcroft et al. 2001). The use of lidar-derived height in this model allowed the model to be initialized with actual vegetation structure, in contrast to potential vegetation conditions, thereby accounting for the effects of land use history and disturbance on standing biomass and C flux.

Collectively, these recent studies combine to illustrate the potential utility of developing lidar technologies to improve estimates of aboveground biomass. Although lidar, like other remote sensing instrumentation, cannot directly measure NPP, the ability to measure vegetation structural attributes, aboveground biomass, and change in biomass over time from airborne and/or spaceborne platforms should help to increase our understanding of trends in NPP at local and regional scales.

## Concluding Remarks

In reviewing approaches for applying remote sensing to the study of terrestrial NPP, it should be clear that a wide variety of methods exists, and each has its own set of inherent strengths and limitations relative to factors such as spatial scale and resolution, frequency of data collection, the nature of the ecological properties that can be captured, and the ability to examine the influence of changing environmental conditions. Methods that rely on broadband vegetation indices, for example, can be applied at regular intervals over continental spatial scales, given the large scene size and regular orbits of sensors such as MODIS and AVHRR. However, their coarse spatial resolution (often ~1 km) and limited spectral detail often don't provide the detail needed to capture fine-scale features of interest to researchers and managers working at subregional scales. Methods involving imaging spectroscopy or active sensors may account for finer-scale variation related to additional vegetation attributes (e.g., biomass or biochemistry), but presently cover smaller portions of the Earth's surface or do not collect data at frequent intervals.

Despite these differences, a common feature of all remote sensing methods is their ability to increase dramatically the amount of land surface that can be sampled, often providing spatially contiguous information at multiple periods in time. As such, remote sensing provides a complement to field measurements by filling gaps between measurements at individual locations and specific points in time. Because nearly all remote sensing methods are in some way or another built up from ground-level measurements—be they measurements of leaf physiology, canopy-light interactions, or stand productivity—remote sensing should be seen as an extension of other methods, rather than divorced from them.

An aspect of remote sensing that is beyond the scope of this chapter, but worth mentioning nonetheless, is the potential to use remotely sensed spatial patterns as a means of examining underlying controls on ecosystem productivity. Whereas field plots typically capture limited samples of the environmental conditions in which ecosystems exist, the large sample sizes provided by remote sensing instruments allow NPP estimates to be contrasted with spatial and/or temporal patterns of important variables, such as climate, topography, and soil properties. Extensive data for variables such as these often exist in the form of geographic information systems or climatological databases, and can be used to shed light on mechanisms controlling variation in NPP (e.g., Braswell et al. 1997). Analyses of this nature should open the minds of researchers to the potential role of remote sensing in basic ecological research.

*Acknowledgments*   This work was supported by NASA through grants from the Terrestrial Ecology Division's Carbon Cycle Sciences Program (CARBON/0000-0243 and CARBON/04-0120-0011), the Interdisciplinary Science Program (NASA IDS/03-0000-0145), and the Earth System Science Fellowship Program (NGT5-30481. We also received support from the U.S. Department of Energy's National Institute of Global Environmental Change (NIGEC grants DE-FC02-03ER63613 and DE-FC03-90ER61010) and the USDA Forest Service Northeastern Research Station. Portions of this research were carried out at the Jet Propulsion Laboratory, California Institute of Technology, under a contract with the

National Aeronautics and Space Administration. We received helpful assistance from Lucie Plourde and Julian Jenkins, and we thank Steve Running for providing us with the image shown in figure 11.3.

References

Aber, J. D., S. V. Ollinger, C. A. Federer, P. B. Reich, M. L. Goulden, D. W. Kicklighter, J. M. Melillo, and R. G. Lathrop, Jr. 1995. Predicting the effects of climate change on water yield and forest production in the northeastern U.S. Climate Research 5:207–222.

Askne, J., M. Santoro, G. Smith, and J. E. S. Fransson. 2003. Multitemporal repeat-pass SAR interferometry of boreal forests. IEEE Transactions on Geoscience and Remote Sensing 41(7):1540–1550.

Asner, G. P., C. Borghi, and R. Ojeda. 2003. Desertification in Central Argentina: Regional changes in ecosystem carbon-nitrogen from imaging spectroscopy. Ecological Applications 13:629–648.

Bergen, K. M., and M. C. Dobson. 1999. Integration of remotely sensed radar imagery in modeling and mapping of forest biomass and net primary production. Ecological Modelling 122:257–274.

Blair, J. B., D. L. Rabine, and M. A. Hofton. 1999. The Laser Vegetation Imaging Sensor (LVIS): A medium-altitude, digitisation-only, airborne laser altimeter for mapping vegetation and topography. ISPRS Journal of Photogrammetry and Remote Sensing 54:115–122.

Braswell, B. H., D. S. Schimel, E. Linder, and B. Moore. 1997. The response of global terrestrial ecosystems to interannual temperature variability. Science 238:870–872.

Cohen, W. B., and C. O. Justice. 1999. Validating MODIS terrestrial ecology products: Linking in situ and satellite measurements. Remote Sensing of Environment 70(1):1–3.

Cohen W. B., T. K. Maiersperger, S. T. Gower, and D. P. Turner. 2003. An improved strategy for regression of biophysical variables and Landsat ETM+ data. Remote Sensing of Environment 84(4): 561–571.

Comins, H. N., and R. E. McMurtrie. 1993. Long-term response of nutrient-limited forests to $CO_2$-enrichment: Equilibrium behaviour of plant-soil models. Ecological Applications 3:666–681.

Dobson, M. C., F. T. Ulaby, L. E. Pierce, T. L. Sharik, K. M. Bergen, J. Kellndorfer, J. R. Kendra, E. Li , C. Lin, A. Nashashibi, K. Sarabandi, and P. Siqueira. 1995. Estimation of forest biophysical characteristics in Northern Michigan with SIR-C/X-SAR. IEEE Transactions on Geoscience and Remote Sensing 33(4):877–892.

Drake, J. B., R. O. Dubayah, D. B. Clark, R. G. Knox, J. B. Blair, M. A. Hofton, R. L. Chazdon, J. F. Weishampel, and S. D. Prince. 2002. Estimation of tropical forest structural characteristics using large-footprint lidar. Remote Sensing of Environment 79: 305–319.

Drake, J. B., R. G. Knox, R. O. Dubayah, D. B. Clark, R. Condit, J. B. Blair, and M. Hofton. 2003. Above-ground biomass estimation in closed canopy neotropical forests using lidar remote sensing: Factors affecting the generality of relationships. Global Ecology and Biogeography 12:147–159.

Dubayah, R. O., and J. B. Drake. 2000. Lidar remote sensing for forestry. Journal of Forestry 98:44–46.

Ehleringer, J. R., and C. B. Field (eds.). 1993. Scaling Physiological Processes: Leaf to Globe. Academic Press, San Diego.

Fahey, T. J., T. G. Siccama, C. T. Driscoll, G. E. Likens, J. Campbell, C. E. Johnson, J. D. Aber, J. J. Cole, M. C. Fisk, P. M. Groffman, S. P. Hamburg, R. T. Holmes, P. A. Schwarz, and R. D. Yanai. 2005. The biogeochemistry of carbon at Hubbard Brook. Biogeochemistry 75 (1): 109–176.

Fassnacht, K. S., and S. T. Gower. 1997. Interrelationships among the edaphic and stand characteristics, leaf area index, and aboveground net primary production of upland forest ecosystems in north central Wisconsin. Canadian Journal of Forest Research 27:1058–1067.

Ferrazzoli, P., S. Paloscia, P. Pampaloni, G. Schiavon, S. Sigismondi, and D. Solimini. 1997. The potential of multifrequency polarimetric SAR in assessing agricultural and arboreous biomass. IEEE Transactions on Geoscience and Remote Sensing 35(1):5–17.

Field, C. 1991. Ecological scaling of carbon gain to stress and resource availability. Pages 35–65 in H. A. Mooney, W. E. Winner, and E. J. Pell (eds.), Response of Plants to Multiple Stresses. Academic Press, San Diego.

Field, C., and H. A. Mooney. 1986. The photosynthesis-nitrogen relationship in wild plants. Pages 25–55 in T. J. Garvish (ed.), On the Economy of Plant Form and Function. Cambridge University Press, Cambridge.

Flood, M. 2001. Lidar activities and research priorities in the commerical sector. Pages 3–7 in Hofton, M.A. (ed), Land Surface Mapping and Characterization Using Laser Altimetry. International Archives of Photogrammetry, Remote Sensing, and Spatial Information Sciences. Vol XXXIV-3/W4. ISPRS Workshop. Annapolis, MD. 22–24 Oct. 2001.

Friedl, M. A., D. K. McIver, J. C. F. Hodges, X. Y. Zhang, D. Muchoney, A. H. Strahler, C. E. Woodcock, S. Gopal, A. Schneider, A. Cooper, A. Baccini, F. Gao, and C. Schaaf. 2002. Global land cover mapping from MODIS: Algorithms and early results. Remote Sensing of Environment 83(1–2):287–302.

Fuentes, D. A., J. A. Gamon, H.-L. Qiu, D. A. Sims, and D. A. Roberts. 2001. Mapping Canadian boreal forest vegetation using pigment and water absorption features derived from the AVIRIS sensor. Journal of Geophysical Research 106:33565–33577.

Gower, S. T., C. J. Kucharik, and J. M. Norman. 1999. Direct and indirect estimation of leaf area index, f(APAR), and net primary production of terrestrial ecosystems. Remote Sensing of Environment 70(1):29–51.

Gower, S. T., P. B. Reich, and Y. Son. 1993. Canopy dynamics and aboveground production of five tree species with different leaf longevities. Tree Physiology 12(4): 327–345.

Gower, S. T., K. A. Vogt, and C. C. Grier. 1992. Carbon dynamics of Rocky Mountain Douglas fir: Influence of water and nutrient availability. Ecological Monographs 62:43–65.

Green, D. S., J. E. Erickson, and E. L. Kruger. 2003. Foliar morphology and canopy nitrogen as predictors of light-use efficiency in terrestrial vegetation. Agricultural and Forest Meteorology 3097:1–9.

Harding, D. J., M. A. Lefsky, G. G. Parker, and J. B. Blair. 2001. Laser altimeter canopy height profiles. Methods and validation for closed-canopy, broadleaf forests. Remote Sensing of Environment 76:283–297.

Haxeltine, A., and I. C. Prentice. 1996. A general model for the light-use efficiency of primary production. Functional Ecology 10:551–561.

Hirose, T., and M. J. A. Werger. 1987. Maximizing daily canopy photosynthesis with respect to the leaf nitrogen allocation pattern in the canopy. Oecologia 72:520–526.

Hofton, M. A., J. B. Minster, and J. B. Blair. 2000. Decomposition of laser altimeter waveforms. IEEE Transactions on Geoscience and Remote Sensing 38:1989–1996.

Hollinger, D. Y. 1989. Canopy organization and foliage photosynthetic capacity in a broad-leaved evergreen montane forest. Functional Ecology 3:53–62.

Hurtt, G. C., R. Dubayah, J. Drake, P. Moorcroft, S. W. Pacala, J. B. Blair, and M. G. Fearon. 2004. Beyond potential vegetation: Combining lidar data and a height-structured model for carbon studies. Ecological Applications 14:873–883.

Imhoff, M. L. 1995. A theoretical analysis of the effect of forest structure on synthetic aperture radar backscatter and the remote sensing of biomass. IEEE Transactions on Geoscience and Remote Sensing 33(2):341–352.

Jacquemoud, S., and F. Baret. 1990. PROSPECT: A model of leaf optical properties spectra. Remote Sensing of Environment 34:75–91.

Kellndorfer, J., W. Walker, L. Pierce, C. Dobson, J. A. Fites, C. Hunsaker, J. Vona, and M. Clutter. 2004. Vegetation height estimation from shuttle radar topography mission and national elevation datasets. Remote Sensing of Environment 93(3):339–358.

Kicklighter, D. W., A. Bondeau, A. L. Schloss, J. Kaduk, and A. D. McGuire. 1999. Comparing global models of terrestrial net primary productivity (NPP): Global pattern and differentiation by major biomes. Global Change Biology 5(suppl. 1):16–24.

Kimball, J. S., A. R. Keyser, S. W. Running, and S. S. Saatchi. 2000. Regional assessment of boreal forest productivity using an ecological process model and remote sensing parameter maps. Tree Physiology 20(11):761–775.

Kotchenova, Y. S., X. Song, N. V. Shabanov, C. S. Potter, Y. Knyazikhin, and R. B. Myneni. 2004. Lidar remote sensing for modeling gross primary production of deciduous forests. Remote Sensing of Environment 92:158–172.

Lefsky, M. A., W. B. Cohen, S. A. Acker, G. Parker, T. A. Spies, and D. Harding. 1999a. Lidar remote sensing of the canopy structure and biophysical properties of Douglas fir-western hemlock forests. Remote Sensing of Environment 70:339–361.

Lefsky, M. A., W. B. Cohen, D. J. Harding, G. G. Parker, S. A. Acker, and S. T. Gower. 2002a. Lidar remote sensing of aboveground biomass in three biomes. Global Ecology and Biogeography 11:393–399.

Lefsky, M. A., W. B. Cohen, G. G. Parker, and D. J. Harding. 2002b. Lidar remote sensing for ecosystem studies. Bioscience 52(1):19–30.

Lefsky, M. A., W. B. Cohen, and T. A. Spies. 2001. An evaluation of alternate remote sensing products for forest inventory, monitoring, and mapping of Douglas-fir forests in western Oregon. Canadian Journal of Forest Research 31:78–87.

Lefsky, M. A., D. Harding, W. B. Cohen, G. Parker, and H. H. Shugart. 1999b. Surface lidar remote sensing of basal area and biomass in deciduous forest of eastern Maryland, USA. Remote Sensing of Environment 67:83–98.

Lefsky, M. A., A. T Hudak, W. B. Cohen, and S. A. Acker. 2005a. Geographic variability in lidar predictions of forest stand structure in the Pacific Northwest. Remote Sensing of Environment 95:532–548.

Lefsky, M. A., D. P. Turner, M. Guzy, and W. B. Cohen. 2005b. Combining lidar estimates of above ground biomass and Landsat estimates of stand age for spatially extensive validation of modeled forest productivity. Remote Sensing of Environment 95: 549–558.

Lim, K., P. Treitz, M. Wulder, B. St-Onge, and M. Flood. 2003. Lidar remote sensing of forest structure. Progress in Physical Geography 27(1):88–106.

Liu, J., J. M. Chen, J. Cihlar, and W. Chen. 1999. Net primary productivity distribution in the BOREAS region from a process model using satellite and surface data. Journal of Geophysical Research—Atmospheres 104:27735–27754.

Luckman, A., J. Baker, M. Honzák, and R. Lucas. 1998. Tropical forest biomass density estimation using JERS-1 SAR: Seasonal variation, confidence limits, and application to image mosaics. Remote Sensing of Environment 63:126–139.

Luckman, A., J. Baker, and U. Wegmüller. 2000. Repeat-pass interferometric coherence measurements of disturbed tropical forest from JERS and ERS satellites. Remote Sensing of Environment 73:350–360.

Luo, T. X., Y. D. Pan, H. Ouyang, P. L. Shi, J. Luo, Z. L. Yu, and Q. Lu. 2004. Leaf area index and net primary productivity along subtropical to alpine gradients in the Tibetan Plateau. Global Ecology and Biogeography 13(4):345–358.

Magnussen, S., and P. Boudewyn. 1998. Derivations of stand heights from airborne laser scanner data with canopy-based quantile estimators. Canadian Journal of Forest Research 28(7):1016–1031.

Marion, J. B. 1965. Classical Electromagnetic Radiation. Academic Press, New York.

Martin, M. E., and J. D. Aber. 1997. Estimation of forest canopy lignin and nitrogen concentration and ecosystem processes by high spectral resolution remote sensing. Ecological Applications 7:441–443.

Martin, M. E., S. D. Newman, J. D. Aber, and R. G. Congalton. 1998. Determining forest species composition using high spectral resolution remote sensing data. Remote Sensing of Environment 65:249–254.

Matson, P., L. Johnson, C. Billow, J. Miller, and R. Pu. 1994. Seasonal patterns and remote spectral estimation of canopy chemistry across the Oregon transect. Ecological Applications 4(2):280–298.

Means, J. E., S. A. Acker, D. J. Harding, J. B. Blair, M. A. Lefsky, W. B. Cohen, M. E. Harmon, and W. A. McKee. 1999. Use of large-footprint scanning airborne lidar to estimate forest stand characteristics in the Western Cascades of Oregon. Remote Sensing of Environment 67:298–308.

Mencuccini, M., and J. Grace. 1996. Hydraulic conductance, light interception and needle nutrient concentration in Scots pine stands and their relations with net primary productivity. Tree Physiology 16:459–468.

Monteith, J. L. 1972. Solar radiation and productivity in tropical ecosystems. Applied Ecology 9:747–766.

Moorcroft, P. R., G. C. Hurtt, and S. W. Pacala. 2001. A method for scaling vegetation dynamics: The ecosystem demography model (ED). Ecological Monographs 71:557–585.

Moreau, S., and T. Le Toan. 2003. Biomass quantification of Andean wetland forages using ERS satellite SAR data for optimizing livestock management. Remote Sensing of Environment 84:477–492.

Myneni, R. B., and D. L. Williams. 1994. On the relationship between FAPAR and NDVI. Remote Sensing of Environment 49:200–211.

Naesset, E. 2002. Predicting forest stand characteristics with airborne scanning laser using a practical two-stage procedure and field data. Remote Sensing of Environment 80:88–99.

Naesset, E. 2004. Effects of different flying altitudes on biophysical stand properties estimated from canopy height and density measured with a small footprint airborne scanning laser. Remote Sensing of Environment 91:243–255.

Nelson, R., M. A. Valenti, A. Short, and C. Keller. 2003. A multiple resource inventory of Delaware using airborne laser data. Bioscience 53:981–992.

Nilsson, M. 1996. Estimation of tree heights and stand volume using an airborne lidar system. Remote Sensing of Environment 56:1–7.

Ollinger, S. V., J. D. Aber, and C. A. Federer. 1998. Estimating regional forest productivity and water balances using an ecosystem model linked to a GIS. Landscape Ecology 13(5):323–334.

Ollinger, S. V., and M. L. Smith. 2005. Net primary production and canopy nitrogen in a temperate forest landscape: An analysis using imaging spectrometry, modeling and field data. Ecosystems 8:760–778.

Ollinger, S. V., M. L. Smith, M. E. Martin, R. A. Hallett, C. L. Goodale, and J. D. Aber. 2002. Regional variation in foliar chemistry and soil nitrogen status among forests of diverse history and composition. Ecology 83:339–355.

Paloscia, S., G. Macelloni, P. Pampaloni, and S. Sigismondi. 1999. The potential of C- and L-band SAR in estimating vegetation biomass: The ERS-1 and JERS-1 experiments. IEEE Transactions on Geoscience and Remote Sensing 37(4):2107–2110.

Pampaloni, P. 2004. Microwave radiometry of forests. Waves in Random Media 14: S275–S298.

Papathanassiou, K. P., and S. R. Cloude. 2001. Single-baseline polarimetric SAR interferometry. IEEE Transactions on Geoscience and Remote Sensing 39:2352–2363.

Parker, G. G., M. A. Lefsky, and D. J. Harding. 2001. Light transmittance in forest canopies determined using airborne laser altimetry and in-canopy quantum measurements. Remote Sensing of Environment 76:298–309.

Patenaude, G., R. A. Hill, R. Milne, D. L. A. Gaveau, B. B. J. Briggs, and T. P. Dawson. 2004. Quantifying forest above ground carbon content using LiDAR remote sensing. Remote Sensing of Environment 93:368–380.

Potter, C., J. T. Randerson, C. B. Field, P. A. Matson, P. M. Vitousek, H. A. Mooney, and S. A. Klooster. 1993. Terrestrial ecosystem production: A process model based on global satellite and surface data. Global Biogeochemical Cycles 7:811–841.

Pregitzer, K. S., and E. S. Euskirchen. 2004. Carbon cycling and storage in world forests: Biome patterns related to forest age. Global Change Biology 10(12):2052–2077.

Pulliainen, J., M. Engdahl, and M. Hallikainen. 2003. Feasibility of multi-temporal interferometric SAR data for stand-level estimation of boreal forest stem volume. Remote Sensing of Environment 85:397–409.

Reich, P., D. Turner, et al. 1999a. An approach to spatially-distributed modelling of NPP at the landscape scale and its application in validation of EOS NPP products. Remote Sensing of Environment 70:69–81.

Reich, P. B., D. S. Ellsworth, M. B. Walters, J. M. Vose, C. Gresham, J. C. Volin, and W. D. Bowman. 1999b. Generality of leaf traits relationships: A test across six biomes. Ecology 80:1955–1969.

Reich, P. B., D. F. Grigal, J. D. Aber, and S. T. Gower. 1997. Nitrogen mineralization and productivity in 50 hardwood and conifer stands on diverse soils. Ecology 78(2): 335–347.

Roberts, D. A., M. Gardner, R. Church, S. Ustin, G. Scheer, and R. O. Green. 1998. Mapping chaparral in the Santa Monica Mountains using multiple end member spectral mixture models. Remote Sensing of Environment 65:267–279.

Running, S. W., and S. T. Gower. 1991. FOREST-BGC: A general model of forest ecosystem processes for regional applications. II. Dynamic carbon allocation and nitrogen budgets. Tree Physiology 9:147–160.

Running, S. W., R. R. Nemani, F. A. Heinsch, M. Zhao, M. Reeves, and H. Hashimoto. 2004. A continuous satellite-derived measure of global terrestrial primary production. Bioscience 54(6):547–560.

Running, S. W., P. E. Thornton, R. R. Nemani, and J. M. Glassy. 2000. Global terrestrial gross and net primary productivity from the Earth Observing System. Pages 44–57 in O. Sala, R. Jackson, and H. Mooney (eds.), Methods in Ecosystem Science. Springer-Verlag, New York.

Salas, W. A., M. J. Ducey, E. Rignot, and D. Skole. 2002. Assessment of JERS-1 SAR for monitoring secondary vegetation in Amazonia: II. Spatial, temporal, and radiometric considerations for operational monitoring. International Journal of Remote Sensing 23 (7):1381–1399.

Santoro, M., J. Askne, G. Smith, and J. E. S. Fransson. 2002. Stem volume retrieval in boreal forests from ERS-1/2 interferometry. Remote Sensing of Environment 81:19–35.

Santos, J. R., C. C. Freitas, L. S. Araujo, L. V. Dutra, J. C. Mura, F. F. Gama, L. S. Soler, and S. J. S. Sant'Anna. 2003. Airborne P-band SAR applied to the aboveground biomass studies in the Brazilian tropical rain forest. Remote Sensing of Environment 87: 482–493.

Sellers, P. J. 1987. Canopy reflectance, photosynthesis, and transpiration. II: The role of biophysics in the linearity of their interdependence. International Journal of Remote Sensing 6:1335–1372.

Sellers, P. J., J. A. Berry, G. J. Collatz, C. B. Field, and F. G. Hall. 1992. Canopy reflectance, photosynthesis, and transpiration. III. A reanalysis using improved leaf models and a new canopy integration scheme. Remote Sensing of Environment 42: 187–216.

Smith, G., L. M. H. Ulander and J. Askne. 1999. VHF Backscatter Sensitivity to Forest Properties: Model Predictions. Geoscience and Remote Sensing Symposium, 1999. IGARSS '99 Proceedings. IEEE 1999 International 4: 1889–1891.

Smith, M. L., S. V. Ollinger, M. E. Martin, J. D. Aber, and C. L. Goodale. 2002. Direct prediction of aboveground forest productivity by remote sensing of canopy nitrogen. Ecological Applications 12(5):1286–1302.

Townsend, P. A., J. R. Foster, and R. A. Chastain, Jr. 2003. Imaging spectroscopy and canopy nitrogen: Application to the forests of the central Appalachian Mountains using Hyperion and AVIRIS. IEEE Transactions on Geoscience and Remote Sensing 41(6): 1347–1354.

Treuhaft, R. N., G. P. Asner, and B. E. Law. 2003. Structure-based forest biomass from fusion of radar and hyperspectral observations. Geophysical Research Letters 30: 1472–1475.

Treuhaft, R. N., B. E. Law, and G. P. Asner. 2004. Forest attributes from radar interferometric structure and its fusion with optical remote sensing. BioScience 54(6): 561–571.

Treuhaft, R. N., S. N. Madsen, M. Moghaddam, and J. J. Van Zyl. 1996. Vegetation characteristics and surface topography from interferometric radar. Radio Science 31: 1449–1485.

Treuhaft, R. N., and P. R. Siqueira. 2004. The calculated performance of forest structure and biomass from interferometric radar. Waves in Random Media 14(2):S345–S358.

Tucker, C. J. 1979. Red and photographic infrared linear combinations for monitoring vegetation. Remote Sensing of Environment 8:27–150.

Turner, D. P., W. B. Cohen, R. E. Kennedy, K. S. Fassnacht, and J. M. Briggs. 1999. Relationships between leaf area index and Landsat TM spectral vegetation indices across three temperate zone sites. Remote Sensing of Environment 70:52–68.

Turner, D. P., S. V. Ollinger, and J. S. Kimball. 2004. Integrating remote sensing and ecosystem process models for landscape to regional scale analysis of the carbon cycle. BioScience 54:573–584.

Turner, D. P., S. Ollinger, M. L. Smith, O. Krankina, and M. Gregory. 2003a. Scaling net primary production to a MODIS footprint in support of Earth observing system product validation. International Journal of Remote Sensing 25:1961–1979.

Turner, D. P., W. D. Ritts, W. B. Cohen, S. T. Gower, M. Zhao, S. W. Running, S. C. Wofsy, S. Urbanski, A. Dunn, and J. W. Munger. 2003b. Scaling gross primary production (GPP) over boreal and deciduous forest landscapes in support of MODIS GPP product validation. Remote Sensing of Environment 88:256–270.

Ustin, S. L., D. A. Roberts, J. A. Gamon, G. P. Asner, and R. O. Green. 2004. Using imaging spectroscopy to study ecosystem processes and properties. Bioscience 54(6): 523–534.

Verhoef, W. 1984. Light scattering by leaf layers with application to canopy reflectance modeling: The SAIL model. Remote Sensing of Environment 16:125–141.

Vose, J. M., and H. L. Allen. 1988. Leaf area, stemwood growth and nutritional relationships in loblolly pine. Forest Science 34:547–563.

Wagner, W., A. Luckman, J. Vietmeier, K. Tansey, H. Balzter, C. Schmullius, M. Davidson, D. Gaveau, M. Gluck, T. Le Toan, S. Quegan, A. Shvidenko, A. Wiesmann, and J. J. Yu. 2003. Large-scale mapping of boreal forest in Siberia using ERS tandem coherence and JERS backscatter data. Remote Sensing of Environment 85: 125–144.

Waring, R. H., and S. W. Running. 1998. Forest Ecosystems: Analysis at Multiple Scales. San Diego, Academic Press.

Webb, W. L., W. Lauenroth, S. R. Szarek, and R. S. Kinerson. 1983. Primary production and abiotic controls in forests, grasslands, and desert ecosystems in the United States. Ecology 64(1):134–151.

Wehr, A., and U. Lohr. 1999. Airborne laser scanning—an introduction and overview. ISPRS Journal of Photogrammetry and Remote Sensing 54:68–82.

Xiao, X., D. Y. Hollinger, J. D. Aber, M. Goltz, E. A. Davidson, Q. Zhang, and B. Moore. 2004. Satellite-based modeling of gross primary productivity in an evergreen needleleaf forest. Remote Sensing of Environment 89:519–534.

Yin, X. W. 1993. Variation in foliar nitrogen concentration by forest type and climatic gradients in North America. Canadian Journal of Forest Research 23(8):1587–1602.

Yoder, B. J., and R. E. Pettigrew-Crosby. 1995. Predicting nitrogen and chlorophyll content and concentrations from reflectance spectra (400–2500 nm) at leaf and canopy scales. Remote Sensing of Environment 53:199–211.

Zagloski, F., V. Pinel, et al. 1996. Forest canopy chemistry with high spectral resolution remote sensing. International Journal of Remote Sensing 17:1107–1128.

Zwally, H. J., B. Schutz, W. Abdalati, J. Abshire, C. Bentley, A. Brenner, J. Bufton, J. Dezio, D. Hancock, D. Harding, T. Herring, B. Minster, K. Quinn, S. Palm, J. Spinhirne, and R. Thomas. 2002. ICESat's laser measurements of polar ice, atmosphere, ocean and land. Journal of Geodynamics 34:405–445.

# 12

# Quantifying Uncertainty in Net Primary Production Measurements

Mark E. Harmon
Donald L. Phillips
John J. Battles
Andrew Rassweiler
Robert O. Hall, Jr.
William K. Lauenroth

Net primary production (NPP; e.g., g m$^{-2}$ yr$^{-1}$), a key ecosystem attribute, is estimated from a combination of other variables, such as standing crop biomass at several points in time, each of which is subject to errors in measurement. These errors propagate as the variables are mathematically combined, and the distribution of these propagated errors reflects the uncertainty in the NPP estimate. While often not reported, quantification of the component error terms and the resultant NPP estimation error is important for several reasons. First, such information allows the user of the data to assess its reliability. A single-point estimate of NPP does not convey any notion of how good the estimate is, but an estimate with an associated confidence interval does. Second, it allows for more meaningful comparisons because no two estimates will ever be exactly the same. The interpretation of a 10% difference in NPP between two forest stands would be viewed differently if the NPP estimate had an uncertainty of 5% rather than 20%. Third, dissection of estimation error into its constituent components allows one to understand what factors are the major contributors to uncertainty and where efforts might best be focused to reduce this uncertainty.

In this chapter we review concepts related to uncertainty, factors that contribute to uncertainty, and how uncertainty can be estimated, as well as provide examples of uncertainty for selected biomes. While uncertainty is often viewed negatively, we encourage the view that it is just another dimension of understanding an ecological system. Although a major goal of science is to reduce the uncertainty of prediction, this is difficult to achieve when uncertainty is not quantified or explicitly expressed. Finally, great progress has been made in methods of estimating uncertainty using statistical error propagation and other methods, such as Monte Carlo analysis. Given the availability of software and computers to perform these estimates, this task has become relatively easy.

## The Dimensions of Uncertainty

We will use the simple definition of uncertainty as a lack of confidence in a single value, and represent uncertainty about a quantity as a range of potential values in the form of a probability density function (Heath and Smith 2000). The term "error" denotes the deviation of a quantity from its true value. Random errors can vary in direction and magnitude but average out to zero, while systematic errors (or bias) represent a consistent tendency toward deviation in one direction. The distribution of errors is also characterized by a probability density function (e.g., normal distribution defined by a mean and standard deviation or variance) that can be used to represent uncertainty.

Uncertainty can arise from either natural variability or imperfect knowledge (figure 12.1 in this volume; Gardner et al. 1990). Natural variability is inherent in most processes, and includes both spatial and temporal variability. For example, the NPP of 1 $m^2$ of grassland varies spatially as one moves across the landscape due to changing microclimate and soils, as well as temporally due to year-to- year changes in weather and other factors. Natural variability cannot be removed as a source of uncertainty through gaining additional knowledge. More intensive sampling, for example, does not reduce the degree of variation among individual members of the sampled population, although it may allow a better characterization of this variation and more precise estimates of the population mean. Knowledge uncertainty includes model error (sometimes called structural error), which results from imperfect representation of processes in a model, and parameter error, which results from imperfect knowledge of the values of parameters associated with these processes. Knowledge uncertainty can be reduced through further measurement or improved models.

## Sources of Systematic Error

Bias, or systematic error, deserves special attention because many forms of bias can be avoided or minimized. Bias may be due to errors introduced by (1) the measurement system (e.g., decomposition of litter in traps); (2) improper selection of statistical samples (e.g., only forest areas without disturbance); or (3) the exclusion of certain NPP components. Whenever possible, the accuracy of NPP-related instruments and methods should be checked, and corrections made. The magnitude of bias and any corrections made during the analysis should be reported in the methods description. In theory, improper selection of samples can be dealt with by introducing stratification and randomization, but often estimates are made for subjective "typical" conditions. Abandonment of long-term series of NPP measures

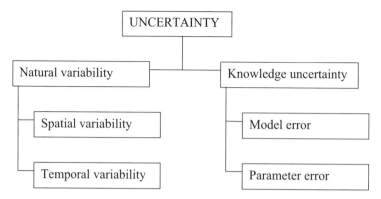

Figure 12.1. Types of uncertainty commonly encountered when making NPP estimates.

from nonrandomly selected sites is often not a practical option. While less than ideal, these estimates can still be useful if the site and conditions are fully described and the measurements can be put in an overall context (e.g., this age class is typically low/high relative to others). The third type of bias is quite common in NPP estimates since certain carbon (C) fluxes are rarely measured, such as C flux to root exudates, mycorrhizae, volatile organic carbons (VOCs), organic leachates, herbivory, and others (Clark et al. 2001). While it is unlikely that all NPP studies can make adequate measurements of all fluxes, values from selected intensive studies can be useful in assigning a likely bias due to omissions.

While having biases is not desirable, the presence of bias does not necessarily make NPP estimates invalid. As with any error, large bias in the estimation of one flux does not mean that the overall estimate is strongly biased. For example, when a small flux is ignored, the overall bias is also small. When estimates are compared over time, bias can be canceled as long as it does not change with time. The same is true for comparisons over space. Bias can be influenced by the time interval between measurements. For example, NPP estimated from increment coring of boles is biased downward because growth of trees that died during the measurement period is not included. For short periods of time (5 yr) this bias is probably <5%; however, for long periods it can become quite sizable, as evidenced by the difference in gross versus net volume growth of forests. For example, by the age of 160 years, the exclusion of mortality in Douglas fir forests led to underestimation of bole-related NPP by 57% (Staebler 1955).

## Sources of Random Error

### Sampling Error

Sampling error reflects the uncertainty in estimates due to the selection of a statistical sample of units from the population for quantification. For example, if NPP were estimated within 10 randomly selected quadrats in a study area, the mean NPP would be different if another set of 10 random quadrats had been selected; sampling error quantifies this variability. For mean estimates, sampling error decreases as the number of samples increases:

$$SE_s = \frac{SD_s}{\sqrt{n}} \qquad (12.1)$$

where $SE_s$ and $SD_s$ are the sampling standard error and standard deviation, respectively, and n is the sample size.

## Measurement Error

Measurement error quantifies the uncertainty in physical measurements that are made in estimating NPP. For example, McRoberts et al. (1994) reported that repeated measurements of the diameter at breast height (DBH) of individual trees in a Michigan forest gave measurement standard deviations of 0.12 cm for 10 cm DBH trees (1.2%), and 0.37 cm for 50 cm DBH trees (0.7%).

## Regression Error

Allometric regression equations are often used for nondestructive estimates of plant biomass, especially when the size of the plants makes destructive measurements impractical. For example, a large number of allometric equations have been developed for tree biomass based on measurements of DBH or DBH and height (e.g., Jenkins et al. 2003; Means et al. 1994). While the regression equations represent the best fit lines through the data points for the harvested and weighed trees, the scatter of points about the lines indicates a random error component (figure 12.2; top). Regression standard deviations for predicted biomass, given a value of the independent variable (e.g., DBH), can easily be derived (Dixon and Massey 1983). If logarithmic transformations are used in the allometric equations, a correction factor should be applied to these estimation standard deviations to avoid introducing a systematic bias (Sprugel 1983). As alluded to above, not only random error but also systematic error may be present if the trees to which the allometric equations are being applied are not representative of the population of trees from which the equations were derived. Since regressions from one area are often used for another area, this type of error is quite common in NPP studies. Clearly, variation in growth form due to differences in genetics or competitive status or climatic/edaphic factors might result in either underestimation or overestimation of biomass. Allometric equations should be chosen which were derived under the conditions most similar to which they are being applied (e.g., same geographic region, similar range of diameters, etc.). In some cases there may be more than one allometric equation that can be used, which reflects additional uncertainty in the derived biomass estimates (figure 12.2; bottom).

## Conversion Factors

Another potential source of uncertainty in production estimates involves the conversion from one type of unit to another. Conversion factors between different units

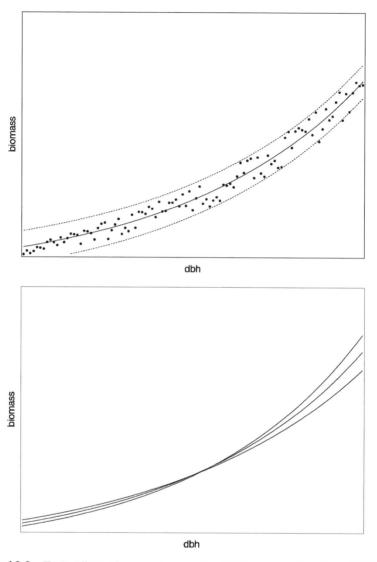

Figure 12.2. (Top): Allometric regression equation for biomass as a function of DBH. Individual data points for equation derivation are shown, along with 95% confidence limits for a biomass estimate given DBH. (Bottom): Multiple allometric equations potentially applicable to a study site. Note that the regression error is highest for larger trees.

that measure the same physical quantity are generally known exactly, and do not add uncertainty to the estimate (e.g., conversion of biomass density from tons acre$^{-1}$ to g m$^{-2}$. However, conversion from one physical quantity to another may entail additional uncertainty; for example, converting NPP from biomass units (g m$^{-2}$ yr$^{-1}$) to C units such as g C m$^{-2}$ yr$^{-1}$. This conversion is made by multiplying biomass NPP by the proportional C content, which may have some random error due to natural variability or measurement error. Systematic error is also possible if conversion factors from an inappropriate source are used, such as a different ecosystem, taxon, or growing conditions.

## Scaling of Error Terms

Ideally all components of NPP are measured on the same spatial extent or time interval. Practically this is often difficult; for example, tree growth is easy to measure on the scale of hectares, whereas litterfall is not. Likewise, changes in some components can be adequately estimated over days and weeks (e.g., litterfall), whereas others (e.g., tree mortality) may require years to estimate adequately. Therefore, it is necessary to put components on a similar spatial and temporal scale when making an overall estimate of uncertainty. This usually involves aggregating the finer measurements up to the level of the broadest measurements. For example, when combining litterfall (measured on $m^2$) and tree growth numbers (measured on the scale of 100–10,000 $m^2$), subsets of litterfall measures should be averaged so that a more reasonable approximation of variation at the larger scale is made (fig. 12.3). This is necessary because it is unlikely that a single small litter trap represents the value at the level of a forest plot. If tree measurements are subdivided spatially, then similar subsets can be used for litter as well. Temporally the process is similar, but may require summation of values within a year and averaging over years. The need to scale error terms does set limits on the spatial and temporal levels for which overall uncertainty can be estimated. Finer-scale measurements (e.g., litterfall measurements at $m^2$ scale) can be aggregated up to coarser scales to match other

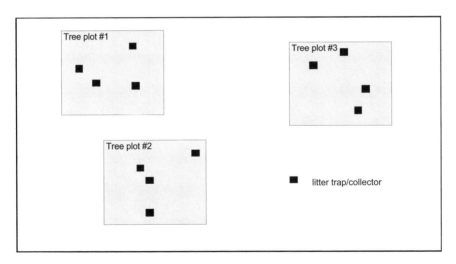

Figure 12.3. Aggregating of samples can be important for calculating uncertainty at the proper scale. Here small litter traps are scattered randomly throughout large tree plots. To combine these terms, an averaging of litter traps within the larger tree plots is desirable.

measurements (e.g., tree growth measurements at ha scale), but the coarser scale measurements cannot be disaggregated down to finer scales because they contain no information on the degree of variability at finer scales. Similarly, on a temporal basis, monthly measurements can be aggregated up to yearly estimates, but not vice versa.

## Error Propagation Approaches

### Statistical Approach

From a statistical point of view, each of the variables used in a calculation of NPP has some uncertainty, due to the sources of random error discussed above, that can be characterized by a variance. The variance of NPP estimates may be easily determined analytically if they are derived from simple combinations of these variables. In the simplest case, if NPP over a time interval t is calculated as

$$NPP = B_t - B_0 \tag{12.2}$$

where $B_t$ and $B_0$ are independent estimates of biomass at time t and time 0, then the variance of NPP is equal to the sum of the variances of the biomass estimates:

$$\sigma^2_{NPP} = \sigma^2_{B_t} + \sigma^2_{B_0} \tag{12.3}$$

Other expressions of NPP may involve a greater number of variables, which may or may not be independent (uncorrelated). The variance of a function of several variables can be approximated by a Taylor series expansion (evaluated at the mean of each of the variables) and dropping the higher-order terms (Stuart and Ord 1987). For NPP as a function of multiple variables $x_i$ this leads to the general expression

$$\sigma^2_{NPP} \approx \sum_i \left[ \left( \frac{\partial NPP}{\partial x_i} \right)^2 \sigma^2_{x_i} \right] + \sum_i \sum_{j \neq i} \frac{\partial NPP}{\partial x_i} \frac{\partial NPP}{\partial x_j} r_{x_i x_j} \sigma_{x_i} \sigma_{x_i} \tag{12.4}$$

where the $\partial NPP / \partial x_i$ terms are partial derivatives and $r_{x_i x_j}$ is the correlation between variables $x_i$ and $x_j$ (Taylor 1997). If the measurements of the variables are independent, and thus have uncorrelated error terms, then the latter term of this equation drops out.

Consider, as an example, estimation of aboveground NPP for a stand of trees according to the equation

$$NPP = (B_t - B_0) + M + L + H + R \tag{12.5}$$

where $B_t$ and $B_0$ are estimated biomass at time t and time 0, and M, L, H, and R are estimates of mortality, litterfall, herbivory, and removals (harvest) of biomass during that time interval. Assume that estimates of M, L, H, and R and their variances are made by various methods for the entire stand, and that estimates of $B_t$ and $B_0$ are made by allometric regression on DBH for a statistical sample of individual trees (the same ones each time) and scaled up to a stand-level estimate.

Since the errors in biomass estimation at two times for a given tree are unlikely to be independent of one another (Phillips et al. 2000), we will hypothesize a correlation coefficient of 0.6 for $B_t$ and $B_0$ ($r_{B_tB_0}$). In this example, the partial derivative of NPP with respect to each variable is 1, except for $B_0$, which has a derivative of -1; thus, by substituting these values into equation (4), the variance of NPP can be approximated as

$$\sigma^2_{NPP} \approx \sigma^2_{B_t} + \sigma^2_{B_0} + \sigma^2_M + \sigma^2_L + \sigma^2_H + \sigma^2_R - 1.2\,\sigma_{B_t}\sigma_{B_0} \qquad (12.6)$$

Note that in this case the positive correlation between $B_t$ and $B_0$ results in a reduction of the NPP variance compared to the case where they are independent of one another, in which case the last term would be zero. This results from a cancellation of errors. If the biomass for an individual tree were underestimated at time 0 (due to growth form, for example), it is also likely to be underestimated at time t, but these errors may largely cancel one another to give a reasonable estimate of the change in biomass over this period.

As an example of terms combined through multiplication or division, consider converting the NPP estimate derived above to C units instead of biomass units. Thus, equation (12.5) could be written as

$$NPP_c = [(B_t - B_0) + M + L + H + R]C \qquad (12.7)$$

where $NPP_c$ is NPP in C units (such as gC m$^{-2}$ yr$^{-1}$) and $C$ is the C content of the biomass (g g$^{-1}$). The variance of $NPP_c$ would be approximated by

$$\sigma^2_{NPP_C} \approx C^2[\sigma^2_{B_t} + \sigma^2_{B_0} + \sigma^2_M + \sigma^2_L + \sigma^2_H + \sigma^2_R - 1.2\,\sigma_{B_t}\sigma_{B_0}]$$
$$+[(B_t - B_0) + M + L + H + R]^2\sigma^2_C \qquad (12.8)$$

where $\sigma^2_C$ is the variance of $C$, the conversion factor for C.

This analytical approach to error propagation can give estimates of the variance of NPP estimates where the equations are tractable and easily differentiated, as may often be the case for computation of NPP. However, there are several other limitations of this method, as was pointed out by Robinson (1989). Error propagation formulas that combine variances for contributing variables, and use variances for calculating specific confidence intervals, are designed for use with normal distributions. If the measured variables have asymmetric, multimodal, or irregular distributions, these formulas are less applicable. Also, first-order Taylor series approximations do not work well when the coefficients of variation (standard deviation/mean) are greater than about 0.3 (Stuart and Ord 1987). Thus, for situations where the equations are mathematically complex, the distributions are irregular, or the coefficients of variation are too high, another method must be used to determine error propagation in NPP estimates.

## Monte Carlo Simulation

As noted above, when the complexity of terms or distributions differs widely from normality, another approach is often needed. Monte Carlo simulation can be a useful tool in this context. Monte Carlo simulations utilize information about the uncertainty of components and their distribution to create an overall distribution

(Rubenstein 1981; Fishman 1996). The method is quite simple and involves repeatedly drawing randomly from the component distributions, combining the components, and then accumulating the results (figure 12.4). In addition to saving the individual combinations of results, the mean, standard deviation, or particular confidence levels can be computed.

Monte Carlo simulations can be developed by using many forms of standard statistical programs or by tools developed specifically for this purpose (e.g., Goodman 2003). The key to this analysis is a random number generator to select values of the components, a function that can define the distributions (so that some values of the components are more likely than others), a function that can combine the components, and a function to save the results. The specific software can further process the results, or routine statistical software can be used instead.

Several issues arise before conducting a Monte Carlo simulation, including the size of the error for each component, the distribution of this error, the possibility of co-variation between the error terms, and the purpose of the estimate. If field samples are taken in different locations, this can be used to estimate the sampling error. Measurement errors should be included if they are nontrivial. In some cases, both types of estimates are unavailable; and in this situation, information from other, more detailed studies can be used to set some bounds on the possible error. The range of values observed can be used when a reasonable statistical sample is not available, using the statistical approximation that the range includes ±2 standard deviations. Multiple types of error distributions can be used, although care must be taken when these are combined (particularly if errors among components are correlated). Commonly used distributions include normal, uniform (all values have equal probability), lognormal, negative exponential, beta, gamma, and triangular (figure 12.5; see McLaughlin 2002 for other possible distributions). The latter distribution is often quite useful because it can approximate many kinds of distributions, yet is easy to parameterize by knowing the minimum, maximum, and modal values. While the exact distribution from which the observed data were drawn is not known, and the use of a best-fit statistical distribution may introduce some unknown degree of error, this trade-off is necessary to provide a quantitative basis for combining sources of variation in the Monte Carlo simulations. Covariance or correlation between components is often not known, but can have an impact on uncertainty estimates (see examples below). In some cases biological theory or understanding of measurements can help determine the degree of correlation. For example, C allocation to one plant part might imply a decreased allocation to another plant part. Or it may be that overinclusion of biomass in one component (e.g., the stem of a tree) might be offset by its exclusion from another component (e.g., the stump of a tree). The final issue to consider is the purpose of the analysis. If, on the one hand, the purpose is to understand the possible uncertainty of individual estimates, then the standard deviations or their analogues should be used. If, on the other hand, the purpose is to understand the possible uncertainty of the estimate of the mean, then the standard errors or their analogues should be used.

In addition to simply combining error components, Monte Carlo simulation can be used to compute the size of individual error components. For example, it can be used for estimating regression error when several allometric equations exist

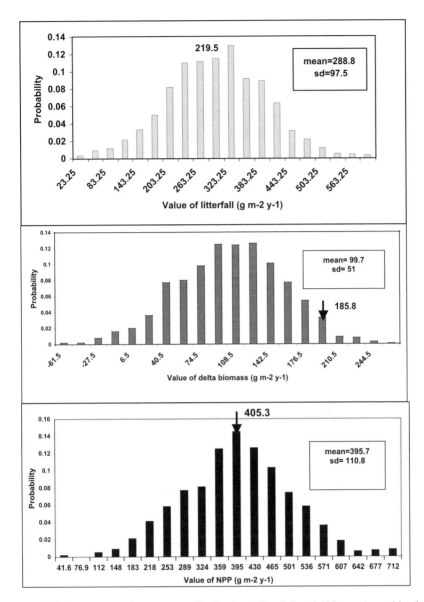

Figure 12.4. Example of component distributions (litterfall and Δbiomass) combined using Monte Carlo simulation to predict a combined distribution using the formula NPP= litterfall ± Δbiomass. A realization value representing one round of calculations is shown with arrows.

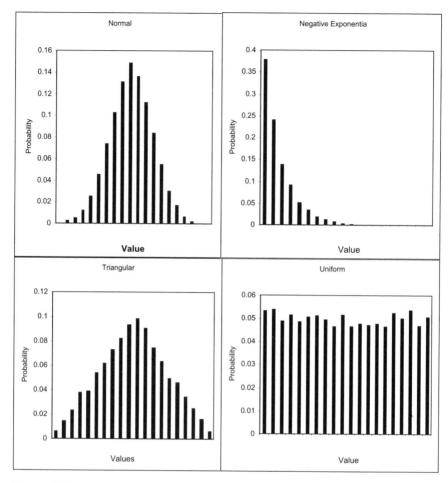

Figure 12.5. Common types of distributions used with uncertainty analysis.

(figure 12.2; bottom). Rather than assuming one equation is correct without comparison against independent data, one could use the range of values predicted by the allometric equations for each size class to limit a random selection of possible solutions. This would be preferable to choosing an equation without any knowledge that it is any better than another possible equation. Another use of Monte Carlo simulation is to estimate growth for plants that were not measured. For example, it is not unusual to use increment cores to estimate growth of trees. While it is ideal that all trees are cored, sometimes the number is too great, and an estimate is determined from a subsample of trees. When trees are subsampled, one needs to apply growth rates from cored trees to those that were not cored. Although one could apply the mean growth increment, this does not provide any estimate of the uncertainty introduced by the subsampling procedure. By using Monte Carlo simulation to repeatedly calculate the possible solutions, such an estimate can be provided.

## Compounding of Errors

While there is often an impression that uncertainty expands exponentially as one considers additional sources of error, the degree to which errors compound in NPP calculations depends on the mathematical operations performed, the relative size of the components, and the degree and sign of the correlation among the variables. In general, errors compound more rapidly when variables are combined by multiplication and division, rather than by addition and subtraction. The effect of correlation is dependent on the sign of the correlation and the mathematical operation used (figure 12.6). Positive correlation among variables reduces the compound NPP error when the variables are combined by division or subtraction, because the errors tend to cancel each other, as was illustrated above. Similarly, negative correlation among variables reduces the compound error when they are combined by multiplication or addition. Conversely, positive correlation among variables increases the compound NPP error when they are combined by multiplication or addition, as does negative correlation for division and subtraction.

As far as NPP measurements are specifically concerned, the mathematical operations usually applied involve addition and subtraction of terms, although conversion to C units from organic matter or estimation of belowground NPP from aboveground NPP involves multiplication. A negative correlation between added components therefore typically reduces compound error, whereas positive correlation typically increases compound error. The opposite is true for subtracted components, such as biomass at time t minus biomass at time 0.

## Examples of Estimating Uncertainty of NPP in Selected Biomes

This section provides examples of selected biomes in which the uncertainty of NPP estimates has been determined. Our goal is to illustrate real-world examples of errors and their consequences on NPP estimates. Additional analysis of uncertainty regarding NPP estimates can be found in chapter 3 of this volume.

### Forest Ecosystems

Though forests contain a range of life-forms, they are dominated by trees that are difficult to measure directly in terms of NPP. An example of the types of measurements taken in forests and their influence on uncertainty of NPP estimates can be found in Harmon et al. (2004). In this study of NPP from a Douglas fir-western hemlock (*Pseudotsuga-Tsuga*)–dominated forest, all trees >5 cm DBH in 12 contiguous, 1-ha plots were measured over a 5-yr period. Mortality of these trees was determined annually. Both changes in tree biomass and mortality were estimated using biomass equations. Litterfall was measured over the course of 3 yr in 4 sets of traps in 4 of the plots, which were averaged to represent a flux at the plot level. Shrub and herb biomass was estimated, and NPP of these life-forms was estimated from leaf longevity and, in the case of shrubs, "rough estimates" of the rate of stem

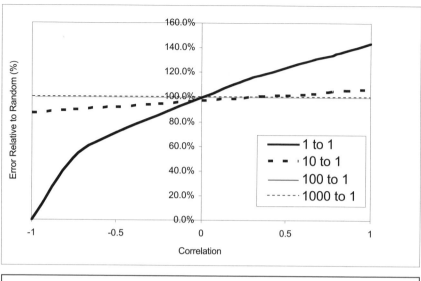

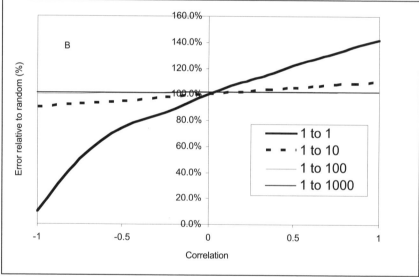

Figure 12.6. Combined error relative to the random error as a function of correlation for different ratios of error between components. The coefficient of variation was set to 10% for both components in the 1:1 case. In the 10:1 case the coefficient of variation was set to 10% for 1 component and 1% for the other, and so on. A = additive terms. B = multiplicative terms.

mortality. Herbivory was estimated on the basis of the amount of leaf area that had been eaten over a 2-yr period. Belowground NPP was estimated from fine root biomass determined from 20 1-m-deep soil cores and fine root mortality observations derived from minirhizotrons. As with the littertraps, the soil cores were placed in 4 sets in 4 of the plots, and were averaged in each plot to approximate the mean for each plot. For each flux that was measured, the mean and standard error of 1-ha plot-level estimated NPP was computed (table 12.1), and Monte Carlo methods were used to estimate the combined uncertainty, employing random realizations drawn from normal distributions defined by these means and standard errors. This uncertainty was reanalyzed using 10,000 iterations, first by assuming there was no correlation in the fluxes and then assuming there was a positive correlation of 1.0 in all the tree woody tissue-related fluxes. The mean for both estimates was 597.1 g m$^{-2}$ year$^{-1}$. As expected, the uncertainty expressed as a standard deviation of estimates for the assumption of positive correlation of tree woody tissues was higher (47.7 g m$^{-2}$ year$^{-1}$) than for assumption of no correlation of fluxes (44.8 g m$^{-2}$ year$^{-1}$). However, expressed as a ratio of the mean, both assumptions gave an uncertainty of NPP of ~8%, which implies that 95% of the estimates were within ±16% of the mean estimate.

Table 12.1. Estimated stores of carbon and rates of production associated with live biomass in a Douglas fir/western hemlock forest

| Pool | Store g C m$^{-2}$ | ΔStore g C m$^{-2}$ y$^{-1}$ | Mortality/litterfall g C m$^{-2}$ y$^{-1}$ | NPP g C m$^{-2}$ y$^{-1}$ |
|---|---|---|---|---|
| Stem sapwood | 6567 (198) | 2 (1) | 30 (9) | 32 (9) |
| Stem heartwood | 15,351 (1151) | 26 (2) | 50 (13) | 76 (13)[d] |
| Stem bark | 3337 (263) | 2 (4) | 12 (2) | 14 (3) |
| Live branches | 4489 (112) | 8 (10) | 100 (34) | 108 (34) |
| Dead branches | 318 (20) | 0 (0)[a] | 3 (1) | 3 (1) |
| Tree foliage | 941 (322) | 0 (0)[a] | 135 (12) | 150 (14)[e] |
| Coarse roots | 8122 (639) | 21 (7) | 30 (5) | 51 (7) |
| Fine roots | 362 (26) | 0 (0)[a] | 91 (16)[f] | 91 (16) |
| Understory shrubs | 144 (37) | 0 (0)[a] | 26 (5)[b] | 26 (5) |
| Understory herbs | 76 (8) | 0 (0)[a] | 40 (8)[c] | 40 (8) |
| Epiphytes | 100 (25) | 0 (0)[a] | 6 (1) | 6 (1) |
| Total random | 39,807 (2479) | 59 (24) | 523 (69) | 597 (44.8) |
| Total positive corr. | | | | 597 (47.7) |

[a] The net change in stores in these pools was assumed to be zero.

[b] It was assumed that all leaves from shrubs died each year and that 0.5% to 1% of the stems died. Litter traps indicate that the value of shrub litterfall may be as low as 1 g C m$^{-2}$ y$^{-1}$.

[c] It was assumed that litterfall from herbs was 40%–60% of live stores to account for the fact that some small, woody, stemmed, evergreen plants are included in the herb category.

[d] It was assumed that no heart rot is present.

[e] This includes grazing of 15 g C m$^{-2}$ y$^{-1}$.

[f] It was assumed that 20% to 30% of fine roots die annually.

Note: Mean (standard error).

Source: After Harmon et al. (2004).

## Savanna Ecosystems

The co-dominance of tree and grass life-forms in savanna ecosystems (House et al. 2003) produces high spatial variability in NPP. Monte Carlo methods were used to quantify the sources of uncertainty of the major aboveground components (live wood, herbaceous plants, and litter) of NPP in blue oak (*Quercus douglasi*) savanna in California (Battles et al. in press). The primary sample unit for this analysis is the plot, in which tree diameters, litterfall, and herbaceous plant growth were measured. Dendrometer bands were installed on a size-stratified subsample of trees within these plots to estimate stem diameter increment. Several plots were established in a stratified random manner (strata based on canopy cover class) to characterize the larger study area, and nested subplots were located within them for litterfall and herbaceous vegetation measurements. The following sources of error (table 12.2) apply to the plot-level estimates: (1) regression errors in the equations predicting mass parameters from tree diameter; (2) model errors in the application of the tree growth model; and (3) sampling errors for the components subsampled within plots (herb and litterfall). Allometric equations for oak tree biomass were developed on-site; published equations were used for pine (Jenkins et al. 2003), which accounted for only 20% of the basal area. Measurement errors for tree diameter and weighed biomass are regarded as minor, and were not included.

A Monte Carlo analysis with 1000 randomizations of the subplot-level data was used to estimate means and standard deviations for each pool and flux. Errors were assumed to be normally distributed, and standard errors (or their equivalents) were used to propagate uncertainty. Biomass pools and fluxes for the strata were based on 1000 randomizations of the plot-level data (i.e., results from subplot-level randomizations), assuming a normal distribution characterized by the mean and variance of each pool or flux. The 2.5 and 97.5 percentiles were used to determine the confidence intervals (95%) of study area means. Of the components of NPP considered here (table 12.2), leaf litterfall had the greatest within-plot variation with a mean coefficient of variation (CV) equal to 34%. The herb component was the next most variable, with a CV equal to 19%. Most of the variation in wood production was due to uncertainty in the growth model (CV = 14%). In comparison, allometric regression model imprecision was 6% CV. Together, the growth and allometric regression models produced plot-level estimates of aboveground wood production with a CV = 15%. In any uncertainty analysis it is instructive to note the sources of these errors. For both litterfall and herbs, plot-level variation stemmed from the high spatial heterogeneity within the plots. For the trees, uncertainty resulted from the growth projections and allometric equations. Propagating these errors to the watershed scale, by area-weighting the results from the four strata, resulted in a 95% confidence interval equivalent to ±18.9% of the mean estimate. For comparison, the watershed estimate of NPP was calculated assuming that plot-level uncertainty was zero. The results were a mean with a small positive bias (+4%) along with the expected smaller estimate of uncertainty (table 12.2).

Table 12.2. Sources of uncertainty in NPP estimates: Small watershed, blue oak savanna of California

| Component | Source of Error | Plot-Level CV (%) | Plot ID | NPP Plots (g m$^{-2}$ ha$^{-1}$) | | Strata (% tree cover) | NPP Strata (g m$^{-2}$ ha$^{-1}$) | | NPP Watershed (g m$^{-2}$ ha$^{-1}$) | | |
|---|---|---|---|---|---|---|---|---|---|---|---|
| | | | | Mean | SE | | Mean | SE | Mean | SE | 95%CI |
| Live Wood | Sampling | — | | | | | | | 371 | 35.1 | 306–441 |
| | Regression | 6 | 215 | 252 | 52.5 | >60 | 399 | 32.0 | (387) | (29.1) | (330–444) |
| | Model | 14 | 216 | 374 | 40.0 | >60 | | | | | |
| | Total | 15 | 221 | 573 | 72.5 | >60 | | | | | |
| Litter | Sampling | 34 | 204 | 242 | 38.2 | 30-60 | | | | | |
| | Regression | — | 206 | 293 | 49.3 | 30-60 | 291 | 32.3 | | | |
| | Model | — | 207 | 337 | 76.8 | 30-60 | | | | | |
| | Total | 34 | 203 | 256 | 33.1 | 15-30 | | | | | |
| | | | 211 | 473 | 56.0 | 15-30 | 364 | 27.2 | | | |
| | | | 213 | 361 | 57.5 | 15-30 | | | | | |
| Herb | Sampling | 19 | 202 | 705 | 213.9 | <15 | | | | | |
| | Regression | — | 208 | 336 | 17.2 | <15 | 442 | 74.0 | | | |
| | Model | — | 220 | 278 | 49.6 | <15 | | | | | |
| | Total | 19 | | | | | | | | | |

Notes: SE = standard error of the mean; CV = coefficient of variation; 95% CI refers to the 95% confidence interval of the mean. Values in parentheses under "NPP watershed" are estimates without plot-level error.

## Grassland Ecosystems

While determining peak biomass of grasslands is a relatively simple, direct measure of ANPP, many have argued that other terms, such as losses to mortality, grazing, leaching, volatilization, and decomposition of recently senesced plants, should be considered to reduce bias (chap. 3 in this volume). An analysis conducted by Lauenroth et al. (in press) examined how bias and uncertainty changed with addition of terms, using data collected from the Chapingo site grassland, which is located approximately 20 km northeast of Mexico City (19° 27' 30" N, 98° 54' 30" W) at 2241 m elevation (Garcia-Moya and Castro 1992). Mean annual precipitation is 579 mm, and mean annual temperature is 15.1 °C at this site. An interesting feature of the study by Garcia-Moya and Castro is that many NPP-related terms were quantified with different methods. This allowed Lauenroth et al. (in press) to use Monte Carlo simulations to assess the relative variability in NPP estimates obtained using six different NPP estimation equations that varied in both the number of parameters and the intricacy of mathematical operations (table 12.3). Lognormal distributions were generated using the mean and standard deviation of each input variable to generate input for a total of 20,000 random estimates for each method. Because many of the means were associated with large standard deviations, the lognormal distribution was used to guarantee that all of the random values were equal to or greater than zero. The results indicated that more complex equations may result in greater uncertainty without reducing the probability of underestimation bias. For example, harvesting the peak live and dead biomass one time (method 2) resulted in a CV of 22% to 27%, whereas summing all the positive increments of live and dead multiple times over the year (method 5) resulted in a CV of 94% to 86% for the 2-yr period considered (table 12.3). The mean NPP estimates for methods 2 and 5 were 1287 and 801 g m$^{-2}$ yr$^{-1}$ and 1031 and 743 g m$^{-2}$ yr$^{-1}$, respectively for the 2 yrs considered. Thus, while it could be argued that some methods, such as peak live biomass, underestimate ANPP (the mean ANPP for this method was the among the lowest: 316 to 254 g m$^{-2}$ yr$^{-1}$), the inclusion of dead biomass in the annual harvest eliminates this bias without increasing uncertainty as much as more complex and time-consuming methods. As could be expected from our general review above, the amount of uncertainty associated with estimates of NPP was influenced by the number of parameters as well as the variability in the data and the nature of the mathematical operations. By standardizing input data to "control" the variability in the data, Lauenroth et al. (in press) were able to demonstrate that equations with product terms (such as the inclusion of decomposition losses) have the potential to magnify the uncertainty in the estimates of NPP. This analysis suggests that more complex NPP estimation equations can increase uncertainty without necessarily reducing risk of underestimation.

## Kelp Forest Ecosystems

Marine forests of the giant kelp *Macrocystis pyrifera* are widely distributed in cool seas of the northern and southern hemispheres (Wormersley 1954) and are among the most productive ecosystems in the world (Mann 1973, 2000). The wave-swept

Table 12.3. Means (g m$^{-2}$ yr$^{-1}$), standard deviations (g m$^{-2}$ yr$^{-1}$), and coefficients of variation (%) for aboveground net primary production

| | Peak Live Biomass[a] | Peak Standing Crop[b] | Max-Min Live Biomass[c] | Sum of Positive Increments Live Biomass[d] | Sum of Positive Increments Standing Crop+Litter[e] | Sum of Changes Live and Dead, Adjusted for Decomposition[f] |
|---|---|---|---|---|---|---|
| *1985* | | | | | | |
| Mean | 316 | 1287 | 265 | 164 | 1031 | 864 |
| Std Dev | 54 | 283 | 56 | 251 | 965 | 371 |
| CV | 17 | 22 | 21 | 153 | 94 | 43 |
| *1987* | | | | | | |
| Mean | 254 | 801 | 239 | 241 | 743 | 661 |
| Std Dev | 110 | 213 | 110 | 101 | 638 | 254 |
| CV | 43 | 27 | 46 | 42 | 86 | 38 |

[a]NPP estimated from peak live biomass.

[b]NPP estimated from peak standing crop of live plus dead, assuming biomass produced during the interval of interest senesced before sampling.

[c]NPP estimated from the difference between the minimum and maximum estimates of live biomass, assuming some live material is carried over from the previous intervals.

[d]NPP estimated by summing all of the positive increments in live biomass, assuming that live material is carried over from the previous interval and that there are multiple peaks in live biomass.

[e]NPP estimated by summing of positive increments in live biomass, standing dead biomass, and litter. An increment in live biomass also corresponds to an increment in total dead (standing dead+litter) with same assumptions as method 4.

[f]NPP estimated by summing the changes in live and dead biomass (positive or negative) and the amount of material that decomposed during the interval.

*Source*: The data are from a grassland in Chapingo, Mexico, and are based on data from Garcia-Moya and Castro

habitat in which giant kelp lives, coupled with its large and complex morphology, rapid growth, and high turnover, introduces many sources of uncertainty when attempting to estimate NPP. Researchers at the Santa Barbara Coastal LTER have been estimating NPP of giant kelp at three sites in southern California since 2002 (Reed, Rassweiler, and Arkema unpubl. data). Monthly estimates of NPP are obtained by measuring biomass changes in fixed plots and accounting for losses of whole plants as well as fronds of surviving plants (each plant consists of many fronds arising from a single basal holdfast). Biomass is estimated from diver measurements of the density of plants and their morphology, which are then converted to dry weight based on empirically derived relationships between morphometric data, such as plant length, and wet weight. Losses are estimated from changes in plant density in the fixed plots and from losses of fronds on a subset of tagged plants. Sources of error include observer error in estimates of biomass and plant loss, sampling error in estimates of frond loss, and regression and model error in the conversion of morphometric data to dry weight (table 12.4). Because errors in estimates of plant density and biomass within a plot are correlated, the error in biomass is treated as a function of the error in density.

Table 12.4. Sources of uncertainty in NPP estimates for a kelp forest in southern California

| Component of NPP | Type of Error | Error Distribution | Scale Applied |
|---|---|---|---|
| Biomass | | | |
| | Observer | Normal (CV: 15 %) | Plot |
| | Sampling | — | |
| | Model | Normal (CV: 21%) | Plant |
| | Regression | Bimodal | Plant |
| Plant loss | | | |
| | Observer | Normal (CV: 15%) | Plot |
| | Sampling | — | |
| | Model | — | |
| | Regression | — | |
| Frond loss | | | |
| | Observer | — | |
| | Sampling | Normal (CV: 12% to 30%, depending on number of plants sampled) | Plant |
| | Model | — | |
| | Regression | — | |

CV = coefficient of variation.

Monte Carlo simulations were used to quantify uncertainty in this system because one of the errors is not normally distributed and because the calculations of annual kelp NPP use particular measurements multiple times. Four of the five sources of error used here can be approximated with a normal distribution, but one cannot: the distribution of length:weight conversions has a bimodal shape. Monte Carlo methods also easily incorporate situations in which particular errors enter into a calculation multiple times. For instance, an observer who overestimates the biomass at a given time will cause an overestimate of NPP for the previous month, because the change in biomass will appear more positive, but an underestimate of biomass in the following month, when the change in biomass will appear more negative. The negative correlation between errors in adjacent months does not affect the uncertainty in NPP for a particular month, but when production is calculated on an annual scale, these monthly errors tend to cancel out, and the total uncertainty is reduced. In the calculations of kelp NPP, the median standard error for monthly estimates of production is ~30% of the production, whereas the standard errors for the annual estimates are less than 10% of the annual production.

## Aquatic Ecosystems

Uncertainty estimation of NPP in aquatic ecosystems with microalgae as the dominant producers involves errors from methods that are very different from those used in the above ecosystems. In fact, due to the inability to separate autotrophic from heterotrophic respiration, NPP is difficult to estimate using $O_2$ flux measurements; therefore, production estimates in aquatic systems consider GPP rather than NPP. Chamber estimates of aquatic ecosystem metabolism and GPP contain both measurement uncertainty and considerable spatial uncertainty when scaled up. Additionally, chambers may not represent the entire ecosystem, which may introduce bias. For example, respiration estimates from surface chambers do not include the hyporheic zone, resulting in underestimation of total stream respiration (Fellows et al. 2001). The number of chambers necessary to characterize metabolism in a particular ecosystem depends upon the spatial heterogeneity relative to the size of the chambers; thus investigators should estimate variability from preliminary data to determine the number of replicate chambers necessary for their particular question.

Within a reach of a river or lake, free-water or open channel estimates of metabolism do not suffer from spatial scaling error as chambers do. However, measurements from one reach of a river or from a lake may not be representative of other reaches or lakes, and temporal variability is often high, so measurement at one time may not represent other times (Uehlinger and Naegeli 1998). Open channel estimates can involve bias such as that caused by groundwater movement into a river. The groundwater usually has a much lower oxygen concentration than the river water, and this will increase the estimate of community respiration (CR) while decreasing the estimate of gross primary productivity (GPP) (Hall and Tank 2005). The level of bias can be quite high, with a 10% increase in groundwater discharge potentially biasing CR estimates 50% too high. It is possible to correct estimates of CR for this bias, provided the concentration of $O_2$ in the groundwater is known (Hall and Tank 2005). Fortunately, the bias in GPP estimates is independent of the $O_2$ concentration of groundwater, and therefore is easily corrected by knowing only groundwater inputs. We suggest that bias of GPP be corrected in all metabolism measurements, and that investigators attempt to correct bias in CR when potential for bias is >30%, since other errors in CR estimation are often that high (see below). Free-water metabolism estimates have several other sources of error (McCutchan et al. 1998). Estimates of $O_2$ concentration contain some error because electrodes may vary by a few percent of saturation. Streams with high rates of reaeration can be problematic because the $O_2$ concentrations in water will approach saturation, causing small variations in probes (e.g., 2%) to amount to large errors in metabolism. In contrast, streams with low reaeration rates have large, easily measured $O_2$ differences from saturation, but the estimate of metabolism is still dependent on the estimate of reaeration rates. We suggest that to minimize error in metabolism, nighttime $O_2$ concentrations should be at least below 95% saturation, and that the reaeration rate is known with high confidence.

There has been little error assessment of whole-ecosystem metabolism derived from free-water methods, although McCutchan et al. (1998) determined that metabolism estimates are sensitive to reaeration and $O_2$ measurement error. We have used data from the Two Ocean Lake Creek, Wyoming, from a single day (29 July 1999) to make a preliminary estimate of uncertainty. Oxygen concentration and temperature were measured at 10-min intervals throughout a 24-hr period at the top and bottom of a stream reach with an average travel time of 16.2 min. The reaeration was estimated to average 0.0257 min⁻¹, when corrected for temperature. We assumed that $O_2$ probes were accurate to within 2% and that a probe deviated from the ideal calibration consistently throughout the day. All the elements of the equation to estimate community metabolism (see chap.10, this volume), except depth, were assumed to vary normally about a mean (table 12.5) and were randomly combined 10,000 times for each time interval. Community respiration was estimated by scaling nighttime estimates to 24 hr; and to calculate GPP, average respiration was subtracted from the daytime metabolism estimates, which were then summed. We therefore assumed a positive correlation among the sampling intervals. Mean CR was estimated to be −9.1 $gO_2$ m⁻² d⁻¹, with a 95% confidence interval ranging from −6.2 to −12.1 g $O_2$ m⁻² d⁻¹. Mean GPP estimate was 1.84 $gO_2$ m⁻² d⁻¹ with lower and upper 95% confidence levels of 1.34 and 2.34 $gO_2$ m⁻² d⁻¹, respectively. This analysis indicates that for this reach and day, GPP was within ±27% and CR within ±32%.

## Summary

We encourage ecological researchers to address and quantify the uncertainty in NPP estimates to the extent possible from known sources of variability in their study systems. These analyses can be useful, not only to provide a sense of the reliability of the production estimates but also to aid in comparisons among sites and to help focus future work where additional efforts will be most fruitful in reducing uncertainty. In the interests of promoting "principles and standards," we have described common contributing factors and outlined several basic approaches to quantifying NPP uncertainty. However, as the examples above demonstrate, their application to particular biomes requires adaptation to particular sets of measurement variables and study designs.

Table 12.5. Estimates of error and uncertainty for GPP for a reach of Two Ocean Lake Creek, using the two-station method

| Source of Error | Mean over 24 Hrs | Standard Deviation | Unit |
|---|---|---|---|
| $O_2$ probe | | | |
|   $O_2$ saturation | 80.19 | 0.5 | % |
|   $O_2$ concentration | 6.67 | 0.05 | mg l⁻¹ |
| Reaeration rate | 0.0257 | 0.00385 | Minute⁻¹ |
| Travel time | 16.2 | 0.01 | Minutes |
| Temperature | 12.66 | 0.1 | C |
| GPP mean | 1.8405 | 0.2513 | $gO_2$ m₋₂ day⁻¹ |
| GPP CV | 13.65 | | % |

*Notes:* Although the parameter value for each 10-min time interval was used, the mean of the value is given to indicate the relative level of variation. The uncertainty was determined by running each of the 191 time intervals 10,000 times, assuming a normal distribution.

*Sources:* Chapter 10, this volume; Hall and Tank (2003).

*Acknowledgments*    This synthesis and the research it was based upon were funded in part by the Pacific Northwest Research Station, the U.S. Environmental Protection Agency, the H. J. Andrews LTER (DEB-0218088), the Shortgrass Steppe LTER (DEB-0217631), the Santa Barbara Channel LTER (OCE-9982105), the Kay and Ward Richardson Endowment, the Bullard Fellowship Program of Harvard University, the Colorado State University Experiment Station (grant no. 1-57661), and the California Integrated Hardwood Range Management Program (project # 00-1). This document has been subjected to EPA's peer and administrative review, and it has been approved for publication as an EPA document. Estimates of error in kelp NPP were developed with K. K. Arkema.

## References

Battles, J. J., R. J. Jackson, A. Shlisky, B. Allen-Diaz, and J. W. Bartolome. In press. Net primary production and biomass distribution in the blue oak savanna: Proceedings of the Sixth California Oak Symposium: Today's Challenges, Tomorrow Opportunities. USDA Forest Service General Technical Report PSW-GTR-19x.

Clark, D. A., S. Brown, D. W. Kicklighter, J. Q. Chambers, J. R. Thomlinson, and J. Ni. 2001. Measuring net primary production in forests: Concepts and field methods. Ecological Applications 11(2):356–370.

Dixon, W. J., and F. J. Massey. 1983. Introduction to Statistical Analysis, 4th ed. McGraw-Hill, New York.

Fellows, C. S., H. M. Valett, and C. N. Dahm. 2001. Whole-stream metabolism in two montane streams: Contributions of the hyporheic zone. Limnology and Oceanography 46:523-531.

Fishman, G. S. 1996. Monte Carlo: Concepts, Algorithms and Applications. Springer-Verlag, New York.

Garcia-Moya, E., and P. M. Castro. 1992. Saline grassland near Mexico City. Pages 70–79 in S. P. Long, M. B. Jones, and M. J. Roberts (eds.), Primary Productivity of Grass Ecosystems of the Tropics and Sub-tropics. Chapman & Hall, London.

Gardner, R. H., V. H. Dale, and R. V. O'Neill. 1990. Error propagation and uncertainty in process modeling. Pages 208–219 in R. K. Dixon, R. S. Meldahl, G. A. Ruark, and W. G. Warren (eds.), Process Modeling of Forest Growth Responses to Environmental Stress. Portland, Ore, Timber Press.

Goodman, J. R. 2003. SUE (stochastic uncertainty estimator)—Users guide. http://www.fsl.orst.edu/lter/pubs/webdocs/models.cfm?topnav=53

Hall, R. O., and J. L. Tank. 2003. Ecosystem metabolism controls nitrogen uptake in streams in Grand Teton National Park, Wyoming. Limnology and Oceanography 48: 1120-1128.

Hall, R. O., and J. L. Tank. 2005. Correcting whole-stream estimates of metabolism for groundwater input. Limnology and Oceanography: Methods 3:222–229.

Harmon, M. E., K. Bible, M. J. Ryan, D. Shaw, H. Chen, J. Klopatek, and X. Li. 2004. Production, respiration, and overall carbon balance in an old-growth *Pseudotsuga/Tsuga* forest ecosystem. Ecosystems 7:498–512.

Heath, L. S., and J. E. Smith. 2000. An assessment of uncertainty in forest carbon budget projections. Environmental Science & Policy 3:73–82.

House, J. I., S. Archer, D. D. Breshears, R. J. Scholes, and N. T.-G. I. Participants. 2003. Conundrums in mixed woody-herbaceous plant systems. Journal of Biogeography 30:1763–1777.

Jenkins, J. C., D. C. Chojnacky, L. S. Heath, and R. A. Birdsey. 2003. National-scale biomass estimators for United States tree species. Forest Science 49(1):12–35.

Lauenroth, W. K., A. A. Wade, M. A. Williamson, B. E. Ross, S. Kumar, and D. P. Cariveau. 2006. Uncertainty in calculations of net primary production for grasslands. Ecosystems 9:843-851.

Mann, K. H. 1973. Seaweeds: Their productivity and strategy for growth. Science 182: 975–981.

Mann, K. H. 2000. Ecology of Coastal Waters, 2nd ed. Blackwell Science, Oxford.

McCutchan, J. H., W. M. Lewis, Jr., and J. F. Saunders 1998. Uncertainty in the estimation of stream metabolism from open-channel oxygen concentrations. Journal of the North American Benthological Society 17: 155-164.

McLaughlin, M. 2002. Common probability distributions. http://www.geocities.com/~mikemclaughlin/math_stat/Dists/Compendium.html.

McRoberts R. E., J. T. Hahn, G. J. Hefty, and J. R. Van Cleve. 1994. Variation in forest inventory field measurements. Canadian Journal of Forest Research 24:1766–1770.

Means, J. E., H. A. Hansen, G. J. Koerper, P. B. Alaback, and M. W. Klopsch. 1994. Software for Computing Plant Biomass—BIOPAK User's Guide. General Technical Report PNW-GTR-340. Portland, Ore, U.S. Dept. of Agriculture, Forest Service, Pacific Northwest Research Station.

Phillips, D. L., S. L. Brown, P. E. Schroeder, and R. A. Birdsey. 2000. Toward error analysis of large-scale forest carbon budgets. Global Ecology & Biogeography 9:305–313.

Robinson, J. M. 1989. On uncertainty in the computation of global emissions from biomass burning. Climatic Change 14:243–262.

Rubenstein, R. Y. 1981. Simulation and the Monte Carlo Method. Wiley, New York.

Sprugel, D. G. 1983. Correcting for bias in log-transformed allometric equations. Ecology 64:209–210.

Staebler, G. R. 1955. Gross yield and mortality tables for fully stocked stands of Douglas-fir. Research paper no. 14. U.S. Dept. of Agriculture, Forest Service, Pacific Northwest Forest and Range Experiment Station, Portland, OR.

Stuart, A., and J. K. Ord. 1987. Kendall's Advanced Theory of Statistics, 5th ed., vol. 1. Oxford University Press, New York.

Taylor, J. R. 1997. An Introduction to Error Analysis: The Study of Uncertainties in Physical Measurements, 2nd ed. University Science Books, Sausalito, CA.

Uehlinger, U., and M. W. Naegeli. 1998. Ecosystem metabolism, disturbance, and stability in a prealpine gravel bed river. Journal of the North American Benthological Society 17:165-178.

Wormersley, H. B. S. 1954. The species of *Macrocystis* with special reference to those on southern Australia coasts. University of California Publications. Botany 27:109–132.

# Index